Polynomial Root-Finding and Polynomiography

To Roger Tobie,
with best wishes.
Bahman

Bahman Kalantari
Rutgers University, USA

Polynomial Root-Finding and Polynomiography

World Scientific
NEW JERSEY • LONDON • SINGAPORE • BEIJING • SHANGHAI • HONG KONG • TAIPEI • CHENNAI

Published by

World Scientific Publishing Co. Pte. Ltd.
5 Toh Tuck Link, Singapore 596224
USA office: 27 Warren Street, Suite 401-402, Hackensack, NJ 07601
UK office: 57 Shelton Street, Covent Garden, London WC2H 9HE

Library of Congress Cataloging-in-Publication Data
Kalantari, Bahman.
Polynomial root-finding and polynomiography / by Bahman Kalantari.
p. cm.
Includes bibliographical references and index.
ISBN-13: 978-981-270-059-9 (hardcover : alk. paper)
ISBN-10: 981-270-059-5 (hardcover : alk. paper)
1. Polynomials. 2. Visualization. 3. Recurrent sequences (Mathematics)
4. Computer graphics. I. Title.
QA161.P59K35 2008
512.9'422--dc22

2008032272

British Library Cataloguing-in-Publication Data
A catalogue record for this book is available from the British Library.

Printed in Singapore by Mainland Press Pte Ltd

To the memory of my parents, and to my wife Azam for her unwavering care and patience rooted in love.

Preface

The genesis of this book dates back to the design of a question for a graduate exam at Rutgers computer science department in the early 1990s. I discovered a method for the approximation of square-root of two, requiring only high school algebra. It coincided with the well-known Newton's method as applied to a corresponding quadratic equation. It appeared interesting and different from the ordinary approach in developing Newton's method. Moreover, it could be easily extended to higher order methods: the third-order method of which for the approximation of square-roots coincided with a method credited to Edmund Halley, the astronomer who has a comet named after him. What I thought would be a matter of a very brief time before I would be sufficiently enlightened about this discovery and then leave it aside has in fact turned out to have taken me beyond my wildest imagination. This book, a by-product of many years of my research, is thus an unexpected fruit of the study of the square-root of two.

I soon came to realize that what stood before me was a mountain. New questions came up and more and more research problems evolved. In short, the polynomial root-finding problem attracted me so much that since then I have never abandoned it. Now I consider it as one of the most fascinating problems of mathematics and science having so much to offer to so many. The mountain that stood before me has no summit, or infinitely many.

Indeed dating back to the ancient civilizations solving algebraic equations has been among the most fascinating and profound intellectual tasks. Even the case of solving quadratic and cubic polynomials - arising naturally in algebraic or geometric settings - it has inspired deep discoveries such as, the irrational numbers, the complex numbers, and even sophisticated algorithms for integer factorization. Perhaps my attraction to the problem during all these years is also due to the very profound nature and gravity of the polynomial root-finding problem, but additionally the beauty of its algorithmic visualization by what I have come to call *polynomiography.*

I came to Rutgers as a computer scientists with expertise in mathematical programming and optimization. Not only the study of polynomial root-finding was not among my research areas at the time, in the eyes of many, the problem was and perhaps still is considered to be old-fashioned and done with. Unless one is specialized in specific aspects such as the complexity of root-finding, or in the context of broader fields of study such as algebra, complexity analysis, computer algebra, dynamical systems, numerical analysis, etc., it is not a common practice to study polynomials in their own right. While scientists and mathematicians use polynomials routinely, no one addresses oneself as a "polynomial theorist." From the academic point of view it was thus very risky to dedicate my research and time into the study of polynomial root-finding with so much history behind it.

Not that I would be abandoning doing research on my other areas of interest, but I knew that by doing so I would diminish such opportunities as receiving grant funding which in turn opens up the way for further explorations, graduate student support, support to attend conferences, and receiving more visibility. I also knew that I would subject myself to unpredictable judgements. Despite these risks and drawbacks I continued to work on the problem because it was simply too beautiful to resist. For several years I was merely interested in the theoretical aspects of polynomial root-finding and related problems I encountered or needed to invent along the way. While I continued to produce publications and brought several collaborators into the field, generally speaking for the most part I found doing research in this field somewhat like mountain climbing where one does it for personal satisfaction of different kinds and simply has to accept the risks that come along.

During the first few years I did not even consider computer visualization of the root-finding process. I knew that images coming from iteration functions generally would fall within the category of images known as *fractal*, a term coined by the famed Mandelbrot. Initially, I did not think that images coming from solving polynomial equations would be much different from images already produced by experts and even many more so by amateurs. For instance, a fractal image coming from the approximation of cubic roots of unity based on Newton's method is very familiar and has even been featured on textbook covers.

However, as I became deeply involved in the root-finding problem, discovering more and more algorithms, it became evident that the computer visualization of the root-finding process through these algorithms would be interesting and worthy of trial. This was partly because of so many choices of algorithms and partly because from the theory I could anticipate the shape of the images coming from some of these algorithms. This was very promising and important since I could sense a reasonable degree of "control" and "design" as opposed to typical fractals.

The initial computer visualization did give rise to striking images and I could foresee that this would just be a beginning of a new set of activities involving visualization. Eventually I felt that this visualization deserved a name of its own. Thus I coined the term *polynomiography* to be the art and science of visualization in the approximation of roots of polynomials (using iteration functions). The word is simply a combination of "polynomial" and the suffix "-graphy." For certain, some would not only find the word hard to pronounce, but perhaps even unnecessary since the term fractal was already so well-known.

However, not only a polynomiography image, called *polynomiograph* is not necessarily a fractal image, even when it is a *fractal polynomiograph* it is very distinct from a typical fractal image and has a precise foundation. Moreover, fractal polynomiographs result in new class of visualizations and thereby dramatically enhance the horizon of fractals. In contrast to polynomiography, the word fractal is so broad and could be interpreted so vaguely that anything from an image of a Julia set coming from the iterations of a quadratic or cubic, to an image from Newton's method, to a complete binary or ternary tree, to a decimal expansion of a rational number, to an ordinary tree, to a mountain, and even to a galaxy, it could all be considered to be fractal. Repetition does not necessarily lead to fractal.

In contrast even in the domain of rational iteration functions on the complex plane, the word polynomiograph is very distinct from a typical fractal image due to such iterations. Indeed even if in such a fractal image we sufficiently zoom into a Fatou component of the underlying rational iteration, there is nothing fractal about the image. In summary, the term polynomiography is quite a logical term and a deserving one, immediately putting a face behind the image, namely a polynomial equation. Moreover, after a few iterations in pronouncing the term, it turns into a memorable and a meaningful one.

After many years of doing polynomial root-finding and polynomiography and numerous experiences that include national and international presentation, I am now convinced that the field of polynomiography will change the view of polynomials and dramatically extend their usage. I feel that polynomiography has the potential to turn into a creative activity that would popularize math among the youth, bring innovations to art and design, bridge art and math, inspire mathematicians and educators, but also engage the general public.

While polynomials remain to be fundamental entities in science and math and in education, never before in their history has there been a systematic development of algorithms to reveal the magnificent visual beauty behind solving polynomial equations. More so to liberate polynomials and to widen their scope of utility to a scale never before imagined. Polynomiography is the algorithmic visualization of polynomial equations. While polynomiography uses sophisticated mathematical algorithms on a computer to create a polynomiograph, with proper software development it turns the polynomial root-finding problem upside down and into a medium of expression, art, design, science, math, education, innovation, discovery, creativity and more.

I have delivered many lectures on polynomiography or the theory that has inspired it, at many levels, at many locations, and to many different audiences. These include, from presentation at theoretical conferences on mathematics or computer science, such as numerical analysis, number theory, computational geometry, computer graphics, art-math conferences, to presentations with audiences that consisted of middle and high school students, university students of science and math or art and graphics, mathematicians and computer scientists, K-12 teachers, general artists, art curators, and even the general public, and in several different countries.

I have had solo or group exhibitions at museums and art galleries. My images have appeared on various covers or inside of magazines and books. Also, articles about polynomiography have appeared in some national and international media. I have even developed a few courses on polynomiography, and with the help of some collaborators have conducted separate teacher and student workshops. During these activities a demo polynomiography software has been tested by the respective middle or high school students, college students, and K-12 teachers. It has never failed to arouse the curiosity and interest of the participants, followed by interesting and deep questions and more importantly much interest in wanting to learn more about polynomiography and its foundation. Because of polynomiography, I have witnessed more interest in polynomials from middle school students than college student who have only been exposed to polynomials abstractly.

Ironically, despite the very multidisciplinary and innovative nature of polynomiography and numerous claimed potentials (that continue to become more and more evident), or the tremendously rich and profound underlying theoretical foundation, at times I have been perceived as an artist when expecting to be considered a computer scientist or a mathematician and educator, while at other times viewed as a scientist or a mathematician when expecting to be considered an artist or an educator. Claiming to be all is unimaginable to some, and a contradiction in terms. Though I had never considered doing art seriously before polynomiography - nor did I try to label myself as an artist after - I am pleased to say that through polynomiography I have come to learn a few things about art and have received the recognition and appreciation of artists of different kinds - from traditional artists, to digital artists, to the general public. Thanks to polynomiography I have also dared to consider myself an artist in addition to being a computer scientist and a mathematician, and given the opportunity I would fill up a large gallery with my personal artwork. Furthermore, I feel that the invention of polynomiography technology for which I hold a U.S. patent will also create many new artists.

As an "algorithmic artist" I consider myself a "polynomiographer," not a "fractal artist." Personally, I enjoy creating a beautiful polynomiograph as much as proving a beautiful theorem, realizing well that beauty is in the eyes of beholder. Indeed I can no longer distinguish between the two kinds of creative activities. Mathematics and art are closely related and computer science as algorithms and technology allows mathematics to unveil its visual beauty. Polynomiography is a set of unique algorithms that unveils the beauty of polynomial equations. I foresee that polynomiography would become well-utilized someday, especially as a powerful medium in education and in art.

Since I make strong claims on the potentials of polynomiography, it is perhaps fair to give the reader a sample of the evaluations and reviews I have received - positive and negative - in order for the reader to judge for himself/herself as to which point of view they would tend to agree with, but also exposing the reader to the vast possibilities in polynomial root-finding and polynomiography.

The reviews to be considered here are limited to the cases of submission of few grant proposals to funding agencies, or submission of articles and proposals to publishers.

On an interdisciplinary proposal to a funding agency regarding polynomiography applications in art, science, math, and education - when the subject was in its earlier stages - a reviewer wrote:

"*Root-finding is indeed a hard field to make a splash in. Since the Sumerians, ancient Greeks, Isaac Newton, through Hermann Weyl and Steve Smale, the best minds have given it a shot. I have taught the material in the classroom and have avoided the glitz associated with a lot of fractal geometry for fear of giving students less than what they need. Well, I am now turning my head! This proposal represents a serious contribution to global root-finding algorithms, computer generated art, and bringing it all to the masses.*"

Ironically, a second anonymous and independent reviewer on the same proposal - though seemingly impressed with the mathematical developments - merely summarized the worth of the proposal as

"*My impression is, 'who cares!'* "

On the journal submission of a theoretical article on the subject of iteration functions for root-finding while a reviewer considered the work among other things "*significant and scholarly,*" another reviewer wrote, "*there is little that would be of value to practising numerical analysts who are already equipped with all the iteration functions they need.*"

A polynomiography proposal focused on applications in K-12 math education with several enthusiastic participating consultants and collaborators from diverse locations in the country, including mathematicians with expertise in K-12 education, and with reasonable initial experimental evidence - to the extent possible without funding - despite the fact that its goals were considered to address a significant problem of national interest was judged to have insufficient experimental evidence!

An anonymous referee of a proposal for a popular or non-technical book on polynomiography suggested that such book could be billed as an elementary introduction to the Riemann Hypothesis. He justified this by saying that though the Riemann Hypothesis (the most famous unsolved problem of mathematics with a one-millon dollar prize for a solution) is not about polynomials, like polynomiography it is about location of complex roots of a function. The reviewer however went on to claim that the availability of computer and specialized software has greatly reduced the level of interest in how one goes about finding roots of polynomials, or solutions of equations in general.

These reviews are revealed here because I would like to raise several important questions regarding the study of polynomial root-finding and polynomiography. Questions that deserve deeper scrutiny than a response based on impulse.

To whom do polynomials belong? Is polynomial root-finding a task exclusive to numerical analysts? Is Newton's method sufficient for root-finding? If the study of iteration functions is pointless why do experts of dynamical systems ("dynamicists") study iterations of rational functions?

A numerical analyst, a mathematician, or a scientist familiar with Newton's method and its quadratic order of convergence may consider the study of higher order iteration functions for root-finding pointless because he/she may argue as follows: if we consider every other iterate in Newton's method we get a sequence which if convergent, it will result in a fourth order method. And if we consider every third iterate, the sequence if convergent, it will result in an eight order method, and so on. Indeed I have heard this argument several times as a quick way to dismiss the study of iteration functions for root-finding. But this view completely misses the point of iteration functions and at best only makes sense when in a small neighborhood of a polynomial root. Moreover, such view should also find pointless the work or Fatou, Julia, and even Halley - a contemporary of Newton - whose iteration function was inspired by Newton's, yet in turn it inspired the celebrated Taylor's Theorem. In summary, such view should also find pointless the entire theory of iteration of rational functions, and polynomiography. Thus it is a superficial view and should be discarded.

Can one ever claim that polynomials are so well understood that need no further research? Can any educator claim that no further worthy curricula can be developed based on polynomials and their applications? What if students acquire a liking of polynomiography and working with polynomials to the extent that they would end up raising deep mathematical questions? What if they would get exposed to deep mathematical concepts through exciting and fun visualizations? How much evidence is needed to convince a reviewer or an educator to approve of the use of a novel medium in the classroom?

Clearly, the utility of polynomials is not restricted to scientists and mathematicians or teachers of math and science. Nor are specialists dealing with polynomials the only ones who can truly appreciate or judge the utility of these fundamental objects of mathematics. Indeed polynomial root-finding through polynomiography has a chance of becoming a subject of interest to many, from K-12 education, to higher education, and even the general public.

Polynomial root-finding is not solely about computing or approximating the roots. It is a process and understanding it. Polynomiography enhances this process and reverses the root-finding problem. One can learn to play with the roots: place them as one pleases and then find the roots of the corresponding polynomial using polynomiography techniques. The Fundamental Theorem of Algebra, one of the most celebrated theorems of mathematics - through polynomiography - suddenly becomes visible, obvious, and as believable as the force of gravity. It becomes discoverable even by children and consequently popular and appreciable. At the same time while the foundation of polynomiography relies on the Fundamental Theorem of Algebra - through polynomiography - it becomes evident that despite many existential proofs, algorithmic attempts in proving the theorem poses a whole new set of mysteries and interesting problems.

In fact suppose that one takes just a few points on the Euclidean plane, say three or more, then forms a polynomial having those as its roots, then selects an arbitrary point. Now consider the question: Does Newton's method gravitate the point toward any of the roots? Physicists may get closer and closer to the understanding of the true nature of gravity, but neither theoretical physicists, nor mathematicians would be able to make a definitive decision on this simple-looking question. It is undecidable!

It is thus fair to say that despite the fact that some "experts" of various fields may have found or may find the root-finding problem old-fashioned, or too specialized, etc., it is in fact an inexhaustible problem and through polynomiography it could become interesting to many different groups and audiences of diverse backgrounds. Polynomials are foundational to math and education. Indeed whenever students are introduced to notions such as functions, derivatives, integrals, solution of equations, graphs and much more, a polynomial is the first example to be considered. Polynomiography opens the way for bringing new views and applications into these topics and even more importantly providing a platform to offer new teaching curricula.

It would of course be tremendously rewarding if polynomiography would ever serve as a medium that would bring any attention or introduction to the Riemann Hypothesis. But it would be at least equally rewarding if polynomiography would help bring recognition and popularity to polynomials themselves. Perhaps like Fermat's Last Theorem, the Riemann Hypothesis too will someday be concurred. But it is plausible that long after a solution to the Riemann Hypothesis is found, polynomiography would still remain to be useful and interesting to many, not just to specialists.

During the years since introducing polynomiography, I have received many interesting questions in one form or another. Here is a sample of such questions from K-12 students. A 7-th grader wrote,

"*I like polynomiography, what are polynomials?*"

An 8-th grader and her classmates wrote,

"*We love your work, are the polynomials you use complex?*"

She and her teacher eventually invited me to their school in New Jersey for a presentation to many students. At a summer math camp for 11–13 year old girls in Illinois who were given the chance to play with a simple demo polynomiography software a student wrote,

"*I love the polynomiography software! I'm into art so to see that math is related to art was really cool. Thank you for that opportunity.*"

She went on to request the camp teachers to allow older girls to attend the following year. The reason: she "*loved it so much*" that she wanted to attend again, but she would be 14 the following year. This type of enthusiasm about working with polynomials is quite novel.

In addition to students, educators from middle schools, high schools, and colleges have expressed excitement and interest in polynomiography. Detailed feedback from both students and teachers will be given in the book. These and many other examples and experiences signify an unprecedented level of interest in polynomiography. And the evidence is mounting. Polynomiography is here to stay and there is really no limit in the extent of its applications.

This book is not intended for the high school student or even a typical undergraduate student - such books are subject of future projects. However, there are chapters or topics in this book that could appeal to an advanced undergraduate student, a high school mathematics teacher, or even an art teacher, or an artist. The general reader of this book could include graduate students of mathematics or science, mathematicians and scientists, science or math teachers. Particular courses where some material from this book can definitely be used include algebra, calculus, numerical analysis, complex analysis, dynamical systems, number theory, computer graphics, and specialized courses dealing with patterns, symmetry, algorithmic art, or honors courses as in interdisciplinary courses. But some parts of this book could perhaps even inspire a high school student to learn about such notions as square-root of minus one, complex numbers, polynomial equations, root-finding algorithms, functions, geometry, Voronoi regions, Fibonacci numbers, homogeneous recurrences, iterations, Newton's method, convergence, limit, the fundamental theorem of algebra, the Gauss-Lucas theorem, the maximum modulus principle, fractals, Julia sets, algorithms and art.

Bahman Kalantari
www.polynomiography.com

Contents

Introduction

Once there was nothing. Not even time. But it seems that 13.7 billion years ago this nothing became everything when a tiny dot of infinite density spontaneously expanded at a phenomenal rate giving birth to the universe, including time.

Michio Kaku, Theoretical Physicist (from TV series "Time")

Solving a polynomial equation could be considered as a game of hide-and-seek with a bunch of tiny dots on a painting canvas. We hide the dots behind a polynomial equation, we then seek them using a formula or an algorithm. *Polynomiography* is the algorithmic visualization of the process of searching for the dots, and painting the canvas along the way.

The above is my informal definition into the fields of polynomial root-finding and polynomiography. It stems from the fact that polynomiography has received a wide range of interest because of which I have had the pleasure and honor of speaking before non-technical as well as technical audiences. I have continually strived to find metaphors that would make audiences get a quick feel of what the underlying foundation is, then when appropriate I would get into more technicalities. But through this kind of informal definition I have also tried to suggest to technical audiences well-familiar with some of the underlying mathematical foundation that there are indeed novelties in polynomiography that in particular could change the way they have viewed or utilized polynomials, and for the better.

Often times I have begun my introductory lectures by explaining first what a polynomial is, using the usual definition: "a linear combination of whole powers of a variable." More formally, one defines a polynomial as:

$$a_n z^n + a_{n-1} z^{n-1} + \cdots + a_1 z + a_0 \qquad (0.1)$$

where $a_n, \ldots, a_0$ are *coefficients*, n is the *degree*, a_n nonzero, and z is a *variable*.

In this book we are interested in the case where the coefficients are *complex numbers* and z a *complex variable*, to be formally defined momentarily. The polynomial is then called a *complex polynomial.* When the coefficients are real numbers and the variable is restricted to the reals, the polynomial may be addressed as a *real polynomial.* In this book we are interested in complex polynomials, whether or not the coefficients are real or complex.

Formally, a polynomial equation is:

$$a_n z^n + a_{n-1} z^{n-1} + \cdots + a_1 z + a_0 = 0. \tag{0.2}$$

A *solution*, or *root*, or *zero* of a polynomial is any specific value of z, say θ that would satisfy the polynomial equation, i.e. when z is assigned the value θ, (0.2) is satisfied.

It would be highly unlikely that in an introductory lecture on polynomials a mathematician, a scientist, or a teacher would give as example a polynomial having a coefficient that is not a real number, even if a complex polynomial is being defined.

Speaking to a non-technical audience it would be a mistake to define a polynomial as in (0.1) and then give as example a polynomial with an imaginary coefficient since in the process this would also force the speaker into defining the square-root of minus one.

Even if one begins by giving an example of a real polynomial, once the corresponding polynomial equation is defined, there is a need to speak of the Fundamental Theorem of Algebra, the complex numbers, and methods to solve the equation. If in between one happen to speak of graphs of real functions, say a quadratic or cubic polynomial and happens to mention Newton's method for finding zeros, undoubtedly one would create a vague and confusing picture of several concepts such as polynomials themselves, polynomial equations, complex numbers, graphs, and Newton's method to say the least.

This route to defining a polynomial equation could sound reasonable if the real roots of real polynomials are being defined. However, from a personal point of view this approach in defining a polynomial equation is not very effective when the main interest lies in polynomiography which deals with the complex plane.

Come to think of which came first - a polynomial or a polynomial equation? Historically, it seems that a polynomial equation came first. Moreover, to introduce complex polynomials it makes more sense to speak first of a polynomial equation than the raw polynomial. The discovery that square-root of two is not a rational number is really a consequence of solving a quadratic equation, as is its approximation. The notion of an abstract function, or a polynomial as a function, is an abstract notion that even college students need much practice to gain a mature feel for. In summary, a polynomial equation and a polynomial do not necessarily provoke very closely related concepts. They could trigger different concepts the realm or domain of which may mean a different thing to different people, even to specialists.

Furthermore, while in general the concept of complex numbers and the corresponding elementary operations are abstract and difficult to comprehend, it is child's play to speak of locations on a map, or points on a canvas. Even with a real polynomial of degree n, according to the Fundamental Theorem of Algebra, we are ensured that what gets hidden behind the corresponding polynomial equation is in fact a set of n points, even if some of these points are placed at the same location.

The following passage from Pan (1997), one of the most authoritative experts in the computational aspects of solving a polynomial equation, eloquently describes the significance of the polynomial root-finding problem:

"*The very ideas of abstract thinking and using mathematical notation are largely due to the study of (0.2). Furthermore, (0.2) has historically motivated the introduction of some fundamental concepts of mathematics (such as rational and complex numbers, algebraic groups, fields, and ideals) and has substantially influenced the earlier development of numerical computing.*"

Since searching for locations on a map from their coordinates is such a familiar task, defining a polynomial equation gives rise to the occasion for defining complex numbers. We may give another informal definition for a polynomial equation:

A polynomial equation is an algebraic encryption of a set of point on a map.

To speak of "hiding points behind a polynomial equation" perhaps also originates or invokes a new perspective at solving a polynomial equation as a task that is not merely that of "seeking the solutions." This together with visualization through polynomiography is the beginning of a new domain of interest and novel applications in the ancient problem of root-finding.

What formalizes the idea of polynomial root-finding as a game of hide-and-seek with dots on a canvas is to convey the fact that a *point* on the *Euclidean plane* is actually a *number*, a *complex number*. This is in the sense that a point on the Euclidean plane with coordinates (a, b), where a stands for the East–West and b for the North–South coordinate, is also a complex number, written as

$$a + ib,$$

where i is the magical number defined as

$$i = \sqrt{-1},$$

i.e. $i * i = -1$. This dual nature is in the sense that a geometric point is actually an algebraic object, a complex number, with respect to which we can perform the four elementary operations on ordinary numbers almost with the same ease. This is a significant discovery behind which lies much brilliance and history due to our forefathers.

Once typical points (a, b) and (c, d) are dressed in their complex number costume, the elementary operations are defined as follows:

$$(a + ib) + (c + id) = (a + c) + i(b + d).$$

$$(a + ib) - (c + id) = (a - c) + i(b - d).$$

$$(a + ib) \times (c + id) = (ac - bd) + i(ad + bc).$$

In fact one can easily discover the geometric effect of multiplication of a general number $a + ib$ by the special number i itself, resulting in the clockwise rotation of the point (a, b) by an angle of 90 degrees, arriving at the point $(-b, a)$.

To discover division, it suffices to realize that the reciprocal of a non-zero complex number can easily be derived by conjugation to be:

$$\frac{1}{c + id} = \frac{1}{(c + id)} \frac{(c - id)}{(c - id)} = \left(\frac{c}{c^2 + d^2}\right) - i\left(\frac{d}{c^2 + d^2}\right),$$

another complex number. Then, the definition of division of two complex numbers follows:

$$\frac{a + ib}{c + id} = \left(\frac{ac + bd}{c^2 + d^2}\right) + i\left(\frac{-ad + bc}{c^2 + d^2}\right).$$

The complex variable z is formally written as

$$z = x + iy$$

with x and y as its *real* and *imaginary* parts, respectively. The *modulus* of z, written as

$$r = |z| = \sqrt{x^2 + y^2}$$

is the Euclidean norm of the point (x, y). The *argument* of z is the angle θ between the point (x, y) as a vector, and the x-axis and by convention and for uniqueness of representation is written as an angle satisfying

$$-\pi \leq \theta \leq \pi.$$

The complex variable/number z can then be represented in the *polar* form. Combining the trigonometric polar form with the exponential representation possible through Euler's formula - which embodies De Moivre's formula for powers - one may write a single formula that combines all these for any integer n:

$$z^n = (x + iy)^n = (re^{i\theta})^n = r^n(\cos\theta + i\sin\theta)^n = r^n(\cos n\theta + i\sin n\theta).$$

A recommended source for further results on complex operations and geometric interpretations is Wikipedia, the free online encyclopedia. Historical development of complex numbers and the number $i = \sqrt{-1}$ and its interesting history can be traced in such books as, "Imagining Numbers: (particularly square-root of minus fifteen)," Mazur (2003) and "The Story of Square Root of Minus One," Nahin (1998).

The problem of solving a polynomial equation is often truly a task that deals with tiny dots having physical width, as opposed to a task that deals with points which are dimensionless geometric object. This view is justified in the sense that one often needs to approximate the roots of a polynomial equation, rather than computing their exact value which could be an impossible task. Indeed even in the course of looking for the dots we need to approximate the intermediate steps due to round off errors.

It is a classic result that closed form formulas for solutions of a general polynomial equation is only possible when the degree is less than five. Even for polynomials of degree less than or equal to four, since the closed formulas involve radicals, from the computational point of view even the case of square-root of a number needs to be approximated through iterative methods. Thus, in general we can only approximate the roots and need to be content with any approximation that would fall within disks of prescribed radii, say ϵ, centered at the exact values.

This book is based on solving a polynomial equation using iteration functions, but with major emphasis on the use of a very special and fundamental family of iteration functions called the *Basic Family*, as well as many iteration functions derived from this family or related to the family. The book reveals many mathematical and algorithmic properties of the Basic Family and many significant connections between the family and different mathematical concepts. The goal of the book is not merely to focus on the computation or approximation of the set of roots of a polynomial equation, but much more, including of course polynomiography, a subject turning the task of solving a polynomial equation upside down, potentially bringing wide range of interest into this ancient problem.

Quoting again from Pan (1997), this time with regard to the scope of today's significance in solving a polynomial equation:

"*In fact, as n grows beyond 10 or 20, the present-day practical needs to solve equation (0.2) become more and more sparse, with one major exception: equation (0.2) retains its major role (both as a research problem and a part of practical computational tasks) in the highly important area of computing called computer algebra, which is widely applied to algebraic optimization and algebraic geometry computations.*"

Not only polynomiography would bring novel interest into the root-finding problem, but in the process it would bring interest in the visualization of much higher degree polynomials than degree 10 or 20. Even middle and high school students exposed to polynomiography always seem to want to go to higher and higher degree polynomials. This says that the nature of "application" effectively changes the need for solving polynomial equations.

Before providing a brief description of chapter contents, also offering a guideline to the reader I wish to acknowledge a number of books on polynomials. Essentially, all these books are directed at specialized or advanced audiences, each a very valuable source of information on polynomials, yet the collection gives evidence on how vast and versatile polynomials are. In a sense it would perhaps be very fair to claim that no one can ever master polynomials or polynomial equations.

The present book neither relies on these books, nor is it intended to be complementary to them. It is a book that is hoped to give a completely novel and popular view of polynomials and polynomial equations. No doubt there may lie many imperfections. Yet optimistically it should broaden the scope of polynomial root-finding to a level much beyond its predecessor books.

"Numerical Methods for Roots of Polynomials," McNamee (2007) a recent book with many results on the particular problem of polynomial root-finding from the iterative point of view. The author's monumental work of gathering an online bibliography of publications on root-finding contains over 8000 items (yet non-exhaustive), of which 50 were published in 2005.

"Polynomial and Matrix Computation," Bini and Pan (1994) deals with computations with polynomials and significant underlying algorithms.

"Polynomials," Barbeau (1989) deals with many topics on polynomials, from very elementary to more advanced.

"Complex Polynomials," Sheil-Small (2002), and "Geometry of Polynomials," Marden (1966) both deal with the geometric theory of polynomials and rational functions in the plane, bringing ideas from algebra, topology, and analysis. In particular they consider the location of zeros of polynomials and those of their derivatives.

"Polynomials," Prasolov (2004) deals with classical and modern algebraic point of viewpoint, including Galois theory.

"Polynomials and Polynomial Inequalities," Borwein and Erdélyi (1995) deals with analytic properties of polynomials as well as such topics as geometric properties, orthogonal polynomials, and inequalities.

"Fundamental Problems of Algorithmic Algebra," Yap (1999) though is not just on polynomials deals with topics from the point of view of computer algebra, the study of efficient algorithms for algebraic operations.

"The Fundamental Theorem of Algebra," Fine and Rosenberger (1997) gives several formal proofs of the theorem in the course of which the reader is introduced to complex analysis.

Finally, there are other books that deal with polynomials from the point of view of dynamical systems. We will refer to these later in the book.

The present book offers modern and novel perspectives into the theory and practice of the historical subject of polynomial root-finding, rejuvenating the field via polynomiography, a creative and novel computer visualization that renders spectacular images of a polynomial equation. Polynomiography will not only pave the way for new applications of polynomials in science and mathematics, but also in art and education. The book presents a thorough development of the Basic Family, arguably the most fundamental family of iteration functions for root-finding, deriving many surprising and novel theoretical results and practical applications such as: algorithms for approximation of roots of polynomials and analytic functions, polynomiography, bounds on zeros of polynomials, formulas for the approximation of π, characterizations of solutions of homogeneous recurrence relation, their polyhedral representation, visualizations associated with a homogeneous linear recurrence relation, connections to Voronoi regions, continued fractions, Fibonacci and Lucas numbers and their generalizations, even novel views into classical theorems such as the Gauss-Lucas theorem and the maximum modulus principle. These discoveries and a set of beautiful images provide new visions and appreciations to polynomials and recurrence relations. The book also describes some polynomiography-related experiences with educators, students, and artists, including middle and high school teachers and students, and a summary of their feedback with respect to the utility of polynomiography.

This book is for all mathematicians, scientists, advanced undergraduates and graduate students, any one who has come to deal with polynomials in formal settings. However, chapters or parts of this book are also for anyone with an appreciation for the connections between a fantastically creative art form and medium and its ancient mathematical foundations.

Many chapters can be read independently of each other, and relevant formulas for the Basic Family are repeated in different chapters. Some chapters include a list of research problems.

Chapter 1, derives algorithms for approximation of square roots and offers their visualization in the complex plane.

Chapter 2, makes use of the Fundamental Theorem of Algebra to derive a special case of Taylor's Theorem that is the genesis of iteration functions for polynomial root-finding algorithms and the Basic Family.

Chapter 3, offers an introduction to the Basic Family of iteration functions and to polynomiography.

Chapter 4, derives several equivalent formulations of the Basic Family.

Chapter 5, deals in much detail with the iterations of the Basic Family from the point of view of dynamical systems, developing the essential parts of the theory of iterations of rational functions, but keeping in view the Basic Family itself. In particular, this chapter should be a useful chapter for anyone interested in the essentials of the dynamics of iterations of rational functions.

Chapter 6, analyzes the properties of the fixed-points of the Basic Family.

Chapter 7, gives an algebraic derivation of the Basic Family and proves many characterizations and optimality results.

Chapter 8, defines and analyzes the Truncated Basic Family and a special case, called Halley Family.

Chapter 9, develops characterizations of solutions of homogeneous linear recurrence relations via the Basic Family and gives polyhedral representations.

Chapter 10, derives a significant determinantal generalization of Taylor's Theorem and Newton's method and its applications in approximation theory.

Chapter 11, develops a multipoint version of the Basic Family and analyzes their order of convergence.

Chapter 12, presents a computational comparison of some of the multipoint Basic Family members.

Chapter 13, develops a general determinantal lower bound and describes its specific applications in root-finding.

Chapter 14, develops new formulas for approximation of pi based on root-finding algorithms, specifically the use of the Basic Family.

Chapter 15, makes use of the Basic Family to derive a unique family of bounds on the roots of polynomials and analytic functions.

Chapter 16, defines a single geometric optimization related to polynomials and derives as its algebraic offsprings the Gauss-Lucas Theorem, the Maximum Modulus Principle, and novelties such as problems in computational geometry, and also introduces the Gauss-Lucas iteration function.

Chapter 17, describes polynomiography as algorithms for visualization of polynomial equations.

Chapter 18, offers techniques for the visualization of homogeneous linear recurrence relations via polynomiography, also develops a whole new family of iteration functions from the Basic Family, the "Induced Basic Family."

Chapter 19, offers applications of polynomiography in art, education, science and mathematics.

Chapter 20, revisits the approximation of square-roots via the Basic Family and develops connections to continued fractions and factorization of integers.

Chapter 21, offers further applications and extensions of the Basic Family and polynomiography.

Whenever possible the book complements the concepts within each chapter via polynomiography images.

Acknowledgements. I would like to thank many people who - in one form or another - have been helpful in the course of completion of this book over several years, or with respect to polynomiography activities addressed in this book, or with respect to the motivation of these activities, or for participation in these activities, or simply for their encouragements. Also, those who have motivated, inspired, collaborated, or have helped me in different ways. All have made me even more determined to advance the topic of polynomial root-finding and polynomiography, hoping to extend the popularity of the latter to the point of bringing it to K-16 education and beyond, perhaps even to the general public. Specifically, I mention the following individuals.

I would like to thank Michio Kaku for permission to quote him.

I would like to thank Curt McMullen for several electronic communications on the subject of dynamical systems and for clarifications of some technical issues, also Tan Lei and Mitsuhiro Shishikura.

I would like to thank Doron Zeilberger for his encouragements, inspirations, and enthusiasm over several years, and for his appreciation of both the technical contents and images in this book.

I would like to thank Yi Jin who as my Ph.D. student wrote a very nice thesis on the combinatorial aspects of polynomial root-finding. This book could only afford to highlight some the results which certainly deserve further attention.

I would like to thank Iraj Kalantari for having been a collaborator in several different ways and for being a significant catalyst and advisor in making possible some very fruitful activities, such as running teacher workshops in New Jersey and in Illinois, and more activities to come.

I would like to thank Fedor Andreev, an ongoing collaborator on some polynomiography-related activities that includes his expert programming of a version of polynomiography software.

I would like to thank Ali Maher for his support that made it possible to organize a very successful teacher workshop at his center CAIT at Rutgers.

I would like to thank several people who in the past few years have made it possible to give special presentations on polynomiography outside of Rutgers. They include Vasek Chvátal, Claude Bruter, Alfred Vendl, Rudolph Taschner, Steven Wolfram, and Iraj Kalantari for presentations at Condordia University (Montreal), Henrie Poincaré Institute (Paris), Angewandte Kunts (Vienna), MuseumsQuartier (Vienna), NKS conference (Boston), and University of Western Illinois (Macomb), respectively.

I also like to thank Reza Emamy-K for several invitations to University of Puerto Rico over the years and for his interest to bring polynomiography to Puerto Rico.

I would like to thank all those at Rutgers who have been supportive of my polynomiography activities over the years. In particular, I thank Kathleen Hull, Victoria Ukachukwu and Sarolta Takács for their encouragements and arrangements in the teaching of polynomiography courses such as First-Year seminars (now Byrne Seminars) and honors courses. Also, Kathleen Hull, Sara Harrington, and Joseph Consoli who made possible a very nice polynomiography exhibition at the Art Library of Rutgers, a photograph of which is featured in the book.

I also thank several K-12 mathematics educators for their invitation to make presentations at their middle and high schools, or their enthusiasm that has brought me closer to K-12 educators and a participant at K-12 teacher conferences in New Jersey, Illinois, Pennsylvania and beyond. They include Carol Ann Altamura (Randolph High School), Jan Gebert (Warren Hills Regional High School), and Mike Stern (Montgomery High School) in New Jersey, Candy Rosene (Western Illinois University) and Beth Shyrock (Macomb High School) in Illinois. It is a pleasure to thank some of their students as well for their enthusiasm which has given me more incentive and determination to proceed with polynomiography in the field of education. In particular, Elizabeth Sergison in New Jersey and Alexandria Munger in Illinois.

I would like to thank Lillian Schwartz with whom I have had many fruitful electronic communications regarding art, digital art, and polynomiography as art.

I would like to thank Hossein Talakoub for his efforts in arranging the production of a beautiful hand-woven carpet based on a polynomiography design featured in this book.

I would like to thank Jihui Zhao for doing a nice job in preparing most of the non-polynomiography figures of this book and for his expert technical assistance with Latex and Beamer.

I would like to thank Ms. Chionh, the Editor of World Scientific for countless communications and for her assistance in bringing this book to publication, and with the level of perfection I had hoped for.

I also thank Aileen Goh of World Scientific who used my desirable foreground and background polynomiography images to come up with the nice cover design.

Finally, I would like to thank my siblings, Iraj, Laleh and Khosrow who have given me much encouragements and moral support in pursuing my goals and obsessions with polynomiography. Even my niece and nephews who ended up encouraging me in their own ways. These include Sara, Leo, and Arshaun who assured me that even a four year old could learn to say "polynomiography."

Chapter 1

Approximation of Square-Roots and Their Visualizations: The Genesis

In this chapter we describe a very simple algebraic method for approximating the square-root of a given positive number. In contrast to geometric approaches which are based on graph of functions and the notion of differentiation, the method can be analyzed with high school algebra. We then extend it to the case of approximation of square roots of the given positive number via sequence of complex numbers. This simple method is in fact the genesis of the entire book. The chapter also considers the case of approximation of cube-roots.

1.1 Introduction

Consider a given positive number α. We construct a rational function $g(x)$, a quotient to two polynomials, such that for any *input* $x_0 > 0$ the *fixed point iteration*:

$$x_k = g(x_{k-1}) = g^2(x_{k-2}) = \cdots = g^k(x_0), \quad k = 0, 1, 2, \ldots \tag{1.1}$$

converges to $\sqrt{\alpha}$ having quadratic-order of convergence. In (1.1) $g^2(x) = g(g(x))$ and thus

$$g^k(x_0) = g(g(\cdots(g(x_0)))),$$

the composition of g with itself k times.

In particular, the convergence of x_k to α, and the continuity of $g(x)$ imply:

$$g(\alpha) = \alpha,$$

i.e. α is a *fixed point* of $g(x)$, justifying the name for the iteration. Later in the chapter we shall consider in detail fixed point iterations and rate of convergence.

Although the algebraic method to be described does not explicitly make use of the notion of derivative, in fact it turns out that $g(x)$ coincides with Newton's iteration function as applied to the quadratic polynomial $p(x) = x^2 - \alpha$. For an arbitrary polynomial $p(x)$, Newton's iteration function is defined as:

$$N(x) = x - \frac{p(x)}{p'(x)}. \tag{1.2}$$

Newton's method refers to the application of fixed point iteration to Newton's iteration function. The algebraic development of this chapter for the approximation of square-roots is in fact the genesis of this book. It leads to the development of a high-order family of iteration function for the approximation of roots of polynomials with real or complex coefficients. The properties of this family, called *Basic Family*, were in fact our inspiration behind the visualization of the root-finding process for polynomials. We call this visualization via the Basic Family *polynomiography*. Later in the book we will justify this definition and its extension to other family of iteration functions inspired by the Basic Family. Although an image coming from polynomiography could turn out to be *fractal*, polynomiography is capable of producing images of polynomials far beyond typical fractal images, including images that cannot be characterized as fractal.

Polynomiography, properly developed, could result in a powerful medium for doing art, for educational applications, and for discovering new mathematical properties on polynomials. The world of polynomials and approximation of their roots is an extremely fascinating world and has played a very significant and influential role in the history of mathematics and science. In this book we shall further explore and examine the polynomial root-finding problem while developing and analyzing the Basic Family and its properties. Furthermore, visualizations through polynomiography brings a new dimension into polynomials and polynomial equations.

Having offered a simple algebraic development of Newton's method for the approximation of square-roots, we will later extend the method to derive iteration functions of order higher than 2. More specifically, for each natural number $m \geq 2$, we derive a rational function $g_m(x)$, so that for any initial approximation $x_0 > \sqrt[2]{\alpha}$ the fixed point iteration

$$x_{k+1} = g_m(x_k), \tag{1.3}$$

enjoys an m-th order monotonic rate of convergence to α. The iterate x_k depends upon m, however for convenience in notation we suppress that dependency. For $m = 2$, $g_2(x)$ coincides with Newton's function $N(x)$ as applied to $p(x) = x^2 - \alpha$.

The convergence properties of these iteration functions are not limited to finding square-root of a positive real number α, but even so the iterations may begin with a complex number as the initial iterate, a *complex input*. In that case we write $p(z) = z^2 - \alpha$, where $z = x + iy$ is the complex variable with x and y as the real and imaginary parts, and $i = \sqrt{-1}$. Then any $z_0 = x_0 + iy_0$ can be selected as the *initial input* as long as it does not lie on the y-axis, i.e. $x_0 \neq 0$. The corresponding fixed point iterates will converge to the roots of the polynomial, namely $\pm\sqrt{\alpha}$, whichever happens to be closer to z_0.

In fact looking at the convergence of the iterates over the complex plane is what makes the problem interesting, giving rise to polynomiography. Polynomiography associates images to a given polynomial equation we call *polynomiographs*. Before describing the iteration functions mentioned above we give two polynomiographs of the process of approximation of the roots of $p(z) = z^2 - \alpha$. Figure 1.1 is a polynomiograph corresponding to the application of Newton's method as applied to this polynomial, within a rectangle centered at the origin. All the points of one color converge to the root closer to these points. In other words the set of all points that converge to a given root are the *Voronoi region* of that root. A point in the plane corresponds to a complex number and conversely. Given any finite set of points in the Euclidean plane, say $S = \{P_1, \ldots, P_t\}$, the Voronoi region of any particular point $P_i \in S$ is the set of all points in the plane that are closer to P_i than to any other point of S. The Voronoi regions turn out to be polygonal regions which may or may not be bounded.

The image in Figure 1.2 is a more interesting polynomiograph corresponding to $p(z) = z^2 - \alpha$. It employs the entire collection of the iteration functions $g_m(x)$, to be described later. The image appeared on the cover of ACM-SIGGRAPH Computer Graphics Quarterly (Kalantari (2004a)).

It may be surprising that the approximation of square-roots should result in such image, but in fact throughout the history of science and mathematics even the simple-looking task of approximating square-root of numbers has resulted in remarkable discoveries. This task was also the inspiration behind polynomiography.

Later in the chapter we will extend our algebraic approach to the case of approximating cube-root of a given positive number α, corresponding to the real root of $p(z) = z^3 - \alpha$. The general case of polynomials will be treated thoroughly throughout the book. Finally, although parallelization

Fig. 1.1 Polynomiograph of $p(z) = z^2 - 2$ based on Newton's iterations.

Fig. 1.2 A polynomiograph of $p(z) = z^2 - 2$ based on the collective use of a family of iteration functions.

is not a theme of this book, in this chapter we will show, in approximation of square-roots via parallel implementation of the high-order methods, it is theoretically possible to gain over Newton's method by an asymptotic speedup factor of 3. This reveals another interesting feature of the high-order methods.

1.2 A Simple Algebraic Method for Approximation of Square-Roots

Consider approximating $\theta = \sqrt{\alpha}$, where α is a natural number. Starting with an initial rational number x_0, we wish to approximate θ with a sequence of numbers x_k having the property that

$$(x_{k+1} - \theta) = \gamma(x_k)(x_k - \theta)^2, \tag{1.4}$$

where $\gamma(x_k)$, if possible, is to be selected so that x_{k+1} remains a rational number.

Expanding the above equation we get

$$x_{k+1} = \gamma(x_k)(x_k^2 + \alpha) - [2\gamma(x_k)x_k - 1]\theta. \tag{1.5}$$

Thus it suffices to choose

$$\gamma(x_k) = \frac{1}{2x_k}. \tag{1.6}$$

Substituting (1.6) into (1.5) gives

$$x_{k+1} = \frac{x_k^2 + \alpha}{2x_k} \tag{1.7}$$

which interestingly happens to coincide with Newton's iterate as applied to the polynomial $p(x) = x^2 - \alpha$.

Since the notion of differentiation is not necessary in deriving the sequence of iterates, the method can possibly be taught to middle school and high school students.

From (1.4) and (1.5) it is easy to prove when $x_0 > \theta$, the sequence of iterates has quadratic-order of converges to θ. More specifically:

$$\lim_{k\to\infty} \frac{(x_{k+1} - \theta)}{(x_k - \theta)^2} = \frac{1}{2\theta}. \tag{1.8}$$

In the next section we will prove this and in more generality. The quadratic-order of convergence implies that when x_0 is sufficiently close to θ, each iteration would roughly double the accuracy in the number of digits.

1.3 High-Order Algebraic Methods for Approximation of Square-Roots

Let α be a given natural number which is not a perfect square. Then $\theta = \sqrt{\alpha}$ is an irrational number we wish to approximate. Let $m \geq 2$ be a given natural number. Given an initial rational number x_0, we wish to obtain a sequence $\{x_k\}_{k=0}^{\infty}$ converging to θ and satisfying

$$x_{k+1} - \theta = \gamma_m(x_k)(x_k - \theta)^m, \tag{1.9}$$

where $\gamma_m(x)$, if possible, is to be selected so that x_{k+1} remains a rational number. If successful, the following gives the desired iteration function

$$g_m(x) \equiv \theta + \gamma_m(x)(x_k - \theta)^m. \tag{1.10}$$

As seen in the previous section, for $m = 2$ this approach gave rise to (1.7) and the corresponding iteration function coincided with Newton's function. We now extend the approach to the general case of $m \geq 2$.

We write

$$(x - \theta)^m = P_0^m(x) + \theta P_1^m(x),$$

where

$$P_0^m(x) = \frac{1}{2}\big[(x+\theta)^m + (x-\theta)^m\big] = \sum_{i\ even} \binom{m}{i} \theta^i x^{m-i}, \tag{1.11}$$

$$P_1^m(x) = -\frac{1}{2\theta}\big[(x+\theta)^m - (x-\theta)^m\big] = -\sum_{i\ odd} \binom{m}{i} \theta^{i-1} x^{m-i}. \tag{1.12}$$

Note that $P_0^m(x)$ and $P_1^m(x)$ are polynomials of degrees m and $m-1$, respectively, having coefficients depend on $\theta^2 = \alpha$. Thus, to have the function

$$g_m(x) \equiv \theta + \gamma_m(x)(P_0^m(x) + \theta P_1^m(x)) \tag{1.13}$$

independent of θ, it suffices to choose

$$\gamma_m(x) = -\frac{1}{P_1^m(x)}, \tag{1.14}$$

from which we get

$$g_m(x) = \gamma_m(x) P_0^m(x) = -\frac{P_0^m(x)}{P_1^m(x)} = \theta\left(\frac{(x+\theta)^m + (x-\theta)^m}{(x+\theta)^m - (x-\theta)^m}\right). \tag{1.15}$$

The convergence of Newton's iterates, the sequence $\{g_2(x_k)\}_{k=0}^{\infty}$, is well-known and is analyzed in practically every numerical analysis textbook.

For $m = 3$ we get

$$g_3(x) = \frac{(x^3 + 3\alpha x)}{(3x^2 + \alpha)}. \tag{1.16}$$

It can easily be verified that θ is a fixed point of $g_3(x)$, and that $g'(\theta) = g''(\theta) = 0$. Then since from Taylor's Theorem we can have

$$g_3(x) = g_3(\theta) + g_3'(\theta)(x-\theta) + \frac{g_3''(\theta)}{2!}(x-\theta)^2 + \frac{g_3'''(\xi)}{3!}(x-\theta)^3,$$

with ξ lying between x and θ, it follows that the iterates of $g_3(x)$ has a cubic-order of convergence to θ.

Indeed as it turns out $g_3(x)$ is another important iteration function, the so-called Halley's method as applied to the polynomial $p(x) = x^2 - \alpha$. Halley's method credited to the astronomer Edmund Halley, is much less known in the literature than Newton's method. It is a very interesting method and will be explored in the book in much more detail later.

In the next section we shall see that for all $m > 2$, the fixed point iterations of $g_m(x)$ enjoy the same global convergence properties as the case of $m = 2$, while having an m-th order rate of convergence.

1.4 Convergence Analysis

In order to analyze the convergence of fixed point iterates of $g_m(x)$ we first derive a closed form for the k-th iterate

$$x_k = g_m(x_{k-1}). \tag{1.17}$$

From (1.15) we may write

$$g_m(x) = \theta\left[\frac{1 + R^m(x)}{1 - R^m(x)}\right], \quad m \geq 2, \tag{1.18}$$

where

$$R(x) = \frac{x - \theta}{x + \theta}. \tag{1.19}$$

Theorem 1.1. *For any $x_0 > 0$ the sequence $\{g_m(x_0) : m = 2, 3, \dots\}$ converges to θ.*

Proof. Since $\theta > 0$, $|R(x_0)| < 1$. Thus $R^m(x_0)$ converges to zero as m approaches infinity. Hence the proof. □

The following result shows that the family of functions $\{g_m(x)\}_{m=2}^{\infty}$, are closed under the operation of composition, denoted by $\circ$.

Lemma 1.1. *Consider the family of functions $\{g_m\}_{m=2}^{\infty}$. For any natural numbers r and s, and any $x \neq 0$ we have*

$$g_r \circ g_s(x) = g_{rs}(x) = g_s \circ g_r(x). \tag{1.20}$$

Proof. Let

$$d = (x+\theta)^s - (x-\theta)^s.$$

We see that

$$g_s(x) \pm \theta = \Big(\frac{2\theta}{d}\Big)(x \pm \theta)^s,$$

and is well-defined. Thus

$$(g_s(x)+\theta)^r \pm (g_s(x)-\theta)^r = \Big(\frac{2\theta}{d}\Big)^r \big[(x+\theta)^{rs} \pm (x-\theta)^{rs}\big],$$

from which the proof of the first equality in (1.20), and clearly the second equality follows. □

By repeated application of the lemma, and since

$$x_k = g_m(x_{k-1}) = g_m \circ g_m \circ \cdots \circ g_m(x_0) = g_m^k(x_0),$$

the k-fold composition of g_m with itself, we conclude:

Theorem 1.2.

$$x_k = g_m^k(x_0) = g_{m^k}(x_0) = \theta\left(\frac{(x_0+\theta)^{m^k} + (x_0-\theta)^{m^k}}{(x_0+\theta)^{m^k} - (x_0-\theta)^{m^k}}\right).$$

□

In the special case where $m = 2$ the theorem gives a closed form for the k-th iterate of Newton's. This special case was proved in Potra and Ptak (1984). Our derivation is more general and very simple, relying only on Lemma 1.1.

In this chapter we will describe other representations for $g_m(x_0)$ and in subsequent chapters will study even more interesting interpretations, relating it to continued fractions.

Theorem 1.3. *Let $x_0 > \theta = \sqrt{\alpha}$. For any $m \geq 2$, consider the fixed point iteration $x_{k+1} = g_m(x_k)$, $k \geq 0$. Then*

$$x_k > x_{k+1}, \quad \forall \ k \geq 0, \quad \text{and} \quad \lim_{k\to\infty} x_k = \theta. \tag{i}$$

$$\lim_{k\to\infty} \frac{(x_{k+1}-\theta)}{(x_k-\theta)^m} = \frac{1}{(2\theta)^{m-1}}. \tag{ii}$$

Proof. From Theorem 1.1 for each *real input* $x_0 > 0$, the sequence $\{g_m(x_0)\}_{m=2}^{\infty}$ converges to θ. But if $x_0 > \theta$, then $0 < R(x_0) < 1$. From this it follows:

$$(1 + R^m(x_0))(1 - R^{m+1}(x_0)) > (1 - R^m(x_0))(1 + R^{m+1}(x_0)). \tag{1.21}$$

From (1.21) and the formula for $g_m(x)$ in (1.18), it is easy to see that for all $m \geq 2$,

$$g_m(x_0) > g_{m+1}(x_0). \tag{1.22}$$

Since $\{x_k = g_m(x_{k-1})\}_{k=1}^{\infty}$ is a subsequence of $\{g_m(x_0)\}_{m=0}^{\infty}$, (1.22) implies of (i). The proof of (ii) follows from (1.9), (1.14), and since

$$\gamma_m(\theta) = \frac{-1}{P_1^m(\theta)} = \frac{1}{(2\theta)^{m-1}}.$$

□

Although we have considered the approximation $\theta = \sqrt{\alpha}$ where α is a natural number, clearly all the convergence properties extend to the case where α is any positive real number.

1.5 Approximation of Square-Roots from Complex Inputs

In the previous sections we have considered approximation of square-roots of a given positive number starting from a real initial input and derived the family $g_m(x)$. In this section we will consider this family in more generality where we replace the real variable x with a complex variable $z = x + iy$ with x and y real, and $i = \sqrt{-1}$. Thus we assume that α is a positive real number, $\theta = \sqrt{\alpha}$, and for any $m \geq 2$ we let

$$g_m(z) = \theta\left(\frac{(z+\theta)^m + (z-\theta)^m}{(z+\theta)^m - (z-\theta)^m}\right) = \theta\left[\frac{1+R^m(z)}{1-R^m(z)}\right], \tag{1.23}$$

where

$$R(z) = \frac{z-\theta}{z+\theta}. \tag{1.24}$$

Theorem 1.4. *Let $z_0 = x_0 + iy_0$ be a complex number with $x_0 \neq 0$. Let θ_0 be the root of $z^2 - \alpha$ closer to z_0. Then*

$$\lim_{m \to \infty} g_m(z_0) = \theta_0. \tag{i}$$

Moreover, for each fixed $m \geq 2$, the sequence of fixed point iterates

$$z_{k+1} = g_m(z_k)$$

converges to θ_0 having order m. More precisely:

$$\lim_{k \to \infty} \frac{(z_{k+1} - \theta_0)}{(z_k - \theta_0)^m} = \frac{1}{(2\theta_0)^{m-1}}. \tag{ii}$$

Proof. Assume $\theta_0 = \theta$. Then $|R(z_0)| < 1$. Thus, $|R(z_0)^m| = |R(z_0)|^m$ converges to zero. Hence, $R^m(z_0)$ converges to zero. This implies the convergence of $g_m(z_0)$ to θ_0. If $|R(z_0)| > 1$, then $\theta_0 = -\theta$. From (1.23) we may write

$$g_m(z_0) = \theta_0 \left[\frac{\frac{1}{R^m(z_0)} + 1}{\frac{1}{R^m(z_0)} - 1} \right]. \tag{1.25}$$

Now since $1/R^m(z_0)$ converges to zero, from (1.25) $g_m(z_0)$ converges to θ. Hence (i) is proved.

To prove (ii), we note that the closed form of g_m in (1.15) as well as Theorem 1.2 remain valid over the complex numbers. Thus

$$z_{k+1} = g_m(z_k) = g_{m^{k+1}}(z_0),$$

implying that the sequence of fixed points iterates is a subsequence of $\{g_m(z_0)\}_{m=2}^{\infty}$. Thus it too converges to θ_0. Moreover, from (1.23) it follows that

$$\frac{(z_{k+1} - \theta_0)}{(z_k - \theta_0)^m} = \frac{2\theta_0}{(z_k + \theta_0)^m - (z_k - \theta_0)^m}.$$

Since z_k converges to θ_0, the above implies (ii). □

Theorem 1.4 in particular justifies the polynomiograph of Figure 1.1 corresponding to the application of Newton's method in solving the polynomial equation $z^2 - 2 = 0$.

For any $\alpha > 0$ the corresponding polynomiograph for $z^2 - \alpha = 0$ will remain similar to the one in Figure 1.1. In particular, the basin of attraction of each root, i.e. the set of all points whose fixed point iterates converge to that root, remains to be its Voronoi region, namely the Euclidean half-plane that contains it, excluding the y-axis.

It is interesting to consider points on the y-axis and the behavior of Newton's method. Since $z = iy$, we get

$$N(iy) = iy - \frac{(iy-2)^2}{2iy} = i\frac{y^2-2}{2y}.$$

Thus if $z_0 = iy_0$ for some $y_0 \neq 0$, all the fixed point iterates are confined to the y-axis, hence non-convergent to square-roots of α. In an intuitive sense one may expect this type of behavior since such initial iterate is equidistant to the square-roots and cannot be attracted to a particular one through any subsequent iterate. Each root attracts a point on the y-axis only with the same gravity as the other root. Hence the fixed point iterates exhibit erratic behavior. Even cyclic behavior could occur. In subsequent chapters we shall examine the dynamics in much more generality.

The theoretical study of the behavior of Newton's method in the complex plane dates back to Cayley (1897) in the late nineteenth century. The observation that Newton's method can be extended to the complex plane is a crucial step in itself. Complex numbers form a field and the four elementary operations: addition, subtraction, multiplication, and division, are all that is needed in extending Newton's method for the approximation of complex roots of real or complex polynomials.

It is fair to assume that Cayley could have visualized the image in Figure 1.1, even though he did not have the luxury of having the computer technology available to him. However, when Cayley considered the behavior of Newton's method for the polynomial equation $z^3 - 1 = 0$, he could not prove similar results. This polynomial has three roots

$$\theta_1 = 1, \quad \theta_2 = -\frac{1}{2} + i\frac{\sqrt{3}}{2}, \quad \theta_3 = -\frac{1}{2} - i\frac{\sqrt{3}}{2}.$$

These are known as cube-roots of unity.

Given the behavior of Newton's for $z^2 - 2$, one might have expected to see the polynomiograph of Figure 1.3, as the result of applying Newton's method to this polynomial, with the basin of attraction of each root is its Voronoi region. However, the actual basins happens to be much more complicated. The corresponding polynomiograph is given in Figure 1.4. The boundaries of the basins happen to exhibit fractal behavior. The term fractal was coined by Mandelbrot (see Mandelbrot (1983)) who popularized the study and visualization of iterative methods, and in particular visualization of the famous *Mandelbrot set*. Later in the chapter we consider the behavior of Newton's method for $z^3 - 1$ in more detail.

Fig. 1.3 A coloring of Voronoi regions of roots of $p(z) = z^3 - 1$. The image is not based of Newton's method.

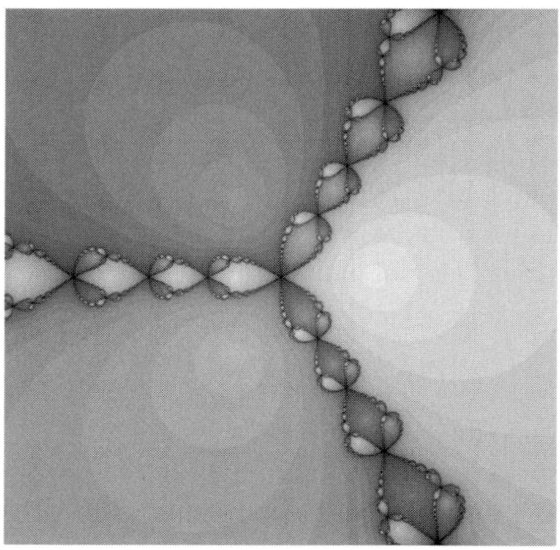

Fig. 1.4 A polynomiograph of $z^3 - 1$ based on Newton's iteration.

1.6 The Basic Sequence and Fixed Point Iterations

Definition 1.1. Given a complex number z_0 we define the sequence $\{g_m(z_0)\}_{m=2}^{\infty}$ as the *Basic Sequence* at z_0.

From the results in previous sections, the basic sequence may be divided into the union of subsequence corresponding to the fixed point iterates of $g_2(z)$, $g_3(z)$, $g_4(z)$, and so on, where all begin at the same initial input z_0. More specifically, we may write this as

$$\{g_m(z_0)\}_{m=2}^{\infty} = \{g_{2^k}(z_0)\}_{k=1}^{\infty} \bigcup \{g_{3^k}(z_0)\}_{k=1}^{\infty} \bigcup \{g_{4^k}(z_0)\}_{k=1}^{\infty} \cdots .$$

Since the iterates of $g_4(z)$ form a subsequence of those of $g_2(z)$, the above may be written as:

$$\{g_m(z_0)\}_{m=2}^{\infty} = \bigcup_{\{p,\ prime\}} \{g_{p^k}(z_0)\}_{k=1}^{\infty}.$$

In particular, Newton's iterates can be viewed as the subsequence $\{g_{2^k}(z_0)\}_{k=1}^{\infty}$. Later in the book we shall define the Basic Sequence in much more generality, and examine the relationship between the particular one above and continued fractions.

1.7 Determinantal Representation of High-Order Iteration Functions and Basic Sequence

In this section we present, without proof, a determinantal representation of the family of iteration functions $g_m(z)$, $m \geq 2$, as well the Basic Sequence. In particular, we will give several representations for $\sqrt{2}$.

Let $p(z) = z^2 - \alpha$, $\alpha > 0$. Set $D_0(z) \equiv 1$, and for each natural number $m \geq 1$, define

$$D_m(z) = \begin{vmatrix} p'(z) & \frac{p''(z)}{2!} & 0 & \cdots & 0 \\ p(z) & p'(z) & \frac{p''(z)}{2!} & \ddots & \vdots \\ 0 & p(z) & p'(z) & \ddots & 0 \\ \vdots & \ddots & \ddots & p'(z) & \frac{p''(z)}{2!} \\ 0 & \cdots & 0 & p(z) & p'(z) \end{vmatrix} = \begin{vmatrix} 2z & 1 & 0 & \cdots & 0 \\ z^2-\alpha & 2z & 1 & \ddots & \vdots \\ 0 & z^2-\alpha & 2z & \ddots & 0 \\ \vdots & & \ddots & 2z & 1 \\ 0 & \cdots & 0 & z^2-\alpha & 2z \end{vmatrix}$$

where $|\cdot|$ denotes the determinant. The $m \times m$ matrix corresponding to $D_m(z)$ is Toeplitz, i.e. having identical entry along each of the diagonals. The matrix is also tridiagonal, i.e. $a_{ij} = 0$, for all i, j with $i > j+1$ or $j > i+1$.

For each $m \geq 2$, define

$$B_m(z) \equiv z - p(z)\, \frac{D_{m-2}(z)}{D_{m-1}(z)}.$$

Theorem 1.5. *For each $m \geq 2$,*

$$B_m(z) = g_m(z) = \frac{1 + R^m(z)}{1 - R^m(z)}.$$

In particular, $\{B_m(z_0)\}_{m=2}^{\infty}$ is the Basic Sequence at z_0.

In subsequent chapters we shall prove the theorem and in much more generality. For now we mention one particular implication, curious representations of $\sqrt{2}$. Let

$$D_m(1) = \begin{vmatrix} 2 & 1 & 0 & \dots & 0 \\ -1 & 2 & 1 & \ddots & \vdots \\ 0 & -1 & 2 & \ddots & 0 \\ \vdots & \ddots & \ddots & 2 & 1 \\ 0 & \dots & 0 & -1 & 2 \end{vmatrix}, \quad D_m(2) = \begin{vmatrix} 4 & 1 & 0 & \dots & 0 \\ 2 & 4 & 1 & \ddots & \vdots \\ 0 & 2 & 4 & \ddots & 0 \\ \vdots & \ddots & \ddots & 4 & 1 \\ 0 & \dots & 0 & 2 & 4 \end{vmatrix}$$

$$D_m(1+i) = \begin{vmatrix} 2+2i & 1 & 0 & \dots & 0 \\ -2+2i & 2+2i & 1 & \ddots & \vdots \\ 0 & -2+2i & 2+2i & \ddots & 0 \\ \vdots & \ddots & \ddots & 2+2i & 1 \\ 0 & \dots & 0 & -2+2i & 2+2i \end{vmatrix}$$

From Theorems 1.4 and 1.5 we get

Corollary 1.1.

$$\sqrt{2} = 1 + \lim_{m \to \infty} \frac{D_{m-2}(1)}{D_{m-1}(1)}$$

$$= 2 - 2 \lim_{m \to \infty} \frac{D_{m-2}(2)}{D_{m-1}(2)} = 1 + i - (-2 + 2i) \lim_{m \to \infty} \frac{D_{m-2}(1+i)}{D_{m-1}(1+i)}.$$

We see that $B_2(1) = B_2(2) = 3/2$ and $B_2(1+i) = 1.0$. The reader can verify that for $m = 3, 4, 5$, and 9 we get the improved approximations given in Table 1.1.

Table 1.1 Approximation of $\sqrt{2}$

m	3	4	5	9
$B_m(1)$	1.4	1.41667	1.41379	1.41421
$B_m(2)$	1.42857	1.41667	1.41463	1.41421
$B_m(1+i)$	$1.4 - .2i$	1.5	$1.41379 + .034482i$	$1.41421 + .00101i$

1.8 Visualizations in Approximation of Square-Roots

One way to visualize the process of approximating square-root of a number $\alpha > 0$ via Newton's method, is through ordinary graphing. We draw the function $p(x) = x^2 - \alpha$ and the iterates on the x-axis. There is a well-known geometric interpretation of the location of the iterates for an arbitrary real polynomial: given x_k, x_{k+1} is the intersection of the tangent line to $p(x)$ at x_k, and the x-axis. The x-intercept of the tangent line is the solution to the equation:

$$p(x_k) + p'(x_k)(x - x_k) = 0.$$

The solution for x gives the next Newton iterate:

$$x \equiv x_{k+1} = x_k - \frac{p(x_k)}{p'(x_k)} = N(x_k).$$

Figure 1.5 offers this visualization for $\alpha = 2$. Depending upon the initial point x_0 being positive or negative, the fixed point converges to $\sqrt{2}$ and $-\sqrt{2}$, respectively. The figure offers insight on the relationship between the iterates, tangent lines, and the nature of convergence.

For general $m \geq 2$, a Newton-like geometric interpretation of the high-order methods for approximation of roots of $p(x) = x^2 - 2$ is not necessarily possible. Viewing the approximation of polynomials roots over the complex plane is a more challenging task, even for such a simple complex polynomial such as $p(z) = z^2 - 2$.

To begin with, we would have to remember that the polynomial $p(z)$ maps the complex number $z = x + iy$, corresponding to the point (x, y) in the complex plane, to

$$z^2 - 2 = (x^2 - y^2 - 2) + i2xy,$$

another point in the complex plane. Thus we need four dimensions to view the graph of this mapping, which is not possible, given that our physical visualization capabilities are limited to three dimensions only.

To remedy this situation we have some options. One option is to replace the image $p(z)$ of a point z by its modulus, $|p(z)|$, a real number. In this

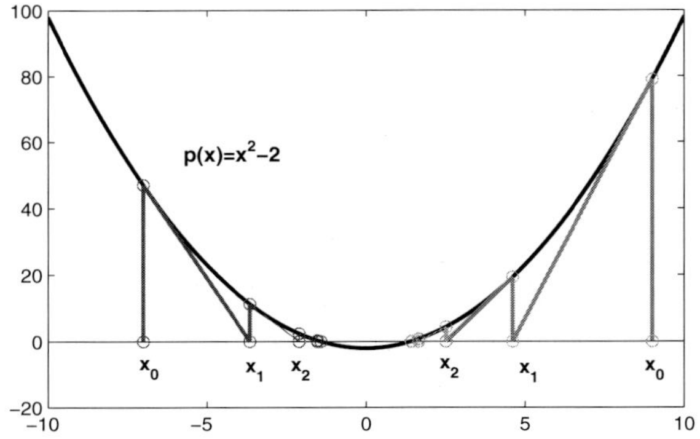

Fig. 1.5 Graph of Newton's iterates for approximation of square-root of two.

fashion we end up with a mapping that takes a complex number $z = x + iy$ (the point (x, y) in the complex plane) to the real number

$$|p(z)| = |(x^2 - y^2 - 2) + i2xy| = [(x^2 - y^2 - 2)^2 + (2xy)^2]^{1/2}.$$

This makes it possible to give 3D views of the corresponding surface. Figure 1.6 gives one such view. Notice that as one might anticipate there are two valleys and two points where the surface would be touching the xy-plane, the points $(1, 0)$ and $(-1, 0)$. Figure 1.7 gives a front view of the surface. When $y = 0$, as we should expect, the graph corresponds to that of $|p(x)| = |x^2 - 2|$.

For a general polynomial, or even the one in consideration, neither a 3D graph such as the one in Figure 1.6, nor the Newton iterates are easy to work with. Moreover, a front or cross-sectional view such as Figure 1.7 does not offer much insight into the behavior of the map $p(z)$, let alone the Newton's iterates.

Instead when dealing with iteration functions such as Newton's, which are maps from the complex plane into itself, we can abandon the range of the map and simply consider the iterates themselves. Then by appropriate coloring of the points in the complex plane we can gain much insight and information about the behavior of the iterates. This type of visualization is the basis of most fractal images, seen in the literature, books, and the Internet in abundance. Figure 1.1, discussed earlier, is one such figure. Figures such as this and Figure 1.2, called polynomiographs, are the basis

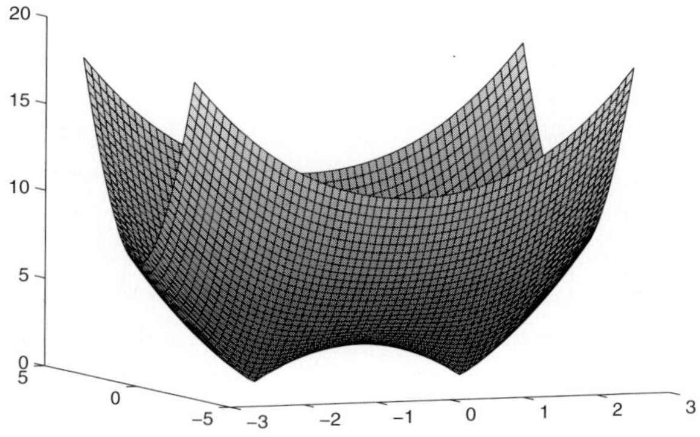

Fig. 1.6 Graph of modulus function corresponding to $p(z) = z^2 - 2$ having two minima at $(1, 0)$ and $(-1, 0)$.

of visualizations we shall be concerned with in this book. The reason to address these images as polynomiographs, as opposed to fractal which is too broad and vague will be addressed in detail.

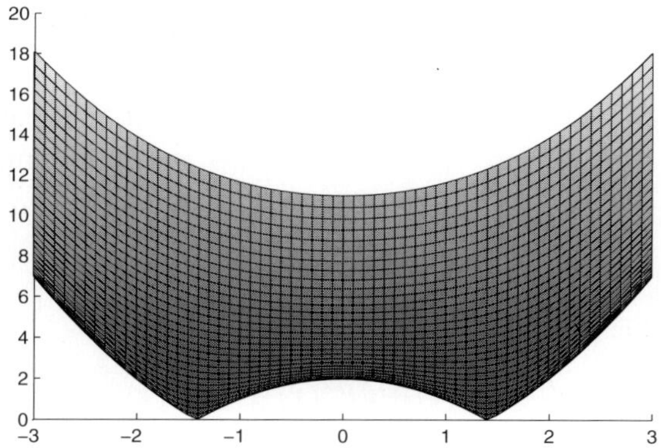

Fig. 1.7 Front view of the surface $|p(z)|$.

1.9 High-Order Methods for Approximation of Cube-Roots

In this section we consider high-order methods for the approximation of cube-roots. Consider the polynomial $p(x) = x^3 - \alpha$, where α is a natural number, not a perfect cube. In other words the real root $\theta = \sqrt[3]{\alpha}$, as can easily be proved, is an irrational number. We are interested in approximation of θ via high-order methods.

Let $m \geq 2$ be a fixed natural number. Analogous to the case of square-root, we ask if it is possible to find a rational function $\gamma_m(x)$ so that the function $\theta + \gamma_m(x)(x - \theta)^m$ depends only on x, and not on θ.

We can easily verify that this is not possible even for $m = 2$. So we consider the existence of rational functions $\gamma_m(x)$ and $\gamma_{m+1}(x)$ so that the function

$$g_m(x) \equiv \theta + \gamma_m(x)(x - \theta)^m + \gamma_{m+1}(x)(x - \theta)^{m+1}, \tag{1.26}$$

depends only on x.

Remark 1.1. We have misused the notation in that $g_m(x)$ here represents a different iteration function than the one in previous sections. But the reader should bear with this since ultimately in subsequent chapters we shall arrive at a general function for all polynomials, not just the simple ones considered here. But the developments for these special polynomials offer much insight into the discovery of the general iteration function we will seek.

To investigate the existence of $g_m(x)$ above, we make use of the fact that $\theta^3 = \alpha$, thus for any exponent $i = 3q + r$ with remainder $r < 3$ we have

$$\theta^i = \alpha^q \theta^r.$$

Using this and expanding we may write

$$(x - \theta)^k = P_0^k(x) + \theta P_1^k(x) + \theta^2 P_2^k(x), \quad k \geq 1 \tag{1.27}$$

in such a way that $P_i^k(x)$ is a polynomial in x with coefficients which are integral multiples of α. Then writing (1.27) for $k = m, m + 1$ and substituting in (1.26), then regrouping the terms as a polynomial in θ, and finally setting the coefficients of θ and θ^2 equal to zero results in the following system of nonlinear equations:

$$\begin{cases} 1 + \gamma_m(x)P_1^m(x) + \gamma_{m+1}(x)P_1^{m+1}(x) = 0, \\ \gamma_m(x)P_2^m(x) + \gamma_{m+1}(x)P_2^{m+1}(x) = 0. \end{cases}$$

Solving the above system of 2×2 equations for $\gamma_m(x)$ and $\gamma_{m+1}(x)$, and substituting the results in (1.26), as the reader can verify, we get

$$g_m(x) = -\frac{P_0^m(x)P_2^{m+1}(x) - P_2^m(x)P_0^{m+1}(x)}{P_1^m(x)P_2^{m+1}(x) - P_2^m(x)P_1^{m+1}(x)}. \tag{1.28}$$

The justification in setting the coefficients of θ and θ^2 equal to zero lies in Theorem 1.6, stated and proved below, which is based on the linear independence of 1, θ, and θ^2. The theorem also makes it possible to derive a more convenient formula for $g_m(x)$ so that it depends only on P_i^m, $i = 0, 1, 2$. We will do that next.

Since $(x-\theta)^{m+1} = (x-\theta)^m(x-\theta)$ and $\theta^3 = \alpha$, from (1.27) we may write

$$(x-\theta)^{m+1} = P_0^{m+1}(x) + \theta P_1^{m+1}(x) + \theta^2 P_2^{m+1}(x)$$

$$= (x-\theta)[P_0^m(x) + \theta P_1^m(x) + \theta^2 P_2^m(x)]$$

$$= [xp_0^m(x) - \alpha P_2^m(x)] + \theta[xP_1^m(x) - P_0^m(x)] + \theta^2[xP_2^m(x) - P_1^m(x)]. \tag{1.29}$$

Equating the coefficient of like powers of θ (see Theorem 1.6), we get

$$P_0^{m+1}(x) = [xP_0^m(x) - \alpha P_2^m(x)],$$

$$P_1^{m+1}(x) = [xP_1^m(x) - P_0^m(x)],$$

$$P_2^{m+1}(x) = [xP_2^m(x) - P_1^m(x)].$$

Substituting these into (1.28), we get the following alternative and more convenient formula

$$g_m(x) = \frac{\left(P_2^m(x)\right)^2\alpha - P_0^m(x)P_1^m(x)}{\left(P_1^m(x)\right)^2 - P_0^m(x)P_2^m(x)}. \tag{1.30}$$

From either formula (1.28) or (1.30) for $g_m(x)$ the reader can readily obtain

$$g_2(x) = \frac{2x^3 + \alpha}{3x^2},$$

$$g_3(x) = \frac{x^4 + 2\alpha x}{2x^3 + \alpha},$$

$$g_4(x) = \frac{4x^7 + 19\alpha x^4 + 4\alpha^2 x}{10x^6 + 16\alpha x^3 + \alpha^2}.$$

As in the case of square-roots, it is easy to see that $g_2(x)$ coincides with the Newton's function $N(x) = x - p(x)/p'(x)$, but as applied to $p(x) = x^3 - \alpha$. As in the case of square-roots, we can prove that for a fixed $m \geq 2$, and a given $x_0 > \theta$, the fixed point iterates $x_{k+1} = g_m(x_k)$, are well-defined and monotonically converge to θ, enjoying an m-th order rate of convergence. However, unlike the case of square-roots, the corresponding functions g_m are not closed under composition. Thus we will need to give a direct proof of convergence properties. In Theorem 1.8 we prove this, but only for the cases where $m = 2, 3, 4$.

Theorem 1.6. *Let $q_0(x)$, $q_1(x)$, $q_2(x)$ be polynomials with rational coefficients. Suppose*

$$q_0(x) + \theta q_1(x) + \theta^2 q_2(x) = 0, \quad \forall \ x.$$

Then the polynomials are identically zero.

Proof. To prove that $q_i(x)$ is identically zero for each i we make use of the fact that the numbers 1, θ and θ^2 are algebraically independent over the field of rational numbers. This means if

$$a_0 + a_1\theta + a_2\theta^2 = 0, \tag{1.31}$$

with each a_i a rational number, then

$$a_0 = a_1 = a_2 = 0.$$

This follows from the general algebraic properties of irreducible polynomials and their roots. However, in this case we give an elementary proof.

Assume (1.31) holds. Without loss of generality we may assume a_i is integer for $i = 0, 1, 2$. If $a_2 = 0$, then θ is rational. If $a_1 = 0$, then θ^2 is rational, hence its square is rational. But this implies that θ is rational. If $a_0 = 0$, then θ is rational. Each leads to a contradiction. Thus no a_i can be zero. Since θ is a root of the quadratic polynomial $a_0 + a_1x + a_2x^2$, we get

$$\theta = -u + \sqrt{v}, \tag{1.32}$$

for some integers u and v. Since θ is irrational, $\sqrt{v}$ must be irrational. But cubing both sides of (1.32), as can readily be verified, implies $v = -3u^2$, making $\sqrt{v}$ an imaginary number, a contradiction. Hence, $a_i = 0$, $i = 0, 1, 2$.

Thus, for each rational number r, $q_i(r) = 0$, for $i = 0, 1, 2$. But since $q_i(x)$ is a polynomial, by the Fundamental Theorem of Algebra it cannot have more roots than its degree, unless identically zero. □

As in the case of square-roots one can consider Newton's iteration in the complex plane. The corresponding polynomiograph, for $z^3 - 1$ is given in Figure 1.4. The complex repeated patterns and self-similarity, justifies why Cayley could not have characterized Newton's behavior. This image is quite familiar in the literature and has appeared in numerous publications. Note that while the basins of attraction are complicated, there is some resemblance to the Voronoi regions of Figure 1.3. Thus, in the case of $p(z) = z^3 - 1$, we may think of Newton's method as a way of approximating the Voronoi regions of the cube-roots of unity.

If in the case $z^3 - 1$ we count every other iterate generated by $N(z)$, we get a 4-th order sequence converging to θ. This sequence however does not coincide with those of $g_4(z)$. In particular, if $z_0 = x_0 = \alpha = 2$, then $g_2(g_2(2)) = 35/27 \neq g_4(2) = 1.28$. In fact comparing the iterates of $g_4(z)$ with every other Newton iterate, it can be verified that $\theta < x_k = g_4(x_{k-1}) < x'_k = N(N(x'_{k-1}))$ for any $x_0 = x'_0 > \theta$, so that x_k is consistently a better approximation to θ than x'_k.

If we were to give the polynomiograph of $p(z) = z^3 - 1$ with respect to the iteration function $N^2(z) = N(N(z))$, the basins of attraction should be identical to the one given in Figure 1.4 (ignoring the hues which are to signify the number of steps in convergence to a root). This can be formally proved and in more generality. We do that next.

Theorem 1.7. *Let θ be a root of a complex polynomial $p(z)$. Then basin of attraction of θ with respect to fixed point iterations of N, N^2, N^3, ... are all identical.*

Proof. Suppose z_0 is in the basin of attraction of θ with respect to N. Thus

$$\{z_k = N^k(z_0)\}_{k=0}^{\infty}$$

converges to θ. For any fixed natural number $m \geq 2$, the fixed point iterates of N^m form a subsequence of those of N, hence convergent to θ. Thus the basin of attraction of θ with respect to N^m is a subset of basin of attraction with respect to N.

To prove the converse we will use the continuity of $N(z)$ at θ. Assume that z_0 is in the basin of attraction of θ with respect to N^m for some $m \geq 2$. In other words the subsequence

$$\{z_{mk} = N^{mk}(z_0)\}_{k=0}^{\infty}$$

converges to θ. We will use this to show that for each $j = 1, 2, \ldots, m-1$ the subsequence

$$\{z_{mk+j} = N^{mk+j}(z_0)\}_{k=0}^{\infty}$$

also converges to θ. But

$$\{z_k = N^k(z_0)\}_{k=m}^{\infty} = \bigcup_{j=0}^{m-1} \{z_{mk+j} = N^{mk+j}(z_0)\}_{k=1}^{\infty}.$$

In other words the fixed point iterates of N at z_0, ignoring the first few terms $z_1, \ldots, z_{m-1}$, consists of the shuffling of the fixed point iterates of N^m, and their images under N, N^2, $\ldots$, N^{m-1}. Thus if for each $j = 0, \ldots, m-1$ the corresponding subsequences converge to θ, so will the entire sequence of $\{z_k\}_{k=0}^{\infty}$.

We now prove the subsequence $\{N^{km+j}(z_0)\}_{k=0}^{\infty}$ converges to θ for all $j = 0, 1, \ldots, m-1$. For $j = 0$ the subsequence corresponds to the fixed point iterates of N^m, hence convergent to θ by assumption. To prove that it is true for $j = 1$, we note

$$N^{km+1}(z_0) = N(N^{mk}(z_0)).$$

Since the sequence of $N^{mk}(z_0)$ converges to θ, continuity of N at θ implies the sequence of images is convergent to $N(\theta)$. But $N(\theta) = \theta$. From this and a similar inductive argument it follows that $\{N^{mk+j}(z_0)\}_{k=0}^{\infty}$ converges to θ for all $j = 0, 1, \ldots, m-1$. Hence the proof is complete. □

Before proving some convergence properties of $g_m(z)$ for $m = 2, 3, 4$, we give two polynomiographs corresponding to $g_3(z)$ and $g_4(z)$ as applied to $p(z) = z^3 - 1$. As may be seen these give better approximation to Voronoi region of the roots than does Newton's method. Indeed it can be shown that as we choose larger and larger values of m, the approximations improve further. In fact the image in Figures 1.3 is a polynomiograph corresponding to some g_m.

Theorem 1.8. *Let $x_0 > \theta = \sqrt[3]{\alpha}$. For $m = 2, 3, 4$, the sequence of fixed point iterates $x_{k+1} = g_m(x_k)$, $k \geq 0$, converges monotonically to θ, having an m-th order rate of convergence.*

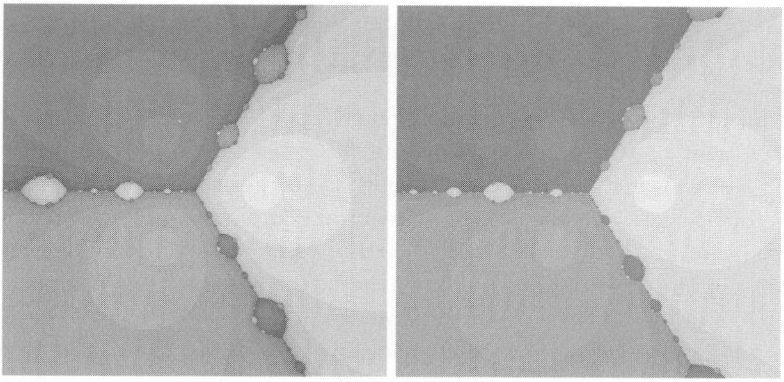

Fig. 1.8 Polynomiographs of $p(z) = z^3 - 1$ corresponding to $g_3(z)$ and $g_4(z)$ from left to right.

Proof. We first show that for $x > \theta$, $x - g_m(x) > 0$. This is readily seen by verifying that $x - g_m(x) = (x^3 - \alpha)u_m(x)$, where

$$u_2(x) = \frac{1}{3x^2}, \quad u_3(x) = \frac{x}{2x^3 + \alpha}, \quad u_4(x) = \frac{6x^4 + 3\alpha x}{10x^6 + 16\alpha x^3 + \alpha^2}.$$

Since $u_m(x) > 0$, for $x > 0$, the fact that $x_0 > \theta$ implies $x_k > x_{k+1}$, for all $k \geq 0$. Next we show that this sequence is bounded below by θ. This follows by verifying that $g_m(x) - \theta = (x - \theta)^m w_m(x)$, where

$$w_2(x) = \frac{(2x + \theta)}{3x^2}, \quad w_3(x) = \frac{(x + \theta)}{(2x^3 + \alpha)}, \quad w_4(x) = \frac{4x^3 + 6\theta x^2 - \alpha}{10x^6 + 16\alpha x^3 + \alpha^2}.$$

Note that $x_0 > \theta$ implies that for all k, $x_k > \theta$. Thus, the sequence $\{x_k\}_{k=0}^{\infty}$ is convergent. Let x^* be its limit. To show $x^* = \theta$ it suffices to take the limit of $x_k - g_m(x_k)$ over k, noting that $u_m(x^*) \neq 0$. Also $\lim_{k\to\infty}(x_{k+1} - \theta)/(x_k - \theta)^m = w_m(\theta)$. □

1.10 Complexity of Sequential Versus Parallel Algorithms

In this final section of the chapter, we consider the approximation of square-root of a positive real number α via sequential and parallel computation of the k-th iterate of the fixed point iteration as applied to $g_m(x)$ given in (1.15), and for different values of m. We will compare them with Newton's iterates. We will consider only real initial input x_0.

Let us denote the k-th iterate of the fixed point iteration, $x_k = g_m(x_{k-1})$, by $x_k^{(m)}$, to indicate its dependency on m. As we have seen $x_k^{(m)}$

can be obtained either recursively, or through its closed form as $g_{m^k}(x_0)$, (1.15). Each of these approaches can in turn be computed sequentially or in parallel.

Consider the sequential computation of $x_k^{(m)}$ via closed form, $g_{m^k}(x_0)$, the quotient of two polynomials of degree m^k and $m^k - 1$, respectively. Using Horner's method the computation can be performed in $O(m^k)$ arithmetic operation. In contrast, the sequential computation of each recursive iteration, $g_m(x_i^{(m)})$, $i = 0, \cdots, k-1$, requires $O(m)$ arithmetic operations (again via Horner's method). Thus, the latter approach for computation of $x_k^{(m)}$ requires $O(mk)$ overall operations. Hence, given that the computations are to be performed sequentially, the use of the closed form is inefficient.

The economical way to compute $x_k^{(2)}$, the k-th iterate of Newton's, is to write

$$g_2(x) = \frac{1}{2}(x + \frac{\alpha}{x}). \tag{1.33}$$

Since each iteration in (1.33) requires 3 operations, the total cost of computing $x_k^{(2)}$ is $3k$.

In what follows we first show using sequential algorithms, the complexity of Newton's is essentially optimal. Next we will show that for $m = 3$, if in each iteration $g_3(x)$ is computed in parallel (using 3 processors), we gain over Newton's by a speedup factor of $\log_2 3 \approx 1.58$. Finally, through existing parallel algorithms for the evaluation of polynomials, we justify that asymptotically, a speedup factor of 3 is possible.

Consider the case where $m > 2$. Let k_m be the number of iterations of the m-th order method needed to obtain an approximation to $\theta = \sqrt{\alpha}$ to within approximately the same accuracy as that of the k-th Newton iterate, i.e. we want

$$x_k^{(2)} \approx x_{k_m}^{(m)}. \tag{1.34}$$

From Theorem 1.1 we must have $2^k \approx m^{k_m}$. Thus, to get at least the same accuracy as that of $x_k^{(2)}$, it suffices to have

$$k_m = \lceil \frac{k}{\log_2 m} \rceil. \tag{1.35}$$

In $O(m)$ arithmetic operations we can compute all binomial coefficients $\binom{m}{i}$, $i = 1, \cdots, m$ as well as the coefficients of the polynomials $P_0^m(x)$, and $P_1^m(x)$ (see (1.11), (1.12)). Ignoring this complexity, for a given x, $g_m(x)$

can be computed in $4m$ arithmetic operations via Horner's method. Thus, the overall complexity to attain at least the same accuracy as that obtained from Newton's method is,

$$4mk_m = 4m\lceil \frac{k}{\log_2 m} \rceil.$$

This number essentially exceeds $3k$ for all $m > 2$. Hence via sequential algorithms, the case of $m = 2$ is essentially optimal.

We now consider parallelization. First consider the case of $m = 3$. An alternative economical formula is

$$g_3(x) = \frac{x^2 + 3\alpha}{3x + \frac{\alpha}{x}}. \tag{1.36}$$

From above it is easy to see that using 3 processors, we can compute $g_3(x)$ in 3 parallel time (as opposed to 7 sequential time). Thus from (1.35) we gain over Newton's by a speedup factor of:

$$\frac{3k}{3k_3} \approx \log_2 3 \approx 1.58.$$

One may observe that the computation of Newton's iterates cannot be accelerated through analogous parallelization.

Next we consider the general case of m and the use of $O(m)$ processors. We will employ some results on the parallel complexity for the evaluation of polynomials. For a given x, $g_m(x)$ can be computed by the evaluation of the polynomials $P_0^m(x)$ and $P_1^m(x)$ in parallel, followed by a division. It can be shown (see Kung (1976)) that for a given input the evaluation of a polynomial of degree m can be achieved in

$$\lceil \log_2 m \rceil + O(\sqrt{\log_2 m})$$

time, using $2m$ processors. In order to apply this result to the evaluation of $P_0^m(x)$ (similarly $P_1^m(x)$), we need to consider the preprocessing time needed to evaluate its coefficients, i.e.

$$\binom{m}{i} \alpha^{i/2}, \quad i = 2, 4, \cdots, \lfloor m/2 \rfloor. \tag{1.37}$$

First using m processors, we compute $\alpha, \alpha^2, \cdots, \alpha^{\lfloor m/2 \rfloor}$ in $\lceil \log_2 m \rceil +$ *constant* parallel time (see e.g. Kung (1976)). The computation of the binomial coefficients $\binom{m}{i}$, $i = 1, 2, \cdots, m$ can also be established in $O(\log_2 m)$ parallel time independently using an additional m processors. Next we compute coefficients in (1.37) in one parallel time. Using an additional $2m$

processors one can independently compute the coefficients of $P_1^m(x)$ in the same amount of parallel time. Hence using $4m$ processors, the j-th iterate of the m-th order method can be obtained in

$$(j+1)\big(\lceil \log_2 m \rceil + O(\sqrt{\log_2 m})\big)$$

parallel time, taking into account the preprocessing time. In particular, if

$$j = k_m = \lceil \frac{k}{\log_2 m} \rceil,$$

there is a gain over Newton's method by a speedup factor of

$$S(k,m) = \frac{3k}{\big[\lceil \frac{k}{\log_2 m} \rceil + 1\big]\big[\log_2 m + O(\sqrt{\log_2 m})\big]}. \tag{1.38}$$

In particular, if we choose $m = 2^k$,

$$\lim_{k \to \infty} S(k, 2^k) = \frac{3}{2}.$$

It is known that any parallel algorithm to compute $x_k^{(2)}$, (regardless of the number of processors used) requires at least k parallel time (see Kung (1976)). Hence the maximum speedup factor any parallel algorithm can gain over Newton's is at most 3.

It is interesting to point out that via the m-th order family $\{g_m(x)\}_{m=2}^{\infty}$ developed in this chapter an asymptotic speedup factor of 3 is theoretically possible. This follows from (1.38) given any $\epsilon > 0$, there exists m_ϵ so that for each fixed $m \geq m_\epsilon$,

$$\lim_{k \to \infty} S(k,m) \leq \frac{3}{(1+\epsilon)}.$$

1.11 Extensions*

Although in this chapter we have only considered the case of square and cube-roots, the iteration functions $\{g_m(x)\}_{m=2}^{\infty}$ can be shown to exist for arbitrary r-th roots, $r > 0$. The case of square-roots was analyzed in (Kalantari and Kalantari (1996)). Furthermore, analogous monotonic m-th order convergence can be established. More importantly, in subsequent chapters we prove that the algebraic approach in constructing high-order methods is also generalizable to the case of approximation of roots of arbitrary polynomials. The study of these iteration functions, their extensions, and their applications in polynomiography are major goal of this book.

*Part of this chapter has been reprinted from High Order Iterative Methods for Approximating Square Roots, *BIT*, Vol. 36 (1996) 395–399, B. Kalantari and I. Kalantari. With kind permission of Springer Science and Business Media.

Chapter 2

The Fundamental Theorem of Algebra and a Special Case of Taylor's Theorem: The Genesis of Iterations Function

In this chapter we employ the Fundamental Theorem of Algebra to derive in a simple algebraic fashion a fundamental family of iteration functions, called *Basic Family*. The first member of the family gives the well-known Newton's method. This family is arguably the most important family of iteration functions for polynomial root-finding. This chapter gives a raw development of the Basic Family. In a sense the entire book is dedicated to the study and development of deep properties of family, its offsprings, as well as its applications, including *polynomiography* a field of study with potentially diverse applications in art, science, and education.

2.1 Introduction

Consider a polynomial

$$p(z) = a_n z^n + \cdots + a_1 z + a_0 \tag{2.1}$$

with complex coefficients. Throughout the history of science the study of solution of polynomial equations and their approximation are among the most fascinating and influential problems. For some aspects of their vast history see e.g. McNamee (1993) and Pan (1997). To find approximation to the roots of $p(z)$ one makes use of iteration functions such as Newton's, $N(z) = z - p(z)/p'(z)$.

A general iteration function $F(z)$ for $p(z)$ is a function with the property that if θ is a root of p, it is a *fixed point* of F, i.e. $F(\theta) = \theta$. The *fixed point iteration* is defined as

$$z_{k+1} = F(z_k), \quad k = 0, 1, \ldots, \tag{2.2}$$

where z_0 is an initial input. The *orbit* of z_0 is the sequence $\{z_k\}_{k=0}^{\infty}$. We are interested in an iteration function $F(z)$ which is a rational function of the coefficients of the polynomial $p(z)$. Then from the continuity of such iteration function it follows that if the fixed point iterates converge at all, they must converge to a fixed point of F.

For a fixed point θ of F the quantity $\lambda = F'(\theta)$ plays a significant role in the local convergence or divergence of the fixed point iteration. If $|\lambda| < 1$ the fixed point iterations converge locally. Moreover, if $\lambda = 0$ the local convergence rate increases, and more generally if the first $m-1$ derivatives are zero at θ, the order of convergence is at least m.

In this chapter we derive a fundamental family of rational iteration functions, the *Basic Family*, to be denoted by

$$\{B_m(z)\}_{m=2}^{\infty}, \tag{2.3}$$

which will have m-th order of convergence for a simple root θ of the underlying polynomial. In this chapter we derive the formula for Basic Family member $B_m(z)$ using a simple algebraic-combinatorial fashion, employing only the Fundamental Theorem of Algebra. The significant properties of the family will be derived and studied in detail in subsequent chapters. To derive the formula for $B_m(z)$ we first derive Newton's method. This leads to a special case of Taylor's Theorem from which we derive a recursive formula to be used for representing the Basic Family.

2.2 Algebraic Derivation of Newton's Method

Let $p(z)$ be a complex polynomial of degree $n \geq 2$ with coefficients in a subfield K of the complex numbers. Let $m \geq 2$ and consider the following problem:

Find a set of n rational functions, $g(z)$, and $\gamma_i(z)$, $i = 2, \ldots, n$ with coefficients in K, so that for any root θ of $p(z)$ we have

$$g(z) = \theta + \sum_{i=2}^{n} \gamma_i(z)(z-\theta)^i. \tag{2.4}$$

The significance of the solvability of the above is as follows. Suppose that for a given z_0 the fixed point iteration

$$z_{k+1} = g(z_k), \quad k \geq 0 \tag{2.5}$$

converges to a root θ of $p(z)$ and suppose that $\gamma_i(z)$ is well-defined for all $i = m, \ldots, m+n-2$. Then, clearly we have

$$\lim_{k\to\infty} \frac{(z_{k+1}-\theta)}{(z_k-\theta)^m} = \gamma_m(\theta), \tag{2.6}$$

i.e. the order of convergence of the sequence of z_k's is m. Furthermore, if z_0 lies in K, then so will all the fixed point iterates.

In this chapter we will prove using elementary techniques that $g(z) = B_2(z) = z - p(z)/p'(z)$, the Newton's iteration function. Here we prove:

Theorem 2.1. *There is a solution to g given by*

$$g(z) = z - \frac{p(z)}{p'(z)} = \theta + \sum_{i=2}^{n} \frac{(-1)^i}{i!} \frac{p^{(i)}(z)}{p(z)}.$$

Before proving the theorem we give an immediate corollary which follows trivially from regrouping of the terms of above:

Corollary 2.1.

$$0 = \sum_{i=0}^{n} \frac{p^{(i)}(z)}{i!}(\theta - z)^i.$$

□

Remark 2.1. Observe that the above is merely a special case of Taylor's Theorem written for $p(z)$ at $z = \theta$. Clearly taking Taylor's Theorem for granted one can easily derive Newton's iteration function. But it is important to note that here we give a direct algebraic-combinatorial proof merely employing the Fundamental Theorem of Algebra.

By applying Corollary 2.1 to the function $q(x) = p(z) - p(a)$ we get the following Taylor's expansion formula:

Corollary 2.2. *For any complex number a we have*

$$p(z) = \sum_{i=0}^{n} \frac{p^i(z)}{i!}(z-a)^i.$$

Remark 2.2. Before proving the theorem we wish to point out that the formula in Corollary 2.1 (or Corollary 2.2) is a special case of Taylor's Theorem in the sense that the remainder term is zero. However, even this special case is sufficiently powerful to derive the raw formula for the Basic Family.

Proof. [Of Theorem 2.1] Let θ be a root of $p(z) = a_n z^n + \cdots + a_0$ which is guaranteed by the Fundamental Theorem of Algebra. Without loss of generality we assume that $a_n = 1$. We wish to determine if there exist $(n-1)$ rational functions $\gamma_k(z)$, $k = 2, \ldots, n$ with coefficient over K so that

$$g(z) = \theta + \sum_{k=2}^{n} \gamma_k(z)(z-\theta)^k \tag{2.7}$$

depends only on z. Using the binomial theorem we write (2.7) as

$$g(z) = \theta + \sum_{k=2}^{n} \gamma_k(z) \sum_{i=0}^{k} (-1)^i \binom{k}{i} z^{k-i}\theta^i. \tag{2.8}$$

Using that $a_n = 1$ and θ is a root, we have

$$\theta^n = -\sum_{i=0}^{n-1} a_i \theta^i. \tag{2.9}$$

Substituting for θ^n in (2.8) we get

$$g(z) = \sum_{i=0}^{n-1} \theta^i h_i(z), \tag{2.10}$$

where

$$h_1(z) = 1 + (-1)^{n-1} a_1 \gamma_n(z) - \sum_{k=2}^{n} \gamma_k(z) \binom{k}{1} z^{k-1}, \tag{2.11}$$

and for $i = 0$ and $i = 2, \ldots, (n-1)$ we have

$$h_i(z) = (-1)^{n-1} a_i \gamma_n(z) + (-1)^i \sum_{k=i}^{n} \gamma_k(z) \binom{k}{i} z^{k-i}. \tag{2.12}$$

If there exist $\gamma_k(z)$, $k = 2, \ldots, n$ such that $h_i(z)$ are identically zero for $i = 1, \ldots, n-1$, then we can define

$$g(z) \equiv h_0(z). \tag{2.13}$$

Setting $h_i(z) = 0$, for $i = 2, \ldots, n-1$ we get

$$\gamma_i(z) = (-1)^{n-i} a_i \gamma_n(z) - \sum_{k=i+1}^{n} \gamma_k(z) \binom{k}{i} z^{k-i}. \tag{2.14}$$

We first prove by induction that for $i = 2, \ldots, n-1$,

$$\gamma_i(z) = (-1)^{n-i} \gamma_n(z) \sum_{l=i}^{n} \binom{l}{i} a_l z^{l-i} = (-1)^{n-i} \gamma_n(z) \frac{p^i(z)}{i!}, \tag{2.15}$$

where the second inequality comes from the definition of formal derivative for a polynomial. From (2.14), for $i = n-1$ we have

$$\gamma_{n-1}(z) = (-1)^{n-(n-1)} a_{n-1}\gamma_n(z) - \gamma_n(z)\binom{n}{n-1}z$$

$$= -\gamma_n(z)\left[\binom{n-1}{n-1}a_{n-1} + \binom{n}{n-1}z\right]. \tag{2.16}$$

Hence (2.15) holds for $i = n-1$. Assuming that it is true for $k = i+1, \ldots, n-1$ we prove that it is true for $k = i$. To do so, using (2.14) it suffices to show that

$$\sum_{k=i+1}^{n} \gamma_k(z)\binom{k}{i} z^{k-i} = -(-1)^{n-i}\gamma_n(z)\sum_{l=i+1}^{n}\binom{l}{i}a_l z^{l-i}. \tag{2.17}$$

From the induction hypothesis we have

$$\sum_{k=i+1}^{n} \gamma_k(z)\binom{k}{i} z^{k-i}$$

$$= \gamma_n(z)\sum_{k=i+1}^{n}\sum_{l=k}^{n}(-1)^{n-k}\binom{l}{k}\binom{k}{i}a_l z^{l-i}. \tag{2.18}$$

Using the identity

$$\binom{t}{r}\binom{r}{s} = \binom{t}{s}\binom{t-s}{r-s}, \tag{2.19}$$

writing $(-1)^{n-k} = (-1)^{n-i}(-1)^{k-i}$, and rearranging terms of the double summation, (2.18) can be written as

$$\sum_{k=i+1}^{n} \gamma_k(z)\binom{k}{i} z^{k-i} = (-1)^{n-i}\gamma_n(z)G_i(z), \tag{2.20}$$

where

$$G_i(z) = \sum_{l=i+1}^{n}\sum_{k=i+1}^{l}(-1)^{k-i}\binom{l}{i}\binom{l-i}{k-i}a_l z^{l-i}$$

$$= \sum_{l=i+1}^{n} a_l z^{l-i}\sum_{k=i+1}^{l}(-1)^{k-i}\binom{l-i}{k-i}. \tag{2.21}$$

But

$$\sum_{k=i+1}^{n}\sum_{l=k}^{n}(-1)^{n-k}\binom{l}{k}\binom{k}{i}a_l z^{l-i}$$

$$= \sum_{j=1}^{l-i} (-1)^j \binom{l-j}{j} = (1-1)^{1-i} - 1 = -1. \tag{2.22}$$

Thus we have shown that for $i = 2, \ldots, n-1$

$$G_i(z) = - \sum_{l=i+1}^{n} \binom{l}{i} a_l z^{l-i}. \tag{2.23}$$

Thus we have proved (2.17). In particular, setting $i = 1$ in (2.18) we get

$$\sum_{k=2}^{n} \gamma_k(z) \binom{k}{1} z^{k-1} = (-1)^{n-1} \gamma_n(z) \sum_{l=2}^{n} \binom{l}{1} a_l z^{l-1}. \tag{2.24}$$

Substituting (2.24) into (2.11), and setting to zero gives

$$1 + (-1)^{n-1} \gamma_n(z) \left[\sum_{l=1}^{n} \binom{l}{1} a_l z^{l-1} \right]$$
$$= 1 + (-1)^n \gamma_n(z) p'(z) = 0. \tag{2.25}$$

Thus $\gamma_n(z) = (-1)^n / p'(z)$ and from (2.15) we have proved

$$\gamma_i(z) = \frac{(-1)^i}{i!} \frac{p^{(i)}(z)}{p'(z)}. \tag{2.26}$$

Next we show

$$g(z) = z - \frac{p(z)}{p'(z)}. \tag{2.27}$$

From (2.15) we get

$$g(z) = h_0(z) = (-1)^{n-1} \gamma_n(z) a_0 + \sum_{k=2}^{n} \gamma_k(z) z^k$$
$$= (-1)^{n-1} \gamma_n(z) a_0 + \gamma_n(z) \sum_{k=2}^{n} \sum_{l=k}^{n} (-1)^{n-k} \binom{l}{k} a_l z^l$$
$$= (-1)^{n-1} \gamma_n(z) a_0 + \gamma + n(z) \sum_{l=2}^{n} \sum_{k=2}^{n} (-1)^{n-k} \binom{l}{k} a_l z^l$$
$$= (-1)^n \gamma_n(z) \left[- a_0 + \sum_{l=2}^{n} a_l z^l \sum_{k=2}^{l} (-1)^k \binom{l}{k} \right]. \tag{2.28}$$

For any $l \geq 2$ we have

$$(l-1) = \sum_{k=2}^{l} (-1)^k \binom{l}{k}. \tag{2.29}$$

Using (2.29) and since

$$-a_0 + \sum_{l=2}^{n} a_l z^l (l-1) = z p'(z) - p(z) \tag{2.30}$$

and $\gamma_n(z) = (-1)^n / p'(z)$ we get

$$g(z) = (-1)^n \gamma_n(z)(z p'(z) - p(z)) = z - \frac{p(z)}{p'(z)}.$$

□

2.3 A Recurrence Relation and the Basic Family

In this section we show that the special Taylor's Theorem in Corollary 2.1 gives rise to the Basic Family of iteration functions. Thus this simple and familiar calculus formula is in fact the source of many families of iteration functions. The formula is a disguised form of a characteristic polynomial corresponding to a homogeneous linear recurrence relation. First we prove a theorem.

Theorem 2.2 (Kalantari (2004b)). *Let a be a complex number, different from a root of $p(z)$. Then, θ is a root of $p(z)$ if and only if $\eta = -p(a)/(\theta - a)$ is a root of the polynomial:*

$$Q_a(z) = z^n - \sum_{i=1}^{n} (-1)^{i-1} p^{i-1}(a) \frac{p^{(i)}(a)}{i!} z^{n-i}. \tag{2.31}$$

Proof. From the formula in Corollary 2.1 we have

$$-p(a) = \sum_{i=1}^{n} \frac{p^{(i)}(a)}{i!} (\theta - a)^i. \tag{2.32}$$

Substituting for $\theta - a = -p(a)/\eta$ in (2.32) we get

$$-p(a) = \sum_{i=1}^{n} \frac{p^{(i)}(a)}{i!} (-1)^i \frac{p^i(a)}{\eta^i}. \tag{2.33}$$

Multiplying (2.33) by $-\eta^n/p(a)$ and right-hand side terms we get

$$\eta^n - \sum_{i=1}^{n} (-1)^{i-1} p^{i-1}(a) \frac{p^{(i)}(a)}{i!} \eta^{n-i} = 0. \tag{2.34}$$

But (2.34) implies that η is a root of $Q_a(z)$. The converse follows by reversing the steps. □

The polynomial $Q_a(z)$ is the characteristic polynomial of the homogeneous linear recurrence relation:

$$D_m(a) = \sum_{i=1}^{n} (-1)^{i-1} p(a)^{i-1} \frac{p^{(i)}(a)}{i!} D_{m-i}(a). \tag{2.35}$$

From this relationship it follows that if $Q_a(z)$ has a unique dominant root say η, then for appropriately selected set of initial conditions, the ratio $D_{m-1}(a)/D_{m-2}(a)$ converges to this root (see e.g. Henrici (1974), Hildebrand (1974), Householder (1970)). Since $\eta = -(\theta - a)/p(a)$ for some root θ of $p(z)$, then if the *Basic Sequence* is defined to be (Kalantari (2004b)) :

$$\left\{ B_m(a) \equiv a - p(a)\frac{D_{m-2}(a)}{D_{m-1}(a)} \right\}_{m=2}^{\infty},$$

then the sequence must converge to θ.

The *Voronoi region* of a root θ is the set of points in $\mathbb{C}$ that are closer to θ than to any other root of $p(z)$. In a subsequent chapter we formally prove the following convergent result:

Theorem 2.3. *Let a be any complex number in the Voronoi region of a root θ of $p(z)$. Then with the set of initial conditions*

$$D_1(a) = 1, \quad D_j(a) = 0, \quad j = -1, \ldots, j = -n+1,$$

the corresponding Basic Sequence $\{B_m(a)\}_{m=2}^{\infty}$ converges to θ. □

The above analysis also justifies the definition of the *Basic Family* of iteration function:

$$B_m(z) = z - p(z)\frac{D_{m-2}(z)}{D_{m-1}(z)}, \quad m \geq 2. \tag{2.36}$$

This family whose first member is Newton's function:

$$B_2(z) = z - \frac{p(z)}{p'(z)}$$

is the main subject of study in the entire book.

2.4 Conclusions

In summary, using the Fundamental Theorem of Algebra we have derived the formula (2.1) which in turn gave rise to the recurrence relation for $D_m(z)$, (2.35), and this in turn gave rise not only to a simple approach for deriving the formula for the Basic Family $B_m(z)$, but for deducing the convergence of the Basic Sequence to a root of $p(z)$, for almost all inputs. It is worth emphasizing that in contrast with an iterative method, such as Newton's method, which repeatedly uses the same iteration function while updating its input, the Basic Sequence makes use of a fixed input while ranging over the entire Basic Family. This pointwise convergence property is a very significant and interesting property. It suggests that if we pick a point at random, with probability one the corresponding Basic Sequence will converge to a root. In contrast, no single rational iteration function can ever achieve that property for all polynomials. The fact that Newton's method could fail even for cubic polynomials is well known, e.g. Smale (1985) where the notion of *general convergence* is defined. This notion will be considered in subsequent chapters. Furthermore, Smale's question on the existence of generally convergent iteration functions led to the negative result of McMullen (1987) that for general polynomials of degree four or higher there is no generally convergent algorithm. These will be analyzed in detail in a subsequent chapter.

We mention that Corollary 2.1 can also give rise to other family of iteration functions. For instance, the Euler-Schröder family, to be discussed in later chapters. However, while this simple Taylor's formula does give rise to the Basic Family and other iteration functions, it cannot be used to prove many significant properties of these iteration functions. The book in fact unveils numerous hidden properties of the Basic Family family of iteration functions, giving evidence that arguably, it is the most significant family of rational functions.

Chapter 3

Introduction to the Basic Family and Polynomiography

In this chapter we describe the Basic Family and give an overview of its fundamental properties and many potential applications in art, education, mathematics and science. In subsequent chapters we will prove the theoretical properties surveyed here and many more results, and will also demonstrate and justify the claimed practical applications.

3.1 Introduction

Newton's method is the best known of the iteration function for finding the roots of polynomials or more general functions. It is taught early on in calculus or even in high school classes. However, only for the real functions or real roots of polynomials with real coefficients. By iteration function we shall mean a rational function with the property that in some neighborhood of each roots the iterates will converge to that root. Cayley (1897) is among the first who considered Newton's method for finding the roots of polynomial with iterations over the complex numbers. Indeed he only considered the case of approximation of square-roots and cube-roots of a number which for simplicity may be assumed to be one. Thus, one can claim Cayley was among the pioneers trying to "understand" or "visualize" the behavior of Newton's method in the complex plane. He was interested in determining the shape of the basins of attraction, the set of points in the complex plane, whose Newton iterates would converge to a root of the underlying polynomial. While the basins of attraction of Newton's method for the square-roots can easily be shown to be their Voronoi regions, for cube-roots of unity, as is now well known, the corresponding basins exhibit fractal boundary behavior. The Mandelbrot set and its complexity together with advancements in computer technology has brought about a great deal of professional and amateur interest in imaging the behavior of iterative methods and some iteration functions for root-finding.

In this chapter we re-introduce the fundamental family of iteration functions for polynomial root-finding, the *Basic Family*. This infinite family of iteration functions possesses many significant properties from the mathematical and algorithmic points of view, as well as from the point of view of visualization. We survey the history of the Basic Family, then describe several modern discoveries and their applications. A particular modern application of the Basic Family we emphasize here is the visualization of a polynomial equation using one or more of the Basic Family members. We call this visualization *polynomiography* and the resulting image a *polynomiograph*. Polynomiography gives rise to a whole set of images, even of a single polynomial, including many novel images that are not necessarily fractal and even if fractals are unlike the typical ones seen in abundance.

Thus the term polynomiography is not merely the assignment of a new term to a familiar fractal image in the visualization of polynomial equations, rather a term that not only emphasizes the origin and characteristics of the image but also suggests novel visualizations beyond the typical and sometimes mechanical imaging of general iterations. In contrast to typical fractal images with their unpredictable behavior, polynomiography places a great deal of creativity, control, and predictability at the hand of the *polynomiographer*. Given an appropriate polynomiography software, a polynomiographer who may or may not know the mathematical detail of iteration functions, can learn to produce images of great diversity. Indeed the theoretical and algorithmic properties of the Basic Family, as uniquely exploited in polynomiography and as supported through exhibited images, could potentially turn polynomiography into a powerful tool with diverse applications in art, science and education, bringing polynomial root-finding to the general public.

More generally, we propose the term polynomiography for the visualization of a polynomial equation with respect to iteration functions, not restricted to be the members of the Basic Family. Although the theoretical properties of the Basic Family make it more interesting in the visualization of a polynomial equation over arbitrary iteration functions, there are good reasons behind our proposition. One reason being that the term polynomiograph brings an identity to an image based on a polynomial equation, even if it happens to be a fractal image. Referring to such image as a polynomiograph or a *fractal polynomiograph* is far more informative than referring to it as fractal. Not only is the term fractal too broad, a fractal image may not have originated from an iteration function for solving a polynomial equation.

A second reason behind our proposition is that interesting visualizations can be obtained from other families of iteration functions, or even individual iteration functions, which although are induced from the Basic Family, or related to it, are driven by their own specific theoretical properties. We will demonstrate such families and their polynomiography later in the book. Yet another reason behind our proposition is that other families of iteration functions may be discovered that although unrelated to the Basic Family, could exhibit their own interesting properties in the visualization of polynomial equations.

We anticipate that the rich theoretical properties of the Basic Family surveyed in this chapter and proved in subsequent chapters, together with applications of polynomiography will initiate the discovery of further properties of this fundamental family of iteration functions for root-finding. Moreover, we have considerable evidence to venture that polynomiography offers a new beginning for the use of polynomials themselves, bringing it to the general domain of appreciation and usage, and changing the way we have viewed these essential objects of science and mathematics, even by expert mathematicians and scientists.

3.2 The Basic Family and its Properties

Consider a fixed polynomial of degree $n \geq 2$

$$p(z) = a_n z^n + \cdots + a_1 z + a_0. \tag{3.1}$$

Assume the coefficients of the polynomial lie in a subfield K of the complex numbers. The following, Problem (P), which is indeed the basis and motivation behind an algebraic development of the Basic Family, was inspired by the simple development of the Basic Family for the approximation of square and cube roots in Kalantari and Kalantari (1996), discussed in Chapter 1.

Problem (P): Given a natural number $m \geq 2$, find a set of n rational functions, $B_m(z)$ and $\gamma_{m,i}(z)$, $i = m, \ldots, m+n-2$ with coefficients in K, if they exist, so that for any root θ of $p(z)$ we have

$$B_m(z) = \theta + \sum_{i=m}^{m+n-2} \gamma_{m,i}(z)(z-\theta)^i. \tag{3.2}$$

If the problem is solvable and for a given z_0 the fixed point iteration

$$z_{k+1} = B_m(z_k)$$

converges to a root θ of $p(z)$, and if $\gamma_{m,i}(z)$ is well defined for each $i = m, \dots, m+n-2$, then we get

$$\lim_{k\to\infty} \frac{(z_{k+1}-\theta)}{(z_k-\theta)^m} = \gamma_{m,m}(\theta), \tag{3.3}$$

implying z_k converges to θ having order m. Furthermore, if z_0 lies in K, then so will z_k for all k.

The existence and uniqueness of the solution to Problem (P) was established in Kalantari *et al.* (1997). In Chapter 1 we proved the existence of solution for the special case of $m = 2$, coinciding with Newton's method and strongly connected with a special case of Taylor's Theorem.

The closed form of $B_m(z)$ above happens to coincide with the recursive form described in Chapter 1, namely the Basic Family. In what follows we give yet another representation of the Basic Family of iteration functions, a determinantal formula. We then state some of its mathematical properties.

To define $B_m(z)$ set $D_0(z) = 1$, and for each $m \geq 1$ let $D_m(z)$ denote the following determinant whose corresponding $m \times m$ matrix is a Toeplitz matrix (i.e. constant along each diagonal), also upper-Hessenberg (i.e. elements below the subdiagonal are zero):

$$D_m(z) = \begin{vmatrix} p'(z) & \frac{p''(z)}{2!} & \dots & \frac{p^{(m-1)}(z)}{(m-1)!} & \frac{p^{(m)}(z)}{(m)!} \\ p(z) & p'(z) & \dots & \frac{p^{(m-2)}(z)}{(m-2)!} & \frac{p^{(m-1)}(z)}{(m-1)!} \\ 0 & p(z) & \ddots & \vdots & \vdots \\ \vdots & \vdots & \ddots & p'(z) & \frac{p''(z)}{2!} \\ 0 & 0 & \dots & p(z) & p'(z) \end{vmatrix}.$$

For each $m \geq 2$ the following defines an iteration function for $p(z)$:

$$B_m(z) = z - p(z)\frac{D_{m-2}(z)}{D_{m-1}(z)}. \tag{3.4}$$

The Basic Family is the entire set of iteration functions $\{B_m(z)\}_{m=2}^{\infty}$. Specific members of the family are

$$B_2(z) = z - \frac{p(z)}{p'(z)},$$

$$B_3(z) = z - \frac{2p'(z)p(z)}{2(p'(z))^2 - p''(z)p(z)},$$

$$B_4(z) = z - \frac{6p'(z)^2p(z) - 3p''(z)p(z)^2}{p'''(z)p(z)^2 + 6p'(z)^3 - 6p''(z)p'(z)p(z)}.$$

The first two members of the Basic Family by themselves possess a very rich and interesting history. For the history of Newton's method see Ypma (1995). The iteration function $B_3(z)$ is credited to the astronomer Halley (1694). Halley's method according to Traub (1964) is the most rediscovered iteration function after Newton's. It apparently inspired the celebrated Taylor's Theorem, see Ypma (1995). Halley's method has been rediscovered and/or derived through various interesting means, see e.g. Alefeld (1981), Bateman (1938), Wall (1948), Bodewig (1949), Hamilton (1950), Stewart (1951), Frame (1944, 1945, 1953), Traub (1964), Hansen and Patrick (1977), Popovski (1980), Gander (1985). For the interesting history of Halley's method see Scavo and Thoo (1995). In Kalantari (1998b) it is shown that Halley's method is only the first member of an infinite family of iteration functions of cubic order.

The determinantal formula for $D_m(z)$ happens to satisfy the homogeneous linear recurrence relation:

$$D_m(z) = \sum_{i=1}^{n}(-1)^{i-1}p(z)^{i-1}\frac{p^{(i)}(z)}{i!}D_{m-i}(z), \tag{3.5}$$

with initial conditions

$$D_0(z) = 1, \quad D_j(z) = 0, \quad \forall \;\; j = -1, \ldots, -n+1. \tag{3.6}$$

This homogeneous linear recurrence relation, also described in Chapter 1, connects $D_m(z)$ to a Taylor expansion of $p(z)$ and leads to the definition of *Basic Sequence* at a given complex number z_0

$$\{B_m(z_0)\}_{m=2}^{\infty}.$$

If z_0 lies in a Voronoi region of a root θ, the Basic Sequence will converge to that root, i.e.

$$\lim_{m \to \infty} B_m(z_0) = \theta. \tag{3.7}$$

We refer to this property as the *pointwise convergence property of the Basic Family.* Aside from Newton and Halley methods, other individual members of the Basic Family have been discovered and rediscovered by different researchers and through different means. In fact in special cases of polynomials, the Basic Family has been discovered and rediscovered several times, but without the global view of their equivalence. For instance, using continued fractions, Yeyios (1992) derives a family of iteration functions for the approximation of square roots. This special case of square root appears to have been known before, e.g. Jamieson (1989). A simple algebraic development for approximation of square roots in Kalantari and Kalantari (1996), described earlier in Chapter 1, in fact leads to proving the equivalence of these special cases as they all coincide with $B_m(z)$ corresponding to $p(z) = z^2 - \theta$.

An earliest derivation of the Basic Family goes back to the late nineteenth century work of Schröder (see English translation by Stewart, Schröder (1870)). Indeed the Basic Family appears to have also been derived by Traub (1966) (p. 130) as a special case of a parameterized family of iteration function for polynomials, with the parameter value equal zero. Traub's motivation is to find high order iterative methods, which for large enough values of the parameter, converge globally to the dominant root of polynomials, assuming such root is unique.

The Basic Family is sometimes known as König's family, but only through a computationally cumbersome formula, see e.g. Vrscay and Gilbert (1988), Buff and Henriksen (2003). The equivalence of König's formula to the determinantal formula of the Basic Family apparently is not always known in literature. In the next chapter we give König's form and a short proof of the equivalence to the Basic Family.

A family of iteration functions often known in the literature as Schröder family, or sometimes as Euler-Schröder family, happens to be another family of iteration functions, see e.g. Henrici (1974), Householder (1970), Shub and Smale (1985), Traub (1964), and Drakopoulos *et al.* (1999). A simple derivation of the Euler-Schröder family is given in Kalantari *et al.* (1997) and Kalantari (2000a).

Schröder's nineteen century article (Schröder (1870)) is an outstanding and classical piece of work on the development of iteration functions for polynomial root-finding and does give rise to the family of iteration functions we call the Basic Family here. However, it certainly does not assert or discover all the underlying properties of this magical family. Indeed in Schröder's work the connections between the Basic Family, the homogeneous linear recurrence relations of $D_m(z)$, and the underlying simple and special case of Taylor's Theorem does not seem to be fully discovered or displayed. Stewart in his scholarly translation of Schröder's article, Schröder (1870), gives his own account of the pointwise convergence property based on what he calls "today's natural approach." Stewart asserts that Schröder could have argued the pointwise convergence property directly, had he known of König's theorem, "if an analytic function has a single, simple pole at the radius of convergence of its power series, then the ratios of the coefficients of its power series converge to that pole."

The pointwise convergence property of the Basic Family for polynomials, can be accomplished by manipulating the special case of Taylor Theorem, Theorem 2.1, which itself is a by-product of the Fundamental Theorem of Algebra. Thus the pointwise convergence property known to Schröder in the late nineteenth century, could have in fact been deduced in the early eighteenth century when the celebrated Taylor's Theorem was proved. In fact as shown in Kalantari (2004b) there is a close relationship between the Basic Sequence and the iterates of Bernoulli method for approximation of dominant roots of polynomials.

Stewart's translation also mentions the point that Bernoulli's method is a special case corresponding to the pointwise evaluation of the family. But even for this special case results in Kalantari (2004b), or via the formula in (3.8), as well as the ones to be presented in subsequent chapters not only offer new findings for the Bernoulli method, but for general homogeneous recurrence relations, in particular for $D_m(z)$.

The pointwise convergence of the Basic Family which can be deduced based on properties of homogeneous linear recurrence relations by itself does not provide any error formula for the convergence of the Basic Family or Basic Sequence. The convergence of the Basic Family or the Basic Sequence takes a precise form through a generalization of Taylor's Theorem proved in Kalantari *et al.* (1997) and in much more generality in Kalantari (2000a) leading to an expansion formula for polynomials and analytic functions that in particular implies the formula (3.8) to be discussed next. For some analytic functions this expansion formula together with other ingredients, a determinantal lower bound proved in Kalantari (1997), also Kalantari and Pate (2001), a result of interest in linear algebra, allows bounding the norm of the gap $|B_m(z) - \theta|$ for a given z and a root θ of the analytic function. As an application, in Kalantari (2000b) this approach is used to give many new formulas for the approximation of the legendary number π.

We now proceed to give several significant properties of the Basic Family member $B_m(z)$. First we need to define additional determinants. Given $m \geq 1$, for $k \geq (m+1)$ set

$$\widehat{D}_{m,k}(z) = \begin{vmatrix} \frac{p''(z)}{2!} & \frac{p'''(z)}{3!} & \dots & \frac{p^{(m)}(z)}{(m)!} & \frac{p^{(k)}(z)}{k!} \\ p'(z) & \frac{p''(z)}{2!} & \dots & \frac{p^{(m-1)}(z)}{(m-1)!} & \frac{p^{(k-1)}(z)}{(k-1)!} \\ p(z) & p'(z) & \ddots & \vdots & \vdots \\ \vdots & \vdots & \ddots & \frac{p''(z)}{2!} & \frac{p^{(k-m+2)}(z)}{(k-m+2)!} \\ 0 & 0 & \dots & p'(z) & \frac{p^{(k-m+1)}(z)}{(k-m+1)!} \end{vmatrix}.$$

Note that the $m \times m$ matrix corresponding to $\widehat{D}_{m,k}(z)$ is like a Toeplitz matrix, but only so for $k = m + 1$. Also note

$$\widehat{D}_{m,k}(z) \equiv 0, \quad k \geq m + n.$$

The following expansion formula can be thought of as a generalization of Taylor's Theorem. Let θ be a root of $p(z)$. For each $m \geq 2$, $B_m(z)$ satisfies the following expansion formula:

$$B_m(z) = \theta + (-1)^m \sum_{k=m}^{m+n-2} \frac{\widehat{D}_{m-1,k}(z)}{D_{m-1}(z)} (\theta - z)^k. \tag{3.8}$$

In particular, from the above expansion it follows that if θ is a root of $p(z)$ there exists a disk centered at θ such that for any z_0 in this disk the fixed point iteration

$$z_{k+1} = B_m(z_k), \tag{3.9}$$

is well defined and converges to θ having order m. More specifically:

$$\lim_{k\to\infty} \frac{(\theta - z_{k+1})}{(\theta - z_k)^m} = (-1)^m \frac{\widehat{D}_{m-1,m}(\theta)}{D_{m-1}(\theta)}, \tag{3.10}$$

where $D_{m-1}(\theta) = p'(\theta)^{m-1}$ and $D_{m-1,m}(\theta)$ satisfies the recurrence relation:

$$D_{m-1,m}(\theta) = \sum_{i=1}^{n} (-1)^{i-1} p'(\theta)^{i-1} \frac{p^{(i+1)}(\theta)}{(i+1)!} D_{m-i-1,m-i}(\theta), \tag{3.11}$$

with

$$D_{0,1}(\theta) = 1, \quad D_{j,j+1}(\theta) = 0, \quad \forall\ j = -1, \dots, -n+1.$$

The proof of (3.8) and its generalization are given in Kalantari *et al.* (1997), and Kalantari (2000a), respectively. The proof of (3.5) and (3.11) and applications are given in Kalantari (2000a), Kalantari (1998a), Kalantari (2000b), and Kalantari (2004b). The asymptotic order (3.10) follows from (3.8). We will prove all these in subsequent chapters.

The algebraic derivation of the Basic Family, Kalantari *et al.* (1997), reveals many interesting minimality and uniqueness properties, in particular the important expansion formula given in (3.8). More generally, a determinantal generalization of Taylor's Theorem, Kalantari (2000a), provides a more general development of the Basic Family, and its multipoint version (multipoint Basic Family) where each member B_m blossoms into m iteration functions,

$$B_m^{(1)}(z_1) = B_m(z), \quad B_2^{(2)}(z_1, z_2), \ldots, B_m^{(m)}(z_1, z_2, \ldots, z_m),$$

where for $k = 1, \ldots, m$, $B_m^{(k)}$ is a k-point iteration function defined in terms of two determinants that depend on the first $m-k$ derivatives of $p(z)$. The order of convergence of these iteration functions is derived in Kalantari (1999).

Figure 3.1 represents the general ascending order of convergence of $B_m^{(k)}$, and the corresponding order for the first few members. The Basic Family corresponds to the first column.

$$\begin{array}{ccccccc}
B_2^{(1)} & \leftarrow & B_2^{(2)} & & & & \\
\downarrow & & \downarrow & \searrow & & & \\
B_3^{(1)} & \leftarrow & B_3^{(2)} & \leftarrow & B_3^{(3)} & & \\
\downarrow & & \downarrow & & \downarrow & \searrow & \\
B_4^{(1)} & \leftarrow & B_4^{(2)} & \leftarrow & B_4^{(3)} & \leftarrow & B_4^{(4)} \\
\downarrow & & \downarrow & & \downarrow & & \downarrow \quad \searrow
\end{array}
\qquad
\begin{array}{ccccccc}
2 & \leftarrow & 1.618 & & & & \\
\downarrow & & \downarrow & \searrow & & & \\
3 & \leftarrow & 2.414 & \leftarrow & 1.839 & & \\
\downarrow & & \downarrow & & \downarrow & \searrow & \\
4 & \leftarrow & 3.302 & \leftarrow & 2.546 & \leftarrow & 1.927 \\
\downarrow & & \downarrow & & \downarrow & & \downarrow \quad \searrow
\end{array}$$

Fig. 3.1 Multipoint Basic Family and corresponding order of convergence.

A computational comparison of the first nine iteration functions in Kalantari and Park (2001) shows that for small degree polynomials $B_m^{(k-1)}$ is more efficient than $B_m^{(k)}$, but as the degree increases, $B_m^{(k)}$ becomes more efficient than $B_m^{(k-1)}$. The most efficient of the nine methods is $B_4^{(4)}$, using only function evaluations, having theoretical order of convergence equal to 1.927. Newton's method, often viewed as the method of choice, was in fact the least efficient of the nine.

In Kalantari and Gerlach (2000), it is shown that each member of the Basic Family can be viewed as Newton's iteration function, but applied to a complicated function that involves $p(z)$. In Kalantari and Jin (2003) it is proved that all extraneous fixed points of the Basic Family are repelling. An extraneous fixed point of an iteration function F is a fixed point that is not a root of p. This establishes another superiority of the Basic Family over other iteration functions. Kalantari (2000b) makes use of the Basic Family to derive new formulas for approximation of π, as well as high-order methods for computing roots of analytic functions. To do so, the polynomial $p(z)$ in the Basic Family is replaced with an appropriate analytic function. Kalantari (2000a) relates the Basic Family to a generalization of Taylor's Theorem which not only gives the Basic Family and the iteration functions in Figure 3.1, but can be used to provide infinitely many approximations to a given function or its inverse.

The Basic Family also gives rise to the *Truncated Basic Family*, defined in Kalantari (2000a), where for each $m \geq 3$ the iteration function $B_m(z)$ is only the first member of an infinite family of m-th order methods. More specifically, the Truncated Basic Family of order t is a family of iteration functions $\{B_{m,t}(z)\}_{m=t+1}^{\infty}$, where $B_{m,t}$ is obtained from B_m by replacing derivatives of $p(z)$ of order higher than t by 0. In particular, for $m = 3$ the Truncated Basic Family is Halley's Family, defined and analyzed in Kalantari (1998b). For general m the order of convergence of the Truncated Basic Family is analyzed in Jin and Kalantari (2005b) using the theory of symmetric function whose application in root-finding is thoroughly investigated in his Ph.D thesis, Jin (2005a). Jin also develops some other very interesting versions of the Basic Family as well as other iteration functions with high order of converge to the poles of analytic functions. A version of the Basic Family for high-order of convergence to multiple roots of the Basic Family is presented in Jin and Kalantari (2005a).

Theoretically it is possible to define the multipoint version of the Truncated Basic Family. If we denote by $B_{m,t}^{(k)}$ the k-point version of the truncated Basic Family, where $m \geq k > t$, we obtain the following 3-dimensional array of iteration functions, see Figure 3.2, where the infinite sequence $\{B_{m,1}^{(1)}\}_{m=2}^{\infty}$ is the ordinary Basic Family, and the vertical 2-dimensional background corresponds to the multipoint Basic Family.

In Kalantari (2005b) (see also Kalantari (2005a)) it is shown that for each $m \geq 2$ we can compute tight upper and lower bounds, U_m and L_m, on the modulus of zeros of a polynomial $p(z)$, computable in terms of its coefficients. These provide a unique infinite family of bounds on the modulus of polynomial zeros, revealing yet a new theoretical application of the Basic Family. Having tight bounds on zeros is not only important and useful from the theoretical or practical point of view, but also from the point of view of polynomiography.

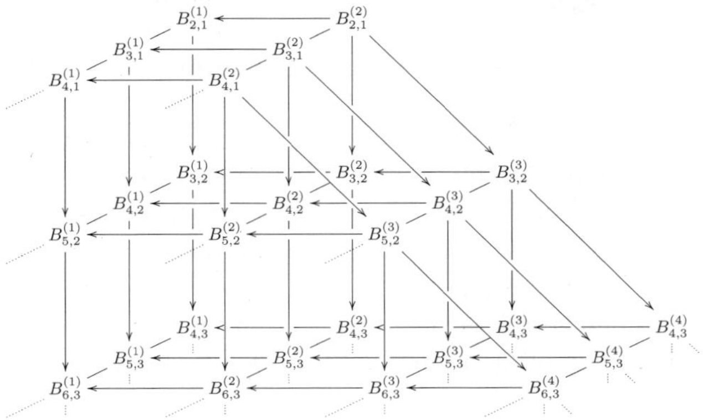

Fig. 3.2 Basic Family and its variants: Truncated, Multipoint.

There are many known individual bounds on roots of polynomials as compiled by McNamee (1993). But to the best of our knowledge no family of such bounds were known previously. In fact McNamee and Olhovsky (2005) (see also McNamee (2007)) make a comparison of 45 different formulas that give upper bounds on the modulus of the roots, including the bounds U_2, U_3, and U_4. They found U_4 to give the most accurate result in comparison to the 45 bounds tested. In Jin (2006) it is shown that as m tends to infinity the upper and lower bounds U_m, L_m, converge to the tightest annulus containing the roots of the given polynomial. Jin has also shown that the computation of the upper bound (or lower bound) for each m can be achieved very efficiently in $O(mn)$ arithmetic operations. Our bounds on zeros are in particular useful in focussing the visualization of polynomial equations to regions that would tightly enclose the roots.

It is worth mentioning here that the mathematical and algorithmic properties of the Basic Family was the inspiration behind the visualization of polynomial equations, leading to polynomiography. Before the introduction of the term polynomiography (see e.g. Kalantari (2002b) and Kalantari (2004c)) there appeared to be no systematic study of this family for the visualization of polynomials, or even a systematic study of its mathematical properties. Polynomiography reveals yet new applications of the Basic

Family. This will be discussed in further detail later in the next section and throughout the book.

3.3 Polynomiography and Its Applications

Formally, we define *polynomiography* to be the art and science of visualization in the approximation of zeros of polynomials using iteration functions. In particular this visualization can be achieved by making use of the Basic Family and its properties. Polynomiography, through a corresponding software, gives rise to an image of a given polynomial, called a *polynomiograph.* A polynomiograph is generated in a given rectangular region, via a given iteration function or a family of iteration functions such as the Basic Family.

A polynomiograph can be produced using a variety of coloring schemes, an art in itself. Producing a polynomiograph of a polynomial via a polynomiography software is somewhat analogous to producing a photograph of an object, where iteration functions and coloring schemes in the polynomiography software, used by the *polynomiographer*, are comparable to a camera, its lenses, and various settings used by a photographer. But polynomiography is also analogous to painting as well, where the polynomiographer can begin coloring an initial polynomiograph using his or her own coloring preferences as supported by the software.

Polynomials which are the most essential of mathematical objects visible in every branch of science, through polynomiography, take a new form that could not only be of interest to mathematicians, scientists, and students, but to artists as well. Polynomiography turns polynomials into an artist's tool that could inspire sophisticated 2D and even 3D artwork. Articles that describe polynomiography and its many applications include Kalantari (2004a), Kalantari (2004c), Kalantari *et al.* (2004), Kalantari (2004d), Kalantari (2007) and Kalantari (2006).

Although a polynomiograph may turn out to be a fractal image, polynomiography is not a subset of fractals, neither as theory nor as application. In particular, polynomiography, viewed as art, is not a subset of fractal art. Fractals have provided an interesting tool of art, see e.g. Mandelbrot (1983), Mandelbrot (1993). As a tool of art, polynomiography is not only a very powerful medium, but is in fact complementary to fractal art. When producing fractal art through a polynomiography software, the polynomiographer is working with a restricted but well-defined class of fractal images, namely those coming from special iteration functions designed for root-finding, such as the Basic Family, whose properties give rise to a powerful tool of art and design. On the one hand, polynomiography provides a basis for producing fractal art with an underlying foundation, as opposed to random fractal art with little or no information on the underlying iteration function or its mission. With polynomiography one can speak of *fractal polynomiographs* and such a term would refer to a well-defined subset of fractal images. Such polynomiographs give rise to many new sets of fractal images, hence broadening the horizon of fractal art. On the other hand, polynomiography also gives rise to a very wide range of interesting polynomiographs which are not at all fractal.

According to the definition of polynomiography, any iteration function can be used to give a polynomiograph. But from the point of view of art and design what makes the use of the Basic Family very appealing is the limiting behavior of the basins of attraction of a root of $p(z)$ as one employs $B_m(z)$ with large values for m. For large m the basins of attraction provide an approximation to the Voronoi regions of the roots. In summary, one visual application of the Basic Family is that it offers methods for approximation of Voronoi regions, whether the roots are given explicitly or implicitly through their polynomial equation. This Voronoi region approximation property of the Basic Family in polynomiography and its potentially vast applications in art and design and even in computational geometry are discussed in Kalantari (2002a) and Kalantari (2005c), respectively. Another interesting property of the Basic Family in art and design is driven by the fact that for any point interior to the Voronoi polygon of a root, the corresponding Basic Sequence converges to that root. This gives rise to polynomiographs with enormous beauty. There are yet many other schemes.

As is well known now, for any iteration function (in particular the Basic Family members) the boundary of the basins of attraction of any of the polynomial roots is the same set, known as a Julia set and often exhibits fractal behavior. Images of the basins of attraction of Newton's method are quite familiar, in particular fractal images of $p(z) = z^3 - 1$ under this method - initially studied in Cayley (1897) (see also Peitgen *et al.* (1984)) - demonstrates well the self-similarity property in typical fractal images. Mandelbrot (1983) popularized the work of Julia (1918) and Fatou (1919) on iterations of rational complex functions. Mathematical analysis of complex iterations may be found in Peitgen *et al.* (1992), Devaney (1986), Falconer (1990). More thorough treatments can be found in the books

Beardon (1991), and Milnor (2006).

It should be mentioned that despite the tremendous research on the theory of iterations of rational complex functions, even the nature of the Julia set corresponding to Newton's method as applied to the simple polynomial $p(z) = z^3 - 1$ is not completely understood, let alone for general roots of unity, the roots of $p(z) = z^n - 1$, and even worse for general polynomials. For instance, given a point in the plane with rational coordinates, can we determine to which basin, if any, it belongs? To the best of our knowledge this question is open even for the polynomial $p(z) = z^3 - 1$. This question is motivated by a question of Wolfram (2004) on whether or not in applying Newton's method to $p(z) = z^3 - 1$ one can quickly decide how to color a given pixel.

For general polynomials the fractal nature of the Julia sets corresponding to an individual member of the Basic Family follows from the general theory on iteration of complex rational functions. But that theory does not predict the behavior of individual Basic Family members on a specific input, or the shape of the basins of attraction. Fractal images based on iteration functions (rational or irrational) such as Newton's method, Laguerre's method, Halley's method, etc. have of course been rendered for a long time. In particular, some members of the Basic Family have been used to give visualizations of polynomial root-finding, see e.g. Varona (2002), or Vrscay and Gilbert (1988) who give images based on a few more members of the Basic Family than Newton or Halley iteration functions. But these studies are limited. In Vrscay and Gilbert (1988) the authors only use the computationally cumbersome König formula, not offering any systematic study of the visualization via this family. The systematic study of the Basic Family or the use of its mathematical properties for the visualization of polynomial root-finding is a novelty that has opened up numerous possibilities. Just as the mathematical properties of the Basic Family are not limited to those discussed in this chapter, neither are their visual applications. In this book we will discover much more about the visualization of polynomial equations using the Basic Family or families that are induced by it.

We now give a brief description of a few sample polynomiographs exhibited in the chapter. All images are created via polynomiography software based on Basic Family and its variants, using an individual member or the Basic Sequence.

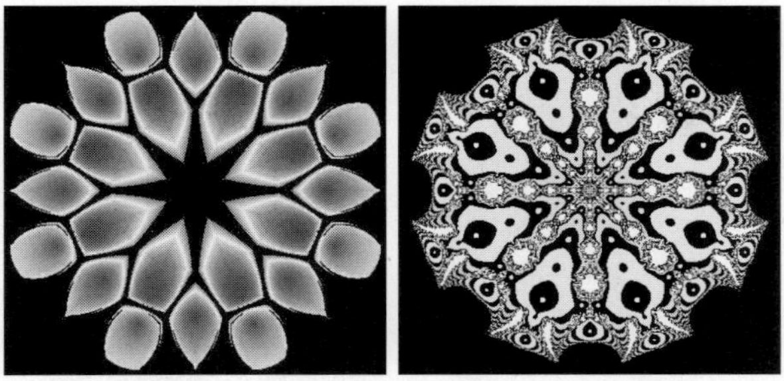

Fig. 3.3 "Jewels."

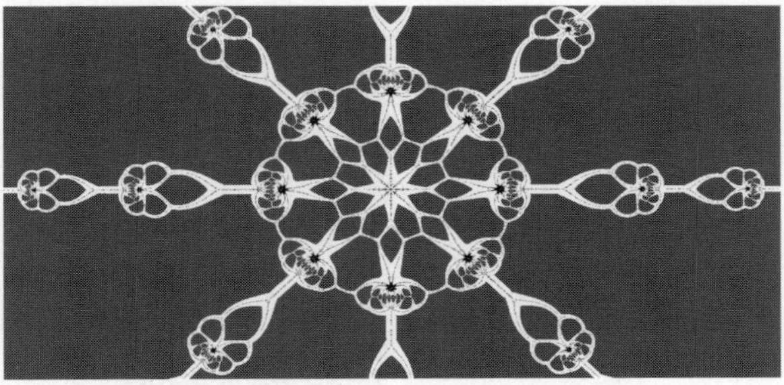

Fig. 3.4 "Acrobats."

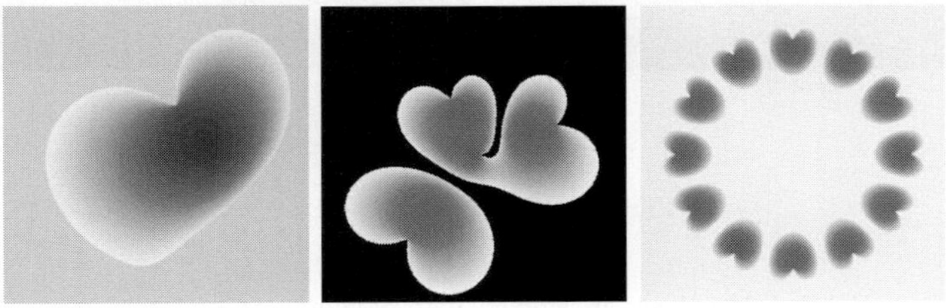

Fig. 3.5 "Hearts."

Some images are presented simply to convey the artistic and design potentials of the field of polynomiography. For instance, the creation of "Acrobats" in Figure 3.4 was a combination of the right choice of an underlying polynomial, the iteration function, as well as the underlying mathematical properties of the Basic Family, and much trial and error.

Fig. 3.6 Playing with hearts.

In Figure 3.7, "Acrobats in Paris" was carved out of "Acrobats." These exhibit the interesting features of polynomiography where the polynomiographer can combine his/her own creativity with the infinity of choices in the selection of the underlying polynomial, the iteration function or functions, as well as coloring choices and personal taste. No collage is used in any of the images. The images in Figure 3.6 shows very clearly that one can create beautiful shapes by design and personal creativity rather than randomness and accident. These qualities should be evident to the reader in the remaining images as well.

Fig. 3.7 "Acrobats in Paris."

In Figures 3.9 and 3.10 the reader may observe the Voronoi region approximation property of the Basic Family. The underlying polynomial is given explicitly via its roots. Several of the figures reveal the power of polynomiography in designing symmetric patterns of great beauty and diversity.

Fig. 3.8 Clockwise from top-left: "Owl," "Symphony," and two symmetric designs.

Next we briefly list several applications of polynomiography. Polynomiography as a tool of education, whether used as images, or as software, can be employed in such college level courses as calculus or numerical analysis, allowing students to tackle important conceptual issues such as the notion of convergence and limits, as well as the idea of iteration functions, and learn about complex numbers and polynomials themselves. It also gives students the ability to understand and appreciate more modern discoveries such as fractals.

Fig. 3.9 "Candy Mosaic," polynomiography giving approximations of Voronoi regions.

Fig. 3.10 More polynomiographs of approximate Voronoi regions.

One may also make use of polynomiographs to study the computational advantage of one iteration function against another by comparing the corresponding times for the generation of polynomiograph of the same polynomial. These times give a measure of average performance of the iteration functions. In general to decide which of two iteration methods is preferable, one can average the times for generating polynomiographs over a set of random polynomials. For such a study which in particular compares some members of the Basic Family among themselves see Andreev *et al.* (2005). Varona (2002) offers other measures for the comparison of iteration function. However, the ones suggested in Andreev *et al.* (2005) (also Kalantari (2002b), Kalantari (2004c)) are probably the most easily checkable, yet meaningful measures of performance.

Fig. 3.11 Cover designs.

Polynomiography through animation allows the visualization of many theorems about polynomials or those of iteration functions. Some examples can be found in Kalantari *et al.* (2004) and their corresponding animations in www.cs.rutgers.edu/~kalantar/Animation.

Polynomiography as a tool of mathematical discovery, gives rise to new conjectures about polynomials and related mathematical properties. We mention one example here concerning the *general convergence* of the Basic Family. The work of McMullen (1987) regarding the general convergence of rational iteration functions in particular implies that no member of the Basic Family can be *generally convergent* for all polynomials. But one may ask if there are classes of polynomials for which the general convergence can be established. Jin and Kalantari (2007) consider the polynomial $p(z) = z^n - 1$, proving partial results for some indices m of the Basic Family members and some indices n. By making an extensive use of polynomiography, it is conjectured that the Basic Family is generally convergent for all n. This conjecture may appear to be intuitively obvious or an intuitively obvious application of polynomiography, but many non-intuitive conjectures can be inspired by polynomiography.

Polynomiography as tool of art and design, has enormous applications. An obvious application is that it turns polynomials into visual objects that an artist who typically would never use polynomials can learn to appreciate and use them in order to generate interesting variety of art. On the other hand a young student, say a middle schooler, by visualization of polynomial equations may develop an interest in the study of their mathematical properties, or math in general. Yet, through polynomiography software a mathematically oriented user could grow an interest in art.

In this chapter we have presented the Basic Family of iteration functions, surveying old and modern discoveries. As seen in Chapter 1 the Fundamental Theorem of Algebra (FTA) gives a simple derivation of the formula for Basic Family members. The world of iteration functions for polynomial root-finding relies on the FTA, perhaps explaining why throughout the history some of these iteration functions have been rediscovered so many times. But just as the FTA proof has a never-ending story so do the iteration functions that are induced by it. In particular despite the many known results about the Basic Family presented in this chapter or even in the entire book, there remains numerous open problems worthy of future investigation. These questions arise with respect to their theoretical and practical applications, such as in designing new root-finding algorithms or with respect to applications that are inspired by polynomiography.

While some theoretical aspects of polynomiography may intersect with those of iteration of complex rational functions, or the theory of fractals and dynamical systems, polynomiography does possess its own independent characteristics. These make polynomiography a subject of its own, offering numerous applications of its own. In particular, we anticipate that not only the study of polynomials through polynomiography will result in a unified and deeper perspective into the theory of root-finding, but it will also inspire the discovery of new properties of polynomials. In subsequent chapters we shall extend polynomiography to the visualization of homogeneous linear recurrence relations and prove mutual relationships between polynomial root-finding and solution of such recurrences. Polynomiography is the most systematic method in the visualization of the polynomial root-finding algorithms, going far beyond the corresponding fractal visualization that preceded it, bringing this visualization into the realm of art and design, but also science and education. As a tool of art or as a tool of education, there are enormous potential applications from middle and high schools to higher education. These will be discussed in detail in later chapters.

Chapter 4

Equivalent Formulations of the Basic Family

Throughout the long history of the root-finding problem iteration functions have been discovered and rediscovered. Often times these equivalences have gone unnoticed even by the researchers in the field. In this chapter we prove the equivalence of several different iteration functions to the Basic Family. They amount to different formulations of the Basic Family. These imply that whatever is proved for one form also applies to the other forms. These different but equivalent formulations also allow the discovery of new properties that may not be evident from other formulation.

4.1 Determinantal Formulation of the Basic Family

Given $p(z)$, a complex polynomial of degree $n \geq 2$, for each $m \geq 2$, the iteration function

$$B_m(z) \equiv z - p(z) \; \frac{D_{m-2}(z)}{D_{m-1}(z)}, \tag{4.1}$$

has been defined in previous chapters, where

$$D_0(z) \equiv 1,$$

and for each $m \geq 1$,

$$D_m(z) = \begin{vmatrix} p'(z) & \frac{p''(z)}{2!} & \dots & \frac{p^{(m-1)}(z)}{(m-1)!} & \frac{p^{(m)}(z)}{(m)!} \\ p(z) & p'(z) & \ddots & \ddots & \frac{p^{(m-1)}(z)}{(m-1)!} \\ 0 & p(z) & \ddots & \ddots & \vdots \\ \vdots & \vdots & \ddots & \ddots & \frac{p''(z)}{2!} \\ 0 & 0 & \dots & p(z) & p'(z) \end{vmatrix},$$

where $|\cdot|$ denotes the determinant. We will prove that D_m satisfies a homogeneous linear recurrence relation.

4.2 Properties of a Determinant

In this section we develop several properties of $D_m(z)$ some of which pertain to the determinant of any matrix whose entries are constant along each diagonal, also having all entries below the subdiagonal entries, equal to zero. The first property makes such a matrix a Toeplitz matrix, while the second property an upper Hessenberg matrix. General Toplitz or upper Hessenburg matrices are discussed in Golub and Loan (1996). However, matrices that satisfy both conditions have their own very special properties. The first result is one such property with significant consequences (see (Kalantari (1998a, 2000b))):

Theorem 4.1. *Let $a_0, \ldots, a_n$ be arbitrary complex numbers, $n \geq 1$. Define $a_i = 0$, for all $i > n$. Given a natural number m, define*

$$d_m = \begin{vmatrix} a_1 & a_2 & \dots & a_{m-1} & a_m \\ a_0 & a_1 & \ddots & \ddots & a_{m-1} \\ 0 & a_0 & \ddots & \ddots & \vdots \\ \vdots & \vdots & \ddots & \ddots & a_2 \\ 0 & 0 & \dots & a_0 & a_1 \end{vmatrix}.$$

Then,

$$d_m = \sum_{i=1}^{n} (-1)^{i-1} a_0^{i-1} a_i d_{m-i},$$

where $a_0^0 \equiv 1$, $d_0 \equiv 1$, and $d_i \equiv 0$, if $i = -1, \ldots, -n+1$.

Proof. Given any pair of natural numbers m and k, define

$$d_m^{(k)} = \begin{vmatrix} a_k & a_{k+1} & \dots & a_{m+k-2} & a_{m+k-1} \\ a_0 & a_1 & \ddots & \ddots & a_{m-1} \\ 0 & a_0 & \ddots & \ddots & \vdots \\ \vdots & \vdots & \ddots & \ddots & a_2 \\ 0 & 0 & \dots & a_0 & a_1 \end{vmatrix}.$$

Note that when $k = 1$, $d_m^{(k)} = d_m$. By expanding along the first column, the determinant of the matrix corresponding to $d_m^{(k)}$, we get

$$d_m^{(k)} = a_k d_{m-1} - a_0 d_{m-1}^{(k+1)}.$$

By expanding along the first column the determinant of the matrix corresponding to d_m, we get

$$d_m = a_1 d_{m-1} - a_0 d^{(2)}_{m-1}.$$

Now by using in the above the previously given formula for various values of m and k repeatedly, the proof of the lemma follows. □

Corollary 4.1. *For all $m \geq 1$ we have,*

$$D_m = \sum_{i=1}^{n} (-1)^{i-1} \frac{p^{i-1} p^{(i)}}{i!} D_{m-i},$$

where $D_1 = 1$, and $D_j = 0$, for $j = -1, \ldots, j = -n+1$.

The following is a key result.

Theorem 4.2 (Kalantari (1998a, 2000b)). *Let $a_0, \ldots, a_n$ be arbitrary complex numbers, $n \geq 1$. Define $a_i = 0$, for all $i > n$. For each i define*

$$b_i = a_0^{\frac{i-1}{2}} a_i,$$

where $a_0^0 \equiv 1$, and we take $a^{1/2}$ to be the principal value. Given a natural number m, let d_m be as in Theorem 4.1, and define

$$c_m = \begin{vmatrix} b_1 & b_2 & \ldots & b_{m-1} & b_m \\ b_0 & b_1 & \ddots & \ddots & b_{m-1} \\ 0 & b_0 & \ddots & \ddots & \vdots \\ \vdots & \vdots & \ddots & \ddots & b_2 \\ 0 & 0 & \ldots & b_0 & b_1 \end{vmatrix}.$$

Then, $d_m = c_m$.

Proof. From Theorem 4.1, c_m satisfies the linear recurrence relation

$$c_m = \sum_{i=1}^{n} (-1)^{i-1} b_0^{i-1} b_i c_{m-i},$$

where, $b_0^0 = 1$, $c_0 \equiv 1$, and $c_i \equiv 0$, if $i < 0$. But from the definition of b_i we have

$$c_m = \sum_{i=1}^{n} (-1)^{i-1} a_0^{\frac{i-1}{2}} a_0^{\frac{i-1}{2}} a_i c_{m-i} = \sum_{i=1}^{n} (-1)^{i-1} a_0^{i-1} a_i c_{m-i}.$$

Thus, $c_m = d_m$, since they satisfy the same recurrence relation and the same initial conditions. □

The following is immediate from Theorem 4.2:

Corollary 4.2. *We have*

$$D_m(z) = \begin{vmatrix} p' & \sqrt{p}\frac{p''}{2!} & \cdots & \sqrt{p}^{m-2}\frac{p^{(m-1)}}{(m-1)!} & \sqrt{p}^{m-1}\frac{p^{(m)}}{(m)!} \\ \sqrt{p} & p' & \ddots & \ddots & \sqrt{p}^{m-2}\frac{p^{(m-1)}}{(m-1)!} \\ 0 & \sqrt{p} & \ddots & \ddots & \vdots \\ \vdots & \vdots & \ddots & \ddots & \sqrt{p}\frac{p''}{2!} \\ 0 & 0 & \cdots & \sqrt{p} & p' \end{vmatrix}.$$

□

The above corollary gives a useful formula for determining lower bounds on the modulus of D_m when $|p(z)|$ is small. In subsequent chapters we will make use of it in bounding the gap $|B_m(z) - \theta|$ when z is close to a root θ.

4.3 Gerlach's Method

It is well-known that Halley's method can be obtained by applying Newton's method to the function $p(z)/\sqrt{p(z)'}$, see Bateman (1938). Gerlach (1994), gives a generalization of this approach, and for each $m \geq 2$, recursively defines an iteration function $G_m(z)$ having order m. In what follows we describe Gerlach's approach and prove $G_m(z) = B_m(z)$. On the one hand, this implies that $G_m(z)$ enjoys the previously derived properties of $B_m(z)$, i.e., the closed formula, its efficient computation, an expansion formula surveyed in Chapter 3 which gives precise asymptotic constant, as well as its multipoint versions and much more. On the other hand, the equivalence gives a new insight on the Basic Family and its connection to Newton's method.

Let $p(z)$ be a polynomial of degree $n \geq 2$ with complex coefficients. The following theorem gives a recipe for constructing high order methods for the approximation of roots of p:

Theorem 4.3 (Gerlach (1994)). *Set $F_1(z) = p(z)$, and for each $m \geq 2$, recursively define*

$$F_m(z) = \frac{F_{m-1}(z)}{F'_{m-1}(z)^{\frac{1}{m}}}.$$

Then, the function

$$G_m(z) = z - \frac{F_{m-1}(z)}{F'_{m-1}(z)}$$

defines an iteration function whose order of convergence for simple roots is m.

For $m = 2$ and $m = 3$ it is easy to verify that

$$G_2(z) = z - \frac{p(z)}{p'(z)}, \quad G_3(z) = z - p(z)\frac{p'(z)}{p'(z)^2 - p(z)p''(z)/2}$$

which coincide with Newton and Halley iteration functions, respectively. Gerlach did not give a closed formula for $G_m(z)$. Indeed it is not even clear that $G_m(z)$ would simplify into a rational function of z, $p(z)$, and its derivatives. The following result in particular proves that $G_m(z)$ is a rational function.

Theorem 4.4 (Ford and Pennline (1996)). *The iteration function* $G_m(z)$ *can be written as*

$$G_m(z) = z - p(z)\frac{Q_m(z)}{Q_{m+1}(z)},$$

where $Q_2(z) \equiv 1$, *and*

$$Q_{m+1}(z) = p'(z)Q_m(z) - \frac{1}{m-1}p(z)Q'_m(z).$$

Ford and Pennline however do not give a closed form for $G_m(z)$.

In the next section we derive a closed formula for $G_m(z)$. We do this by proving the equivalence of the family $\{G_m(z)\}_{m=2}^{\infty}$ to $\{B_m(z)\}_{m=2}^{\infty}$.

4.4 Equivalence to the Basic Family

Let $p(z)$ be a polynomial of degree n with complex coefficients. We prove that for each $m \geq 2$, $G_m(z) = B_m(z)$. The equivalence of these two iteration functions for polynomials implies their equivalence for more general functions: suppose that $P_n(z) = \sum_{i=0}^{n} f^{(i)}(z_0)/i!$ is the n-th degree Taylor polynomial of an analytic function $f(z)$ at a given input z_0. Since $P_n^{(i)}(z_0) = f^{(i)}(z_0)$, for $i = 0, \ldots, n$, then $G_m(z_0)$ and $B_m(z_0)$ can be viewed as the value of these iteration functions as applied to $P_n(z)$ at z_0. Hence, the two iteration functions are equivalent for general analytic functions if and only if they are identical for all polynomials.

The following is a key result. For simplicity of notation throughout the rest of the chapter we will suppress the variable z in $p^{(j)}(z)$, $D_j(z)$, etc.

Theorem 4.5. *For each $m \geq 1$, we have*

$$D'_m = \frac{m+1}{p}\Big(p'D_m - D_{m+1}\Big) = (m+1)\sum_{i=2}^{n}(-1)^i\frac{p^{i-2}p^{(i)}}{i!}D_{m+1-i}.$$

Proof. It is easy to see that the second equality in the theorem follows from Corollary 4.2 as written for $m+1$.

We prove the first equality by induction on m. For $m = 1$, $D_1 = p'$, and $D_2 = p'^2 - pp''/2$. We have,

$$\frac{2}{p}(p'D_1 - D_2) = \frac{2}{p}(p'^2 - p'^2 + \frac{pp''}{2}) = p'' = D'_1.$$

Hence, the theorem is true for $m = 1$. Now assume that $m > 1$. From Corollary 4.1 we have

$$D_m = \sum_{i=1}^{n}(-1)^{i-1}\frac{p^{i-1}p^{(i)}}{i!}D_{m-i}.$$

Differentiating D_m, and grouping terms corresponding to all those that differentiate p^{i-1} $(i > 1)$ into one group, those that differentiate $p^{(i)}$ into a second group, and those that differentiate D_{m-i} into a third group we get,

$$D'_m = A + B + C,$$

where

$$A = \sum_{i=2}^{n}(-1)^{i-1}(i-1)\frac{p^{i-2}p'p^{(i)}}{i!}D_{m-i},$$

$$B = \sum_{i=1}^{n}(-1)^{i-1}\frac{p^{i-1}p^{(i+1)}}{i!}D_{m-i},$$

$$C = \sum_{i=1}^{n}(-1)^{i-1}\frac{p^{i-1}p^{(i)}}{i!}D'_{m-i}.$$

By the induction hypothesis, we have

$$D'_{m-i} = \frac{m+1-i}{p}\Big(f'D_{m-i} - D_{m+1-i}\Big).$$

Thus $C = C_1 + C_2$, where

$$C_1 = \sum_{i=1}^{n}(-1)^{i-1}(m+1-i)\frac{p^{i-2}p'p^{(i)}}{i!}D_{m-i},$$

$$C_2 = -\frac{mp'}{p} D_m + \sum_{i=2}^{n} (-1)^i (m+1-i) \frac{p^{i-2} p^{(i)}}{i!} D_{m+1-i}.$$

Note that

$$A + C_1 = \frac{m(p')^2}{p} D_{m-1} + m \sum_{i=2}^{n} (-1)^{i-1} \frac{p^{i-2} p' p^{(i)}}{i!} D_{m-i}.$$

Since in B, $p^{(n+1)} \equiv 0$, and by changing the summation index, we get

$$B = \sum_{i=1}^{n-1} (-1)^{i-1} \frac{p^{i-1} p^{(i+1)}}{i!} D_{m-i} = \sum_{i=2}^{n} (-1)^{i-2} \frac{p^{i-2} p^{(i)}}{(i-1)!} D_{m+1-i}.$$

Thus,

$$B + C_2 = -\frac{mp'}{p} D_m + (m+1) \sum_{i=2}^{n} (-1)^i \frac{p^{i-2} f^{(i)}}{i!} D_{m+1-i}.$$

From Corollary 4.1 we also have

$$\frac{m(p')^2}{p} D_{m-1} - \frac{mp'}{p} D_m = p' \left(\frac{m}{p}\right) (p' D_{m-1} - D_m) = m \sum_{i=2}^{n} (-1)^i \frac{p^{i-2} p' p^{(i)}}{i!} D_{m-i}.$$

Thus,

$$D'_m = A + C_1 + B + C_2 = (m+1) \sum_{i=2}^{n} (-1)^i \frac{p^{i-2} p^{(i)}}{i!} D_{m+1-i}.$$

Hence the proof. □

Corollary 4.3. *For each $m \geq 1$, we have*

$$D_{m+1}(z) = p'(z) D_m(z) - \frac{1}{m+1} p(z) D'_m(z).$$

Theorem 4.6 (Kalantari and Gerlach (2000)). *For all $m \geq 2$, $G_m(z) = B_m(z)$.*

Proof. From Theorem 4.4 and Corollary 4.3 it follows that Q_m and D_{m-2} satisfy the same recurrence relationship. Since $Q_2 = D_0 = 1$ (see definitions of Q_m and D_m), it follows that $Q_m = D_{m-2}$. Thus, $F_{m-1}/F'_{m-1} = f D_{m-2}/D_{m-1}$, i.e., $G_m = B_m$, for all $m \geq 2$. □

In what follows we give a direct proof of the above theorem using Theorem 4.5, also directly relating F_m and D_m.

Theorem 4.7. *Define $\widehat{F}_1(z) = p(z)$, and for each $m \geq 2$, define $\widehat{F}_m(z) = p(z) D_{m-1}^{-\frac{1}{m}}(z)$. Then,*

$$B_m(z) = z - \frac{\widehat{F}_{m-1}(z)}{\widehat{F}'_{m-1}(z)} = G_m(z).$$

Proof. From differentiation of $\widehat{F}_m$ and the identity $D'_{m-1} = \frac{m}{p}(p'D_{m-1} - D_m)$ derived in Theorem 4.5, we get

$$\widehat{F}'_m = p'D_{m-1}^{-\frac{1}{m}} - \frac{1}{m}pD_{m-1}^{-\frac{m+1}{m}}D'_{m-1} = D_{m-1}^{-\frac{m+1}{m}}D_m.$$

It follows that

$$\frac{\widehat{F}_{m-1}}{\widehat{F}'_{m-1}} = p\frac{D_{m-2}}{D_{m-1}}.$$

To prove that $G_m = B_m$ and to directly relate F_m and D_m, it suffices to show $\widehat{F}_m = F_m$. But they have the same initial condition and satisfy the same recursion since we have

$$\widehat{F}_{m-1}(\widehat{F}'_{m-1})^{-\frac{1}{m}} = pD_{m-2}^{-\frac{1}{m-1}}D_{m-2}^{\frac{1}{m-1}}D_{m-1}^{-\frac{1}{m}} = fD_{m-1}^{-\frac{1}{m}} = \widehat{F}_m.$$ □

4.5 König's Family and Equivalence to the Basic Family

König family is defined as follows, where $(\cdot)^{(m)}$ stands for the m-th derivative:

$$K_m(z) = z - \frac{\Delta_{m-2}(z)}{\Delta_{m-1}(z)}, \quad \Delta_m(z) = \frac{(-1)^m p(z)}{m!}\left(\frac{1}{p(z)}\right)^{(m)}. \tag{4.2}$$

Theorem 4.8. *$K_m(z) = B_m(z)$ for all $m \geq 2$.*

Proof. It suffices to show the following (suppressing the dependency on z for convenience)

$$D_m = p^m\Delta_m = \frac{(-1)^m p^{m+1}}{m!}\left(\frac{1}{p}\right)^{(m)}. \tag{4.3}$$

We prove (4.3) by induction on m. For $m = 0$ this holds since $D_0 = 1$ and $\Delta_0 = p'/p$. Assuming (4.3) is true for $m-1$ and differentiating it we get:

$$D_{m-1} = \frac{(-1)^{m-1}p^m}{(m-1)!}\left(\frac{1}{p}\right)^{(m-1)}, \tag{4.4}$$

$$D'_{m-1} = \frac{(-1)^{m-1}}{(m-1)!}\left\{mp^{m-1}p'\left(\frac{1}{p}\right)^{(m-1)} + p^m\left(\frac{1}{p}\right)^{(m)}\right\}. \tag{4.5}$$

But from Theorem 4.5 we have

$$D_m = p'D_{m-1} - \frac{pD'_{m-1}}{m}. \tag{4.6}$$

Substituting in (4.6) for D_{m-1}, D'_{m-1} from (4.4), (4.5) and simplifying we get (4.3). □

4.6 Notes and Remarks

In this chapter we have proved several equivalent forms of the Basic Family. A different proof of the equivalence of König's Family and the Basic Family using the theory of symmetric function is given in Jin and Kalantari (2005b). In his Ph.D thesis, Jin (2005a), makes a thorough analysis of the theory of symmetric functions in root-finding and connections to the Basic Family.

The equivalence of Ford and Pennline formulation of Gerlach's to the Basic Family was proven in Kalantari and Gerlach (2000). About the same time as the publication of this equivalence, Petković and Hereceg (1999) article independently also proved the equivalence of some iteration functions to Gerlach's family. Curiously, the authors refer to the Basic Family with determinantal formula as Wang's family (apparently an article in Chinese) but not as Schröder's family nor König's family. Yet in Wang (1994) the König formulation of the family is refereed as Halley's family.

As we saw in Chapter 2 the raw form of the Basic Family is easily derivable from a Taylor's formula which in turn is a consequence of the Fundamental Theorem of Algebra. It is thus not surprising that many authors may have discovered or rediscovered these iteration functions since all roads on iteration functions have to initiate with the Fundamental Theorem of Algebra. But the mere formula for an iteration function is as good as Newton's method without Taylor's Theorem. It is in fact Taylor's Theorem which allows the analysis of the order of convergence of Newton's and many other results on iteration functions. For instance, the important one-point theory of Smale (1986) relies heavily on Taylor's Theorem and not just the raw form of Newton's formula.

This book is testimonial to the depth of the Basic Family, a name that as we shall see is quite fitting and is based on an algebraic derivation that reveals much about this most fundamental family of iteration functions for polynomial root-finding. In this book we will develop many variations of the Basic Family and reveal significant connections to many other subjects.

We end the chapter by pointing out an incidental fact that the formulation of the Basic Family as Newton's method, though with respect to a much more complicated function, does result in an automatic Smale one-point theory for the Basic Family as well, albeit through an algebraic formula whose closed form or efficient computation would require further research.

Chapter 5

Basic Family as Dynamical System

In this chapter we are interested in the theoretical study of the behavior of iterations of individual members of the Basic Family, both as rational functions as well as iteration functions for polynomial root-finding. There is much theory on the dynamics of iterations of rational functions. The main motivation behind the study of iterations of rational functions was the analysis of the behavior of Newton's method for finding roots of polynomials. Cayley and Schröder in the late nineteen century were among the first who investigated the behavior of Newton's method in the complex plane. In particular, Cayley examined Newton's method for $z^2 - 1$ and $z^3 - 1$. Later Julia and Fatou considered rational functions in more generality, proving many significant results and laying the foundation of the theory of complex dynamical systems. It is of course fair to say that the Fundamental Theorem of Algebra whose first proof is credited to Gauss is the foundation behind this general theory. The work of Montel which preceded those of Fatou and Julia plays an important role in proving some of the magical and significant properties of the sets that have become known as Fatou and Julia sets. The collective visualization of these two sets, in the popular literature, has become known as *fractal*, a term coined by Mandelbrot to mean a rough or fragmented geometric shape that reveals repeated self-similarity as one zooms in finer and finer scales.

These theoretical results, the work of Mandelbrot and the interest he generated in the visualization of quadratic iterations, as complemented by advancements in computer technology, computer graphics, as well as algorithms that made the visualization of dynamics possible, all renewed interest in the theoretical study of iterations of general rational functions. The modern study of the dynamical behavior of Newton's method for complex polynomials and more general iteration functions was also motivated by theoretical results as well as questions that have inspired new insights, most prominently by the work of Smale.

It would obviously be impossible to cover the entire theory of complex dynamical systems in a single chapter. However, the goal of the chapter is to give highlight of the theory, not just as a survey, but with as much proof as possible, trying also to give as simple proofs as possible. We will strive to give a clear and intuitive picture into the general theory of complex dynamical systems. However, at times we will relate the general theory to polynomial root-finding and the Basic Family.

In order to review and prove the material in the chapter we will rely primarily on the books by Beardon (1991) and Milnor (2006), though our proofs may differ from theirs. There are also excellent survey articles such as Blanchard (1984), and Bergweiler (1993). Other related publication include books such as Devaney (1986), Falconer (1990), and conference proceedings such as Devaney (1994), Lei (2000). There are of course many popular books on general fractals, Mandelbrot (1983), Barnsley (1988), Peitgen *et al.* (1992), Peitgen and Richter (1992).

5.1 Introduction

In order to motivate the entire theory of the dynamics of rational functions over the complex plane we will consider Newton's method,

$$N(z) = z - \frac{p(z)}{p'(z)}$$

for finding roots of a complex polynomial $p(z)$. Let us raise a natural question that would come to mind when trying to analyze the behavior of Newton iterations

$$z_k = N(z_{k-1}),$$

given an arbitrary initial point z_0. It is well known and trivial to prove that for any root θ of $p(z)$, there exists a disk

$$D = \{z \in \mathbb{C} : |z - \theta| < \epsilon\}$$

such that for any z_0 in D the iterates converge to θ.

The question can be posed: What is the maximal open set U that contains a given root θ of the underlying polynomial $p(z)$, such that U is also connected and starting with any point in U Newton's iterates converge to θ?

We will see that the above question with some elementary analysis and detective work gives reasonable insight into what might happen, and even leads into some of the deepest questions about the theory of dynamical systems.

First, we must ask if the set U above is even well-defined? To justify that it is well-defined, note that if Newton iterates starting at an arbitrary point z_0 converge to θ, then it must be the case that within a finite number of iterations, say n, all the subsequent iterates fall into the open disk D around θ. Thus by continuity, the inverse image of the disk D is open, and consequently the repeated inverse images, i.e.

$$N^{-k}(D) \equiv N^{-1}(N^{-k+1}(D))$$

is open for any $k \geq 1$ and will eventually contain z_0.

We can thus conclude that the set of all points whose iterates converge to θ, called the *basin of attraction* of θ, is an open set and hence it is the countable union of open sets, one of which is the *immediate basin of attraction*, the set we called U. Hence U is well-defined.

One may ask: What happens to Newton iterates if we start with a point on the boundary of U? Clearly by maximality of U no point on its boundary can converge to any root of $p(z)$. Nor can such a point converge to any point in the complex plane because, by continuity, such a point must necessarily be a fixed point of $N(z)$. But any finite fixed point of $N(z)$ is necessarily a root of $p(z)$.

One plausible case for Newton iterates starting at a boundary point of U is that their norm converges to infinity. But if we can argue that infinity is *repelling*, then the only other possibility is that the iterates either cycle or come close to being cyclic. With this elementary analysis and detective work, we are forced to conclude that for points on the boundary of the immediate basin of attraction of roots of $p(z)$ the behavior of Newton iterates should be erratic.

Once we have a feel of the behavior of Newton's for an immediate basin of attraction U of a root θ of $p(z)$, we can extend that to the entire basin of attraction. We know that the basin of attraction of θ is merely the union of repeated inverse image of U under Newton's. Moreover, we would anticipate strange behavior on the boundary of the basin of attraction. If we repeat this argument for all the roots, we get a finite collection of basin of attractions and their boundaries.

The set U is a connected component of the Fatou set associated with Newton's iteration function for this polynomial as are all the other components of the basin of attraction. The boundary of U as well as the basin of attraction is a part of the Julia set. These sets will be formally defined and analyzed later.

Some natural questions that arise now are: Do these basins of attraction and their boundaries always exhaust the entire complex plane? If not what does the left-over look like? How many connected components can the basin of attraction of a root contain? Can a boundary point of a basin of attraction of one root have a neighborhood disjoint from the basin of attraction of another root?

From the above analysis and without the need for any visualization, we can deduce that Newton's method in the case of $z^2 - 1$ will give rise to two immediate basin of attractions and for $z^3 - 1$ will give three immediate basin of attractions. With elementary analysis - as was done by Cayley - one can prove that in the case of $z^2 - 1$ the immediate basin of attractions does coincide with the basin of attractions and that the Newton iterates fail to be convergent for the boundary points, namely the y-axis. While the boundary is a straight line, the behavior of Newton's method is expected to be erratic on this boundary. In the case of $z^3 - 1$, without visualization it would be very difficult to describe the behavior of Newton's method. But even with a little bit of computation that would simply consider the repeated inverse images of a root, say $\theta = 1$ would have revealed the fact that the basin of attractions of the three roots would not merely be their Voronoi regions.

Figure 5.1 gives the well-known images - polynomiographs - corresponding to $z^2 - 1$ and $z^3 - 1$, as well as three other polynomials showing the basins of attraction and their boundaries. The fact that the boundary is smooth in the first example is no accident. It can be attributed to the fact that the Newton iterates applied to a point not on the y-axis actually form a subset of the Basic Sequence corresponding to that point, and as the general theory predicts, it should converge to the closets root to this point. This was proved in Chapter3 1 and the Basic Family described in Chapter 3. In the case of $z^3 - 1$, the boundary of basins of attractions, the Julia set in this case, obeys a fractal nature, turning the analysis of Newton's method into a much more difficult problem. In fact even for this simple cubic polynomial - despite the tremendous literature on the iterations of rational functions - one can state numerous open questions. The other examples are important examples and will be discussed in the chapter.

Fig. 5.1 Polynomiographs of $z^2 - 1$ and $z^3 - 1$ based on Newton's (first row); polynomiographs of $z^3 - 2z + 2$ and $z^3 - 3z + 3$ under Newton's (second row), of the same two under Halley's (third row); and polynomiographs of a $p(z) = 3z^5 - 10z^3 + 23z$ (a polynomial due to Barna) under Newton's and of $z^3 - 2z + 2$ under a McMullen's iteration function.

From the point of view of iteration functions, the Basic Family in particular, one can view this chapter as the study of the general theory of dynamical systems for answering many questions such as the ones stated above. In the process we shall learn about some very fascinating results from the theory of dynamical systems, as well as general theory of iteration functions for polynomial root-finding.

5.2 Iterations of a Rational Function

A *rational function* is a function of the form

$$R(z) = \frac{P(z)}{Q(z)}$$

where $P(z)$ and $Q(z)$ are complex polynomials with coefficients in $\mathbb{C}$ (the complex plane) assumed to be coprime, i.e. having no common zeros. The *degree* of R is defined as

$$d = deg(R) = \max\{deg(P), deg(Q)\}.$$

To analyze the properties of a rational map R in the complex plane $\mathbb{C}$, it is necessary to consider the *Riemann* sphere

$$\widehat{\mathbb{C}} = \mathbb{C} \cup \{\infty\},$$

also known as the *extended complex plane*. More precisely, we use *stereographic projection* to map $\mathbb{C}$ into the unit sphere S in three-dimensional Euclidean space centered at the origin:

$$S = \{(x_1, x_2, x_3) : x_1^2 + x_2^2 + x_3^2 = 1\}.$$

More precisely, given a complex number $z = x_1 + ix_2$, we use the north pole, the point $\xi = (0, 0, 1)$, to project z into S where the projection is the point of intersection of the ray through ξ and z with S. The projection point will be denoted by z^*. The ray will intersect the sphere either in the Northern Hemisphere, or the Southern Hemisphere, or the equator, depending upon z being outside the unit circle $\{(x_1, x_2) : x_1^2 + x_2^2 = 1\}$, inside the unit circle, or on its boundary (the equator).

Note that the origin gets mapped to the South pole $(0, 0, -1)$. The points of the complex plane which are very far from the origin get mapped to points very near ξ. This means we can associate the North pole with ∞ in the extended complex plane.

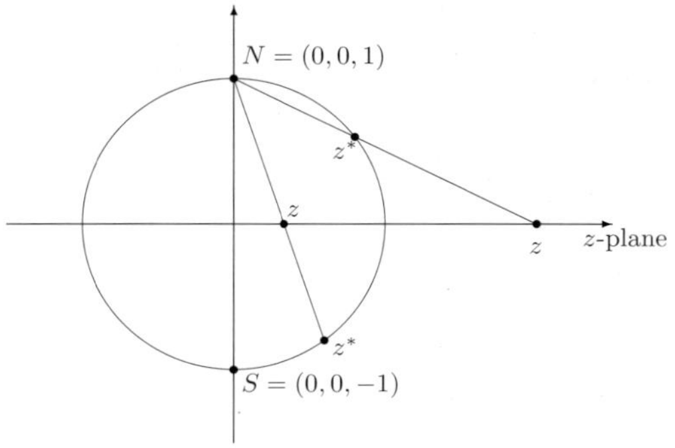

Fig. 5.2 Vertical cross-sectional depiction of stereographic projection.

The *chordal metric* on $\widehat{\mathbb{C}}$ defines a new metric on the points in the extended complex plane: the Euclidean distance between their stereographic projections. Mathematically,

$$\sigma(z, w) = |z^* - w^*| = \frac{2|z - w|}{\sqrt{(1 + |z|^2)(1 + |w|^2)}}.$$

Note that for fixed z, we have

$$\sigma(z, \infty) = \lim_{w \to \infty} \sigma(z, w) = \frac{2}{\sqrt{(1 + |z|^2)}}.$$

In particular, we get $\sigma(0, \infty) = 2$, as we should.

On the Riemann sphere we no longer need to worry about ∞, and use the chordal metric to handle convergence properties.

The *spherical metric* σ_0 is defined as the shortest length path between z^* and w^*, but restricted to S. Clearly σ and σ_0 are within a constant factor of each other.

A rational map $R(z)$ satisfies the Lipschitz condition on the Riemann sphere with respect to the metric σ (and σ_0), i.e.

$$\sigma(R(z), R(w)) \leq M\sigma(z, w),$$

where M is a constant. Now we think of $R(z)$ as a map

$$R : \widehat{\mathbb{C}} \to \widehat{\mathbb{C}}$$

and begin analyzing the dynamics assuming it is of degree d. An important property is that except for possibly a finite number of points w in $\widehat{\mathbb{C}}$ the set

$$R^{-1}(w) = \{w : R(z) = w\} = \{z_1, \dots, z_d\}$$

consists of d distinct elements in $\widehat{\mathbb{C}}$. This property is a consequence of the Fundamental Theorem of Algebra as applied to the polynomial

$$F_w(z) = P(z) - wQ(z).$$

Except for $w = 0$ and another $w \in \mathbb{C}$ which could make the leading coefficient to be zero, the degree of $F_w(z)$ is d. Then, if we have $F_w(z) = 0$, and $F'_w(z) = P'(z) - wQ'(z) = 0$, it implies

$$\frac{P(z)}{Q(z)} = \frac{P'(z)}{Q'(z)}.$$

But this equation has only finitely many solutions. In other words, except possibly for finitely many values of z, the polynomial $F_w(z)$ has no multiple zeros, and thus has d distinct solutions.

A point $w \in \widehat{\mathbb{C}}$ where the cardinality of $R^{-1}(w)$ is less than d is said to be a *critical value* of R. Thus at least one $z_i \in R^{-1}(w)$, when considered as root of $R(z) - w$, will have multiplicity greater than one. Such z_i is said to be a *critical point* of R. For example, consider

$$R(z) = \frac{(z-1)^2(z+1)}{(z-2)^3(z+3)}.$$

The points $z = 1$ and $z = 2$ are critical points corresponding to critical values 0 and ∞ respectively. Note that $R^{-1}(0) = \{1, \infty\}$. But in this case ∞ is not a critical point.

In general it turns out that the critical points will consist of the zeros of $R'(z)$, poles of $R(z)$ having multiplicity greater than one (i.e. multiple roots of $Q(z)$), and possibly ∞. A *pole* of R is a point $z_0 \in \mathbb{C}$ such that $R(z_0) = \infty$.

A critical point happens to be a point where R fails to be injective in every neighborhood of the point. It is easy to argue that the number of critical points is at most $2d - 1$. However, a more careful analysis gives the bound of $2d - 2$. We will prove this later.

Each Basic Family member

$$B_m(z) = z - p(z)\frac{D_{m-2}(z)}{D_{m-1}(z)},$$

is clearly a rational function. If $p(z)$ is a polynomial of degree n having only simple roots, then we may write

$$B_m(z) = \frac{P_m(z)}{Q_m(z)},$$

where P_m and Q_m have no common factors and

$$deg(B_m) = deg(P_m) = deg(Q_m) + 1 = n(n-1)^{m-2}.$$

What makes rational functions interesting is the dynamics of the fixed point iteration,

$$z_n = R(z_{n-1}), \quad n \geq 0,$$

where z_0 is a given initial complex number. We can alternatively write

$$z_n = R^n(z_0),$$

where R^n is the n-fold composition of R with itself.

The *orbit* of R at z_0 is the sequence

$$O^+(z_0) = \{z_n = R^n(z_0)\}_{n=0}^{\infty}.$$

This is also referred to as the *forward orbit*. The *backward orbit* at z_0 is

$$O^-(z_0) = \cup_{n=0}^{\infty} R^{-n}(z_0),$$

where

$$R^{-n}(z_0) = \{w : R^n(w) = z_0\}.$$

Note that $O^-(z_0)$ contains z_0.

Suppose S is a rational map of degree d'. We write RS for composition of R with S. Similarly with SR. Now $(RS)^{-1} = S^{-1}R^{-1}$. From the analysis of the finiteness of the number of critical points of a rational map it follows that except for a finite set of points z, $(RS)^{-1}(z)$ has dd' distinct elements. Thus we can argue:

Lemma 5.1.

$$deg(RS) = deg(SR) = dd'.$$

□

From this and induction we have:

Theorem 5.1. $\deg(R^n) = deg(R)^n = d^n$. □

Hence for all but finitely many z_0 the set $R^{-n}(z_0)$ contains d^n elements. As we shall see both the forward and backward orbits play important role in the analysis of the dynamics of rational function.

Definition 5.1. A *fixed point* of R is a point ξ such that $R(\xi) = \xi$.

Suppose for a given complex z_0 the corresponding orbit converges to a point ξ. Then from continuity of $R(z)$ it follows that ξ is a fixed point:

$$\xi = \lim_{n\to\infty} z_{n+1} = \lim_{n\to\infty} R(z_n) = R(\lim_{n\to\infty} z_n) = R(\xi).$$

Historically, the above property is the most important motivation behind fixed point iterations, the best known of which is for polynomial root-finding via Newton's method. Fixed points are special cases of periodic points:

Definition 5.2. A *periodic point* of R of period $p \geq 1$ is a point ξ such that $R^p(\xi) = \xi$ and $R^k(\xi) \neq \xi$ for any $k < p$. Thus a fixed point is a periodic point of period 1.

For instance, if we consider Newton's $N(z) = z - p(z)/p'(z)$ where $p(z) = z^3 - 2z + 2$, then the roots of $p(z)$ are the fixed points of $N(z)$. It can be easily checked that $N(0) = 1$ and $N(1) = 0$. Hence $N(0) = 0$ and $N(1) = 1$. Thus 0 and 1 are both periodic points of period 2. The orbit at either point gets trapped between the two points.

To each periodic point ξ we can attribute three quantities each of which plays a significant role in the dynamics of $R(z)$: period p, multiplier λ, and multiplicity m to be defined next. We will first consider fixed points.

Definition 5.3. Given a fixed point $\xi \in \mathbb{C}$ the quantity $\lambda = R'(\xi)$ is a well-defined point in $\mathbb{C}$ and is called its *multiplier*.

There are four different basic types of fixed points:

$$\xi : \begin{cases} \textit{superattractive}, & \text{if } \lambda = 0; \\ \textit{attractive}, & \text{if } 0 < |\lambda| < 1; \\ \textit{repelling}, & \text{if } |\lambda| > 1; \\ \textit{indifferent}, & \text{if } |\lambda| = 1. \end{cases}$$

An indifferent fixed point is said to be *rationally indifferent* or *parabolic* if λ is a root of unity, i.e. there exists a natural number n such that $\lambda^n = 1$. Otherwise, *irrationally indifferent*.

The same characterization applies to a periodic point ξ of period p where its multiplier is $\lambda = (R^p)'(\xi)$.

As an example in $N(z)$ for $p(z) = z^3 - 2z + 2$, the roots are superattractive. Using the chain rule it is easy to verify that $(N^2)'(0) = N'(0)N'(1) =$

0. Hence 0 is a superattractive periodic point of $N(z)$. Later we will consider this example in more detail.

Definition 5.4. A fixed $\xi \in \mathbb{C}$ of R is said to be of multiplicity m if ξ is a root of $F(x) = R(z) - z$ with multiplicity m, i.e. $F^{(j)}(\xi) = 0$ for all $j = 1, \ldots, m-1$. If $m = 1$ we will say that ξ is a simple fixed point of R.

It may be possible that ∞ is also a fixed point of R. In order to deal with ∞ as a fixed point it is useful to make use of Möbius transformations. We will consider these in detail. The function

$$g(z) = \frac{1}{z}$$

is a special case of a Möbius transformation and maps ∞ to zero. The inverse of $g(z)$ coincides with itself. The function

$$S(z) = gRg^{-1}(z) = g(R(g^{-1}(z))) = \frac{1}{R(1/z)} = \frac{z^d Q(1/z)}{z^d P(1/z)}$$

is the *conjugate* map of $R(z)$. If $R(\infty) = \infty$, then

$$S(0) = \frac{1}{R(\infty)} = \frac{1}{\infty} = 0,$$

so that zero is a fixed point of S. The multiplicity of ∞ as a fixed point of R is the multiplicity of 0 as a fixed point of S. In fact we have:

$$S'(0) = \frac{1}{\lim_{z \to \infty} R'(z)} = \frac{1}{R'(\infty)}.$$

In summary, to examine the nature of ∞ as a fixed point R is equivalent to considering the nature of 0 as a fixed point of S. In particular, we have:

Proposition 5.1. *If $p(z)$ is a polynomial of degree $d \geq 2$, then ∞ is a superattractive periodic point.*

A polynomial of degree d has $d+1$ fixed points in $\mathbb{C}$. For a rational map $R(z)$ if $d = deg(R(z)) = deg(P(z)) > deg(Q(z))$, then it is easy to see that R has d fixed points in $\mathbb{C}$ as well as ∞ so that it will have $d+1$ fixed points (counting with multiplicity). For instance, $R(z) = z + z^3$ has 0 as a fixed point with 3 as its multiplicity. More generally we have

Proposition 5.2. *A rational map R of degree $d \geq 1$ has precisely $d+1$ fixed points, counted with multiplicity.*

Proof. If $deg(Q) = d$, then $R(z) = z$ has $d+1$ fixed points in $\mathbb{C}$. Suppose $deg(Q) = r < d$. Then ∞ is a fixed point. The finite fixed points of R are solutions to $R(z) = z$. This equation may have no solutions, or as many as d solutions.

First consider the case when the equation has no solution. Then $P(z) - zQ(z)$ must be a nonzero constant c . In this case we must have $deg(P(z)) = d$, $deg(Q(z)) = d - 1$, and

$$P(z) = a_d z^d + \cdots + a_0, \quad Q(z) = a_d z^{d-1} + \cdots + a_1, \quad c = a_0.$$

We claim that zero is a fixed point of multiplicity $d + 1$ in the conjugate map $S(z)$. To show this, note that the finite fixed points of S must be the solutions to the equation

$$z^d Q(\frac{1}{z}) - z^{d+1} P(\frac{1}{z}) = -a_0 z^{d+1}.$$

But this proves the claim.

For the case where the equation $P(z) - zQ(z)$ has $r \geq 1$ solutions, again it is easy to show using the conjugate map $S(z)$ that ∞ as a fixed point of R has to have multiplicity equal to $(d + 1 - r)$. We omit this because analogous proof holds. □

As an example if $p(z) = z^2 - 1$, then

$$R(z) = B_2(z) = z - \frac{z^2 - 1}{2z} = \frac{z^2 + 1}{2z}.$$

The fixed points are the roots of $p(z)$, i.e. ± 1. It is easy to check that $R'(\pm 1) = 0$ so that these are attractive fixed points. But note that $R(\infty) = \infty$ so that ∞ is also a fixed point. Using the conjugate map it is easy to verify that $S'(0) = 2$ and thus ∞ is a repulsive fixed point of R.

As another example consider $R(z) = z^2$. Then ∞ is an attractive fixed point. Indeed for any z_0 with $|z_0| > 1$ the orbit converges to ∞. Alternatively, the conjugate map $S(z) = 1/R(1/z) = z^2$. This implies $S'(0) = 0$, thus ∞ is indeed a superattractive fixed point of R.

5.3 Newton's Method and Connections to Mandelbrot Set

For a general polynomial of degree n, consider Newton's function

$$R(z) = B_2(z) = z - \frac{p(z)}{p'(z)}.$$

Then

$$R'(z) = \frac{p(z)p''(z)}{p'(z)^2}.$$

Each multiple root of $p(z)$ is an attractive fixed point of $R(z)$, and each simple root is a superattractive fixed point. Moreover, ∞ is a fixed point of $R(z)$ and it can be shown to be repelling. There are no other fixed points.

In order to give computer visualization of the dynamics of Newton's method, or more general iteration functions for polynomial root-finding, or even general rational functions, the general idea is simple: for each point z_0 within a selected square, we assign a color or hue to a corresponding pixel of the computer monitor depending upon the number of iterations it would take to converge to a fixed point (including infinity) as measured with respect to a preassigned tolerance, or the number of iterations to detect lack of convergence. As a result a whole region may get assigned a single color and the boundaries of the regions, possibly within the same basin of attraction of a fixed point, will resemble curves. This can be seen in the typical polynomiographs corresponding to Newton's such as in Figure 5.1.

In Figure 5.1 the regions where the iterates converge are clearly seen and except for the boundaries all points converge to a root, except for the bright areas. If in contrast we consider the polynomial $p(z) = z^3 - 2z + 2$ and its corresponding polynomiograph we witness some white region as shown in Figure 5.3.

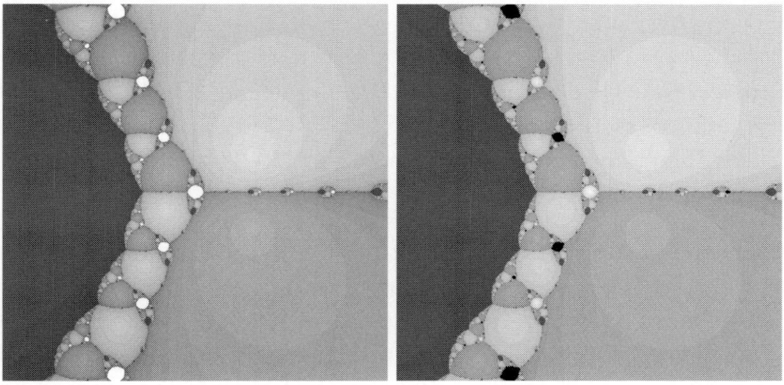

Fig. 5.3 Polynomiograph of $p(z) = z^3 - 2z + 2$ under Newton's (left), and polynomiograph of $p(z) = z^3 - 2z + 2$ under $R(z) = N^2(z)$.

How can we explain the white area around the origin? We know that it is a fixed point of $N^2(z)$. So if we iterate $R(z) = N^2(z)$ we might expect that white areas may decompose. Indeed as we see in polynomiograph in Figure 5.3 the white areas in the original polynomiograph decompose into white and black, white being the basin of attraction of 0 and 1 the basin of attraction of 1.

Near the origin, $N^2(z)$ behaves like its quadratic Taylor polynomial

$$q(z) = N^2(0) + (N^2)'(0)z + \frac{(N^2)''(0)}{2}z^2 = \frac{3}{2}z^2.$$

Let us now consider a slight variation of the polynomial and Newton's

$$p_\epsilon(z) = z^3 - (2+\epsilon)z + 2, \quad N_\epsilon(z) = z - \frac{p_\epsilon(z)}{p'_\epsilon(z)},$$

where ϵ is a complex number. We are actually interested in the cases where the norm of ϵ is small. Since again $N'_\epsilon(0) = 0$, the origin is a critical point. One may ask what happens to the orbits of Newton's near the origin as we vary ϵ. Figure 5.4 gives polynomiographs for some values of ϵ, as well as for more general perturbations of the form

$$z^3 + \epsilon_1 z^2 - (2+\epsilon_2)z + (2+\epsilon_3),$$

where $|\epsilon_i|$ is small.

At this point it is appropriate to give a definition that allows us to speak of the Mandelbrot set and its natural kind of connection to Newton's method and in this very same example of cubic polynomial and its perturbation.

Definition 5.5. Given a polynomial $P(z)$ of degree $d \geq 2$, the *filled Julia set* is defined as the set of all points $z \in \mathbb{C}$ whose orbit $O^+(z)$ stays bounded. The *Julia set* of $P(z)$, denoted by $J(P(z))$ is defined to be the boundary of the filled Julia set.

Since ∞ is a superattractive point, we can alternatively define the filled Julia set to be the set of points whose orbit does not converge to ∞, i.e. the complement of the basin of attraction of infinity. As we shall see the Julia set defined above will coincide with the Julia set as to be defined for an arbitrary rational map $R(z)$. Because the basin of attraction of ∞ is a non-empty open subset of $\widehat{\mathbb{C}}$ it follows that $J(P(z))$ is compact.

We would expect that near the origin, $N^2_\epsilon(z)$ would behave like its quadratic Taylor polynomial:

$$q_\epsilon(z) = N^2_\epsilon(0) + \frac{(N^2_\epsilon)''(0)}{2}z^2.$$

Thus we would expect that the filled Julia set of the quadratic Taylor polynomial would resemble divergence area near the origin corresponding to the iterations of Newton's method. In particular for $\epsilon = 0$ a closeup of the white area near the origin is the upper-left image in Figure 5.4. The corresponding Taylor polynomial $q_0(z) = 1.5z^2$. It is easy to see that the filled Julia set is merely a disk.

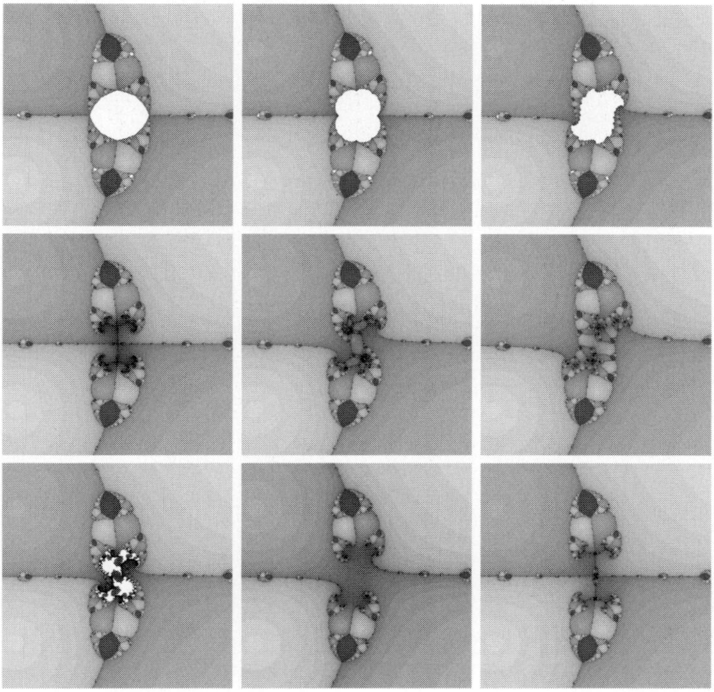

Fig. 5.4 Polynomiographs under Newton's corresponding to $p_\epsilon(z) = z^3 + (2+\epsilon)z + 2$ with $|\epsilon|$ small (top three rows); as well as polynomiographs corresponding to $z^3 + \epsilon_1 z^2 - (2+\epsilon_2)z + (2+\epsilon_3)$.

Let us consider the set

$$\widehat{\mathcal{M}} = \{\epsilon \in \mathbb{C} : \{N_\epsilon^n(0)\}_{n=1}^\infty \text{ does not converge to a root of } p_\epsilon(z)\}.$$

We would expect that the set $\widehat{\mathcal{M}}$ would give resemblance part of the *Mandelbrot set* defined as follows:

$$\mathcal{M} = \{c \in \mathbb{C} : \text{if } P_c(z) = z^2 + c, \{P_c^n(0)\}_{n=1}^\infty \text{ is bounded}\}.$$

The images of the Mandelbrot set are well familiar and will not be considered here.

In Figures 5.5 we give some familiar images corresponding to certain values of c. These seem to resemble what we have seen in Newton's iterations considered earlier.

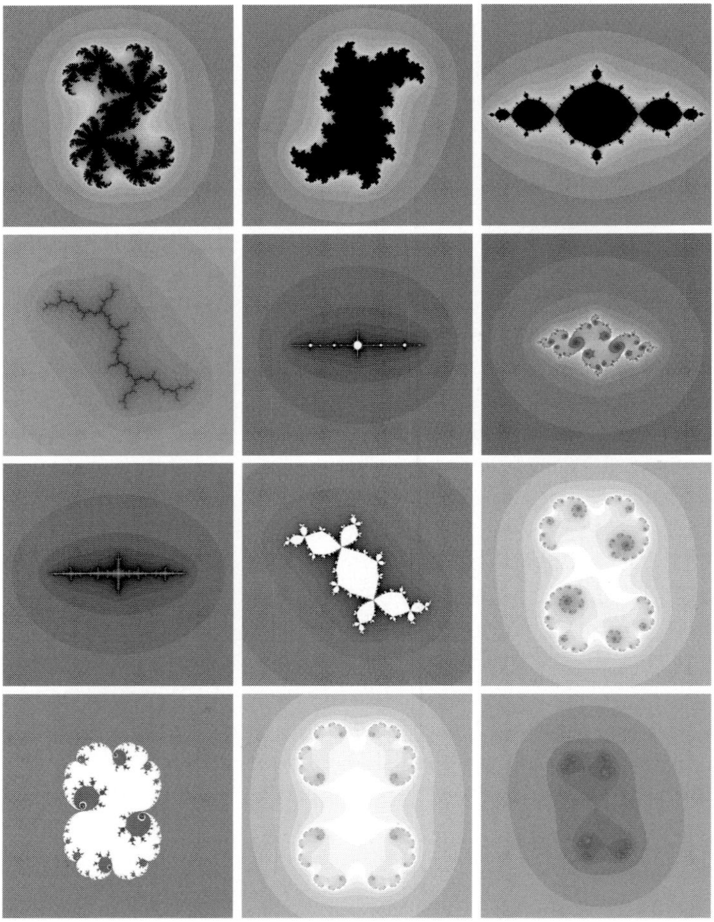

Fig. 5.5 Familiar fractal images of $P_c(z) = z^2 + c$ for some values of c.

It is not difficult to prove that the Mandelbrot set can equivalently be defined as the set of all $c \in \mathbb{C}$ such that the Julia set $J(P_c(z))$ is connected, see e.g. Falconer (1990). An auxiliary result to prove this is that given $P_c(z) = z^2 + c$ and a loop L in the complex plane, the inverse image of L, $P_c^{-1}(L)$ is a loop if c is inside the loop, and a figure-eight image if c lies on the boundary of L. The latter is a consequence of the fact that c is a critical value of $P_c(z)$ (since the origin is a critical point and $P_c(0) = c$).

It can be shown that when $|c| > 2$ the filled-Julia set of $P_c(z)$ coincides with the Julia set itself, it is a Cantor set and consisted of nested figure-eight shapes, see e.g. Devaney (1992). The image in the bottom right in Figure 5.5 shows the nesting of figure-eight loops.

Many properties of Mandelbrot set have been discovered and there is vast literature surrounding it. The case of dynamics of cubic polynomials has been a motivating problem for the definition of several significant discoveries as well as notions. The particular example $p(z) = z^3 - 2z + 2$ was considered by Smale (1985). Smale raised a question on the existence of generally convergent algorithms to be defined later. Douady and Hubbard (1985) studied the dynamics Newton's for cubic polynomials in connection with their relationship to the Mandelbrot set and more generally studied polynomial-like rational behavior of rational functions. The parameterized polynomials analyzed via Newton's are $p_a(z) = z^3 + (a-1)z - a$ and their analysis is given in Haesler and Peitgen (1989). See also Curry *et al.* (1983).

Having seen the appearance of the Mandelbrot set in a parameterized family of cubic polynomials in Newton's method one can justify the appearance of more general Mandelbrot sets in Newton's method and more generally in the Basic Family. We first give a definition:

Definition 5.6. Given a natural number $d \geq 2$ the *generalized Mandelbrot set* is defined to be

$$\mathcal{M}_d = \{c \in \mathbb{C} : \text{if } P_c(z) = z^d + c, \{P_c^n(0)\}_{n=1}^{\infty} \text{ is bounded}\}.$$

Equivalently, it is the set of all $c \in \mathbb{C}$ such that the Julia set of $P_n(z)$ is connected.

Proposition 5.3. *Let $p(z) = z^{d+1} - dz + d$, $d \geq 2$. Then the origin is a periodic point of $N(z) = z - p(z)/p'(z)$. Moreover, the Taylor polynomial of degree d of $N(z)$ is*

$$q(z) = \frac{N^{(d)}(0)}{d!} z^d.$$

Proof. The case of $d = 2$ has already been considered and it is easy to show that $\{0, 1\}$ forms a superattractive cycle. We shall return to this example again later in the chapter. The fact that $p^{(k)}(0) = 0$ for $k = 2, \ldots, d$ and repeated differentiation of $N'(z) = p(z)p''(z)/p'(z)^2$ implies the claimed Taylor polynomial at the origin. □

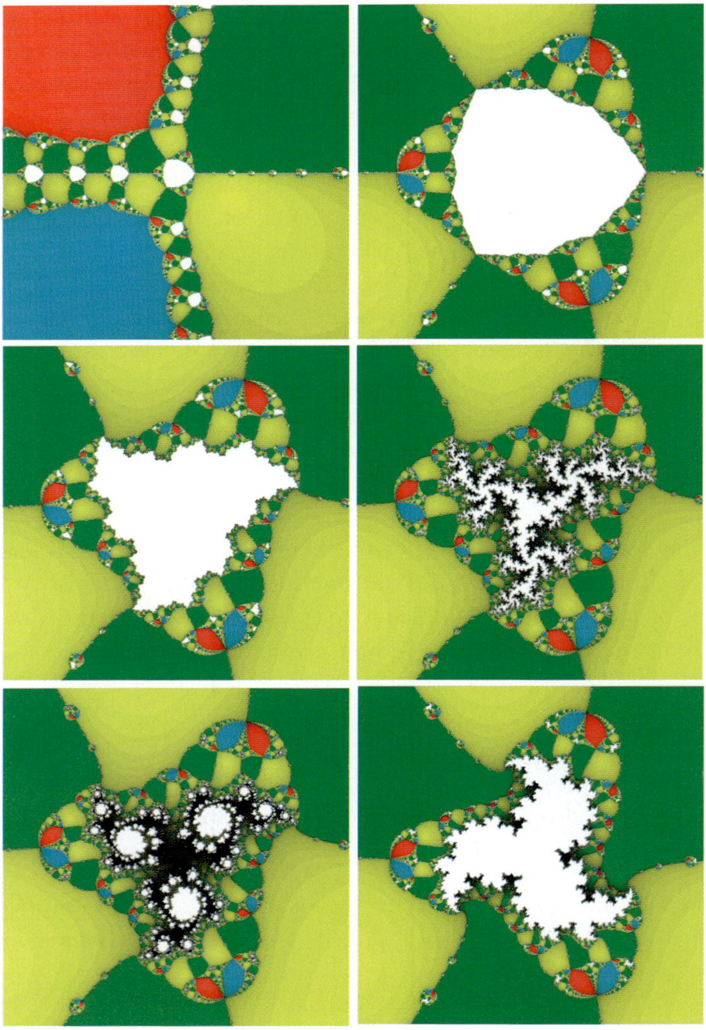

Fig. 5.6 Polynomiographs corresponding to $p(z) = z^4 - 3z + 3$ (top-left), its closeup near the origin (top-right); and closeup of $p_\epsilon(z) = z^4 + (3 + \epsilon)z + 3$ with $|\epsilon|$ small.

As in the case of $d = 2$ for general d we expect that if

$$p_\epsilon(z) = z^{d+1} - (d+\epsilon)z + d, \quad N_\epsilon(z) = z - \frac{p_\epsilon(z)}{p'_\epsilon(z)},$$

near the origin, $N_\epsilon^2(z)$ would behave like its Taylor polynomial of degree d. Thus we would expect that the set

$$\widehat{\mathcal{M}_d} = \{\epsilon \in \mathbb{C} : \{N_\epsilon^n(0)\}_{n=1}^{\infty} \text{ does not converge to a root of } p_\epsilon(z)\}$$

would give resemblance to part of the generalized Mandelbrot set $\mathcal{M}_d$. In Figures 5.6 we give the polynomiograph of $p(z) = z^4 - 3z + 3$ and those of $p_\epsilon(z) = z^4 - (3 \pm \epsilon) + 3$ depicting Newton's behavior near the origin.

In Figure 5.7 we give some images from iteration of $P_c(z) = z^3 + c$ for some value of c.

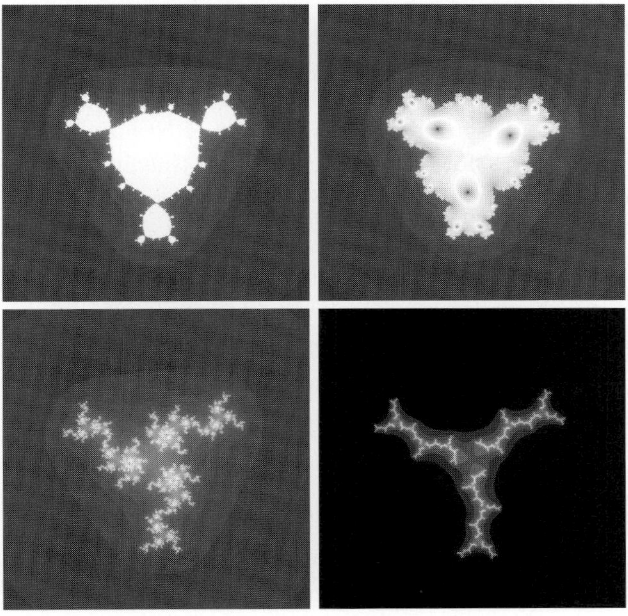

Fig. 5.7 Fractal images of $P_c(z) = z^3 + c$ for some values of c.

Again we see resemblance between the polynomiographs from Newton's method and the ones from the iterations of cubic polynomial. We could have observed this similarity for larger values of d. One would anticipate that the Mandelbort and generalized Mandelbrot sets would also appear in the case of any member of the Basic Family for parameterized polynomials.

McMullen (2007) has proved the versatility of generalized Mandelbrot in more general setting of parameterized rational functions.

Before resuming the study of general rational functions, we emphasize the fact that convergence properties are very much dependent on the particular choice of iteration functions. Figure 5.1 considers the polynomiography of $p(z) = z^3 - 2z + 2$ under the iterations of $B_3(z)$ (Halley's method), McMullen's method to be treated later in the chapter, as well as the polynomiography of $p(z) = z^4 - 3z + 3$ under Halley's method. As can be seen from the figure, the white areas have vanished in all three cases. Although it would take a formal proof that no such white area will exist, the examples emphasize the fact that the existence of white areas is not inherent in the polynomial, rather the particular iteration function.

5.4 Analysis of Infinity as Fixed Point

We have seen that ∞ is a repelling fixed point of Newton's method for the examples considered. More generally, we prove that in Newton's function ∞ is a repelling fixed point. Indeed we do this in more generality.

Theorem 5.2. *Suppose*

$$R(z) = \frac{a_n z^n + a_{n-1} z^{n-1} + \cdots + a_0}{b_m z^m + b_{m-1} z^{m-1} + \cdots + b_0}, \qquad n > m.$$

Then ∞ is a fixed point of R satisfying

$$\infty : \begin{cases} \textit{superattractive}, & \textit{if } m < n-1; \\ \textit{attractive}, & \textit{if } m = n-1,\ |a_n| > |b_m|; \\ \textit{repelling}, & \textit{if } m = n-1,\ |a_n| < |b_m|; \\ \textit{indifferent}, & \textit{if } m = n-1,\ |a_n| = |b_m|. \end{cases}$$

Proof. Clearly ∞ is a fixed point of R. We have

$$R'(z) = \frac{(na_n z^{n-1} + \dots)(b_m z^m + \dots) - (a_n z^n + \dots)(mb_m z^{m-1} + \dots)}{(b_m z^m + \dots)^2}$$
$$= \frac{a_n b_m (n-m) z^{n+m-1} + \dots}{b_m^2 z^{2m} + \dots}.$$

Thus

$$R'(\infty) = \begin{cases} \infty, & \text{if } m < n-1; \\ a_n/b_n, & \text{if } m = n-1. \end{cases}$$

Thus to complete the proof using the conjugate map $S(z)$ we simply need to evaluate $S'(0)$. □

Corollary 5.1. *If $p(z)$ is a polynomial of degree n, then ∞ is a repelling fixed point of Newton function, $N(z) = z - p(z)/p'(z)$.*

Proof. It is straightforward to show that $N'(\infty) = (n-1)/n$. □

Next we state a generalization of the corollary as stated for the members of the Basic Family. The complete proof will follow from results in subsequent chapters.

Theorem 5.3. *Given $p(z)$ a polynomial of degree n, each root is an attractive fixed point of $B_m(z)$ if it is a multiple root, and a superattractive fixed point if it is a simple root. Moreover, ∞ is a repelling fixed point.* □

In the case of $B_m(z)$, other than infinity there are many other fixed points than the roots of $p(z)$. The case of $m = 2$ is the only case where the roots of $p(z)$ are the only finite fixed point of $B_2(z)$.

5.5 Möbius Transformations and Conjugacy

A Möbius transformation (or Möbius map) is a rational map of degree one, thus of the form

$$R(z) = \frac{az+b}{cz+d}, \quad ad - bc \neq 0.$$

Except for the identity map $R(z) = z$, any Möbius map has either one or two fixed points which are easily computable. To understand the behavior of orbits it suffices to understand the two special cases:

$$R(z) = \begin{cases} z + \beta, & \beta \neq 0; \\ \lambda z, & \lambda \neq 0,\, \lambda \neq 1. \end{cases} \tag{5.1}$$

Every other case can be converted into one of the above through conjugacy. The first case is a translation and ∞ is the only fixed point. Trivially

$$R^n(z) = z + n\beta,$$

and as $\beta \neq 0$, for every z_0 the corresponding orbit $O^+(z_0)$ converges to ∞.

In case $R(z) = \lambda z$, $\lambda \neq 1$, $z = 0$ and $z = \infty$ are the only fixed points. Thus

$$R^n(z) = \lambda^n z.$$

If $|\lambda| < 1$, for each z_0, $O^+(z_0)$ converges to 0. If $|\lambda| > 1$, $O^+(z_0)$ converges to ∞. If $|\lambda| = 1$, two cases need to be considered.

If λ is an n-th root of unity, for each z_0 the orbit will cycle on the n points

$$\xi_k = z_0 e^{\frac{2\pi k}{n} i}, \quad k = 1, \ldots, n.$$

When $|\lambda| = 1$, but λ is not a root of unity, the orbit of z_0 will consist of a dense subset of the circle of radius $|z_0|$, i.e. given any point w on the circle, and $\epsilon > 0$, there exists an n such that $|w - R^n(z_0)| < \epsilon$. Here R is said to be rotation of infinite order on the unit disk centered at the origin.

These two cases explain the behavior of general Möbius maps. To prove this, first we define the notion of conjugacy which is very useful in the context of general rational maps.

Definition 5.7. Two rational maps R and S are said to be *conjugate* if there is a Möbius map g such that

$$S = gRg^{-1}.$$

The following theorem groups together the most significant properties of Möbius transformations through conjugacy as used throughout the chapter.

Theorem 5.4. *Suppose R and S are conjugate maps.*

(1) *R and S have the same degree.*

(2) *For any natural number n $S^n = gR^ng^{-1}$.*

(3) *If $R(\xi) = \xi$, then $S(g(\xi)) = g(\xi)$. Conversely, if $S(u) = u$, then $R(g^{-1}(u)) = g^{-1}(u)$.*

(4) *If ξ is a fixed point of R in $\mathbb{C}$ and $g'(\xi)$ is not 0 or ∞, then*

$$R'(\xi) = S'(g(\xi)).$$

In particular, if $g(z) = z - \xi$, then $S(z) = R(z + \xi) - \xi$ and $R'(\xi) = S'(0)$.

(5) *w is a critical value of R if and only if $g(w)$ is a critical value of S.*

Proof. To prove (1), from Lemma 5.1 we have

$$deg(S) = deg(g)deg(R)deg(g^{-1}).$$

But $deg(g) = deg(g^{-1}) = 1$.

To prove (2), by induction:

$$gR^ng^{-1} = g(g^{-1}S^ng)g^{-1} = S^n.$$

To prove (3), suppose ξ is a fixed point of R, then

$$S(g(\xi)) = gRg^{-1}(g(\xi)) = gR(\xi) = g(\xi),$$

and if $S(u) = u$, since $R = g^{-1}Sg$, we have

$$R(g^{-1})(u) = g^{-1}Sgg^{-1}(u) = g^{-1}S(u) = g^{-1}(u).$$

To prove (4), we have $Sg(z) = gR(z)$. From the Chain Rule we get

$$S'(g(z))g'(z) = g'(R(z))R'(z).$$

Substituting $z = \xi$ and canceling $g'(\xi)$ gives the desired result. The special case of $g(z) = z - \xi$ follows trivially.

To prove (5), taking the inverse in $S = gRg^{-1}$ we get $g^{-1}S^{-1}g = R^{-1}$. Since g^{-1} is one-to-one, the set $R^{-1}(w)$ has the same cardinality as the set $S^{-1}(g(w))$. The conclusion then follows from the definition of critical values. □

We now return to a general Möbius transformation R, and using conjugacy show that it is reducible to one of the two special cases in (5.1).

Theorem 5.5. *Suppose $R(z)$ is a non-identity Möbius map, then $R(z)$ is conjugate to one of the two cases in (5.1). If $R(z)$ is conjugate to the first case, it has one attractive fixed point. If conjugate to the second case, it has two fixed points, one attractive and one repulsive.*

Proof. We consider three cases.

(i) Suppose that R has two distinct roots ξ_1 and ξ_2. Assume one of these is ∞. Then $R(z) = az + b$ for some a, b with $a \neq 0$. If $b = 0$, $R(z)$ is the first case in (5.1).

If $b \neq 0$, and $a = 1$, $R(z)$ is the second case in (5.1). If $b \neq 0$ and $a \neq 1$, then the finite fixed point of R is $\xi = b/(1 - a) \neq 0$. Let $g(z) = 1/(z - \xi)$, then $g^{-1}(z) = (1/z) + \xi$. Letting $\lambda = 1/a$ we have

$$S(z) = gRg^{-1}(z) = \lambda z.$$

Thus

$$S^n(z) = gR^n g^{-1}(z) = \lambda^n z.$$

Applying g^{-1} to the above and evaluating at $z = g(z)$ we get

$$R^n(z) = \frac{z - \xi}{\lambda^n} + \xi.$$

Hence $R^n(z)$ converges to ξ if $|\lambda| > 1$, and to infinity if $|\lambda| < 1$.

(ii) Suppose that R has distinct fixed points ξ_1 and ξ_2, both in $\mathbb{C}$. Then the Möbius map

$$g(z) = \frac{z - \xi_1}{z - \xi_2},$$

satisfies $g(\xi_1) = 0$ and $g(\xi_2) = \infty$. Thus $g^{-1}(\xi_1) = \infty$ and $g^{-1}(\xi_2) = 0$ so that the Möbius map $S = gRg^{-1}$ satisfies $S(0) = 0$ and $S(\infty) = \infty$. Hence $S(z) = \alpha z$. Because of condition (4) of Theorem 5.4 $|\alpha| \neq 1$. If $|\alpha| < 1$, 0 is attractive, and if $|\alpha| > 1$, 0 is repelling. In particular, if a Möbius transformation has two fixed points one must be attractive, the other repelling.

(iii) Suppose $R(z) = (az+b)/(cz+d)$ has a single fixed point $\xi \in \mathbb{C}$. Then R is conjugate to the second case in (5.1). In particular, ξ is attractive. In this case c must be nonzero and we may assume without loss of generality that it is 1, i.e. the fixed point of R is the solution to $(az+b)/(z+d) = z$. This implies

$$z^2 + (d-a)z - b = (z-\xi)^2 = z^2 - 2\xi z + \xi^2 = 0,$$

which implies

$$d + \xi = a - \xi, \qquad -b = \xi^2, \qquad \Delta = (d-a)^2 + 4b^2 = 0.$$

Using the above and $g(z) = 1/(z-\xi)$, it is straightforward to show that if $d + \xi \neq 0$, then

$$S(z) = gRg^{-1}(z) = \frac{(d+\xi)z + 1}{a - \xi} = z + \beta,$$

where $\beta \neq 0$ and hence reduces to the first case of (5.1) having only ∞ as its fixed point.

Suppose $\xi = 0$, then $d = a \neq 0$ and hence $d + \xi \neq 0$. We claim if $\xi \neq 0$, then again $d + \xi$ is nonzero. Otherwise, $d = -\xi = -a$. Using this, $b = -\xi^2 = ad$. It follows that $R(z) = a$, contradicting $ad - bc \neq 0$. □

As we have seen Möbius maps can be completely characterized and the iterates $R^n(z)$ can be computed explicitly for any given z. The general case of rational functions is much more complex.

5.6 Periodic Points and Cycles of a Rational Function

In order to analyze the fixed point iterations of R at a given point z_0 we must keep in mind that in reality there are many iteration functions at work. More specifically, for each natural number p the orbit of R at z_0 contains the orbit of R^p at the same point:

$$\{R^{pn}(z_0)\}_{n=1}^{\infty} = \{(R^p)^n(z_0)\}_{n=1}^{\infty} \equiv O^+_{R^p}(z_0) \subset O^+(z_0).$$

If the orbit of R^p at z_0 is convergent, by continuity it converges to a fixed point of R^p. However it may or may not be a fixed point of R.

We recall a periodic point ξ is a fixed point of R^n for some $p \geq 1$. The period of a periodic point is the least natural number p such that it is a fixed point of R^p. In particular, a fixed point is a periodic point of period one.

Definition 5.8. If ξ is a periodic point of period p, then the set

$$\xi^{(p)} = \{\xi, R(\xi), R^2(\xi), \ldots, R^{p-1}(\xi)\}$$

consists of p distinct elements called, *cycle* or *periodic cycle* (see Figure 5.8).

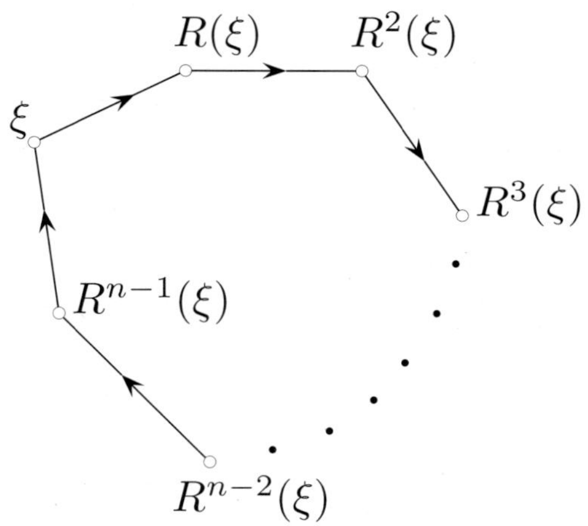

Fig. 5.8 Periodic cycle.

Setting $\xi_i = R^i(\xi)$, $i = 0, \ldots, p-1$, we have

$$R^p(\xi_i) = \xi_i,$$

so that each member on the cycle is a periodic point of R^p. Interestingly, from repeated application of the Chain rule we get

$$\lambda^{(p)} \equiv (R^p)'(\xi_i) = \prod_{j=0}^{p-1} R'(\xi_j), \quad \forall\, i = 0, \ldots, p-1. \tag{5.2}$$

This implies that R^p has the same derivative at each of its fixed points on the cycle. This common value can be considered as the *multiplier* of periodic cycle. For this reason, and analogous to the case of fixed points of R, the cycle $\xi^{(p)}$ will be classified as follows

$$\xi^{(p)} : \begin{cases} \textit{superattractive}, & \text{if } \lambda^{(p)} = 0; \\ \textit{attractive}, & \text{if } 0 < |\lambda^{(p)}| < 1; \\ \textit{repelling}, & \text{if } |\lambda^{(p)}| > 1; \\ \textit{indifferent}, & \text{if } |\lambda^{(p)}| = 1. \end{cases}$$

Here we have assumed that the cycle does not include ∞ but even if it does we can make use of a Möbius transformation to map the cycle so that none of the periodic points is ∞. For instance if the cycle contains ∞, but not zero, we can use the map $1/z$.

It turns out that R has infinitely many periodic points, however only finitely many of them are repelling. These are not obvious results. In fact it is not even obvious that R should have infinitely many periodic points. A more precise statement on the number of non-repelling periodic point is the following important theorem which we will state without proof for now but will return to it later.

Theorem 5.6 (Bound on Cycle, Shishikura (1987)). *Let R be a rational map of degree $d \geq 2$, then there are at most $2d - 2$ periodic cycles which are either attractive, superattractive, or indifferent.* □

It is not difficult to see that the number of superattractive periodic points, including ∞, is finite. This follows because at a finite periodic point, ξ, $R'(\xi) = 0$, i.e. ξ must be a critical point of R. But it turns out that the number of critical points of R, whether periodic points or not, plays a significant role in the bounding of the number of non-repelling cycles, as well as in the dynamics of rational iterations. In fact the bound $2d - 2$ in Theorem 5.6 is a bound on the number of critical points of R which is considerably easier to prove. We will consider these in the upcoming sections.

We will not prove the bound $2d - 2$ of Theorem 5.6 but will give partial proof. It is simpler to derive a larger bound, $6d - 6$, see Milnor (2006).

We will next analyze critical points and aim at proving the infinity of their cardinality and their types as (super)attractive, indifferent, and repelling.

5.7 Critical Points and Their Cardinality

Previously we have defined critical points and values. Here define them formally and derive some of their properties.

Definition 5.9. Assume $z_0 \in \mathbb{C}$, and not a pole of $R(z) = P(z)/Q(z)$. The *valency* of R at z_0, denote by $v_R(z_0)$, is its multiplicity as a zero of $R(z) - R(z_0)$. In other words, it is the natural number k such that

$$R(z) - R(z_0) = (z - z_0)^k g(z), \quad g(z_0) \neq 0.$$

If z_0 is a pole of R we define its valency to be its multiplicity as a root of $Q(z)$. When $z_0 = \infty$ its valency is defined to be the valency of 0 in $S(z) = 1/R(1/z)$.

If $z \in \mathbb{C}$ is zero of $R'(z)$, then it is a *critical point* of R. Other critical points are from among the poles of R, or infinity.

Remark 5.1. We point out that valency and multiplicity are related. The use of multiplicity with respect to roots of an equation is well-known. The word valency is used with respect to an arbitrary point z_0. When z_0 is a fixed point, we will use the word multiplicity for its valency.

Definition 5.10. A critical point of R is a point z at which R fails to be injective in any neighborhood of z. Equivalently, a critical point is a point $z \in \widehat{\mathbb{C}}$ at which $v_R(z) > 1$. Moreover, $R(z)$ is a critical value.

Remark 5.2. Thus if z is a critical point, then $w = R(z)$ is a critical value. However, not every point in $R^{-1}(w)$ is necessarily a critical point. If R has degree d and w is not a critical value, then

$$R^{-1}(w) = \{z_1, \ldots, z_d\}$$

with distinct elements. Clearly if z is not a fixed point it is distinct from z_j. Since z is not a critical point, there exists a neighborhood N of w, and neighborhoods $N_j, j = 1, \ldots, d$ corresponding to $z_j, j = 1, \ldots, d$ such that if R_j is the restriction of R into N_j, then it is one-to-one and onto. By the Inverse Function Theorem there exists inverse analytic map R_j^{-1} from N into N_j, $j = 1, \ldots, d$. These are called *branches* of R^{-1} at w. In other words at a non-critical value w, there is a neighborhood N so that R^{-1} maps the neighborhood bijectively into d neighborhoods of the points z_j. Since these points are distinct, N can be taken so that the d neighborhoods will also be disjoint. For the reason above a rational map of degree, except for its critical points, is considered as a d-fold mapping of the Riemann sphere into itself.

The *critical set* of R, denoted by C, is the set of critical points of R. It is a significant set and so is the dynamics of the set under the iterates of R.

The remaining part of this section is devoted to the proof of the following important theorem.

Theorem 5.7. *A rational map R of degree $d \geq 2$ has $2d-2$ critical points, counting with multiplicity.*

Remark 5.3. There is a topological proof of the above theorem which follows from a formula credited to Riemann-Hurwitz. Our proof is simple and direct.

We first state and prove some auxiliary results.

Lemma 5.2. *Suppose $R(z) = P(z)/Q(z)$ is a rational map of degree $d \geq 2$. The finite critical points of R are the distinct roots of*

$$P(z)Q'(z) - P'(z)Q(z).$$

In particular, there are at most $2d-2$ critical points, counted with multiplicity.

Proof. Consider

$$R'(z) = \frac{P'(z)Q(z) - P(z)Q'(z)}{Q(z)^2}.$$

If R has no pole of multiplicity greater than one (equivalently Q having no multiple roots), then the equation $R'(z) = 0$ corresponds to

$$P'(z)Q(z) - P(z)Q'(z) = 0.$$

However, for each multiple root α of Q the numerator and denominator of $R'(z)$ have $(z - \alpha)$ as a common factor. Hence the number of solutions of $R'(z) = 0$ decreases. To account for these poles and their multiplicity we consider the following approach.

Let α be a nonzero point in $\mathbb{C}$ that is not a critical point of R. Set $S(z) = gRg^{-1}(z)$, where

$$g(z) = \frac{1}{z - \alpha}.$$

Then

$$T(z) = Sg(z) = \frac{Q(z)}{P(z) - \alpha Q(z)}$$

has the property that it has no multiple poles. Moreover, it follows from properties of conjugate maps (Theorem 5.4, (4)) that a point $c \in \mathbb{C}$ is a critical point of R if and only if $g(c)$ is a critical point of $T(z)$. But the finite critical points of $T(z)$ are precisely the roots of $P'Q - Q'P$.

The claimed bound on the number of critical points is trivial when $deg(P)$ or $deg(Q)$ is less than d, and since when $deg(P) = deg(Q) = d$, there is a cancelation of the highest term in $P'Q - Q'P$. □

Lemma 5.3. *The only cases when ∞ is not a critical point of R is when $deg(P) + deg(q) = 2d - 1$, as well as the case when $deg(P) = deg(Q) = d$ and $deg[P(z) - Q(z)R(\infty)] < d - 1$. Moreover, in these cases*

$$m_R(\infty) = |deg(P) - deg(Q)| = (2d - 2) - deg(P'Q - PQ').$$

Proof. As in the proof of Theorem 5.2 it follows that when $deg(P) = d = deg(Q) + 1$, ∞ is a non-superattractive fixed point of R. Hence, zero is a non-superattractive fixed point of $S(z) = 1/R(1/z)$. It follows that ∞ is not a critical point.

If $deg(P) = d \geq deg(Q) + 2$, ∞ is a superattractive fixed point of multiplicity $deg(P) - deg(Q)$. This follows because in this case

$$S(z) = \frac{1}{R(\frac{1}{z})} = z^{deg(P) - deg(Q)} G(z)$$

where $G(z)$ is a rational function with $G(0)$ finite and nonzero, so that 0 has multiplicity $deg(P) - deg(Q) \geq 2$. On the other hand, if $deg(Q) = d \geq deg(P) + 2$, then $R(\infty) = 0$. In this case

$$S(z) = z^{deg(P) - deg(Q)} H(z),$$

for some rational function $H(z)$ such that $H(0)$ is finite and nonzero. In this case 0 is a pole of S of multiplicity $deg(Q) - deg(P) \geq 2$.

Suppose $deg(P) = deg(Q) = d$. Then

$$R(z) = \frac{a_d z^d + \cdots + a_r z^r + \cdots + a_s z^s}{b_d z^d + \cdots + b_u z^u + \cdots + a_v z^v}.$$

Here at least one of the indices s or v must be zero since P and Q must be coprime. Then $R(\infty) = a_d/b_d$. Let us assume that r is the first nonzero coefficient of P so that $a_r/b_r \neq a_d/b_d$ (with $b_r = 0$ allowed). Likewise, let u be the first nonzero coefficient of $Q(z)$ such that $a_u/b_u \neq a_d/b_d$. Then it easily follows that

$$deg\left(P(z) - \frac{a_d}{b_d} Q(z)\right) = \max\{r, u\}.$$

We have

$$S(z) = \frac{b_d + \cdots + b_u z^{n-u} + \cdots + a_v z^{n-v}}{a_d + \cdots + a_r z^{n-r} + \cdots + a_s z^{n-s}}.$$

Thus

$$S(0) = \frac{1}{R(\infty)} = \frac{b_d}{a_d},$$

and 0 is a critical point of multiplicity k if $S^{(j)}(0) = 0$ for all $j = 1, \ldots, k-1$. To find k it suffices to show that

$$S'(z) = z^{k-1}H(z)$$

for some rational function $H(z)$ which is not 0 or infinity at $z = 0$. But from the expression for $S(z)$ it follows that

$$S'(z) = z^{d-1-\max\{r,u\}}H(z)$$

for some such $H(z)$. Thus the multiplicity of ∞ is $d - \max\{r, u\}$. In particular, if $\max\{r, u\} = d - 1$, it implies that ∞ is not a critical point.

To complete the proof we need to show that the critical points with multiplicity again add up to $2d-2$. We avoid doing this since it is a matter of similar analysis as the first case. □

Finally, the two lemmas imply Theorem 5.7. We conclude the section with a property of the critical points and values of R^n.

Theorem 5.8. *Let C be the set of critical points of R. Then*
(i) The set of critical points of R^n is

$$C \cup R^{-1}(C) \cup \cdots \cup R^{-n}(C).$$

(ii) The set of critical values of R^n is

$$R(C) \cup R^2(C) \cup \cdots \cup R^n(C).$$

Proof. Without loss of generality we assume R^n has only finite critical points. Then by the Chain Rule we have

$$(R^n)'(z) = R'(R^{n-1}(z)) \cdots R'(R(z))R'(z).$$

Thus the left-hand side is zero at a given z if and only if

$$R^j(z) \in C, \quad j = 1, \ldots, n-1.$$

This precisely implies that z belongs to the set in (i). The image of the critical set under R^n is then precisely the set in (ii). □

5.8 Cardinality of Periodic Points of Different Types

In this section we prove the infinity of periodic points for rational maps of degree higher than one and analyze the cardinality of various types. The latter issue plays an important role in the dynamics of R^n and we will return to it later. A periodic point ξ is a fixed point of R^p for some $p \geq 1$, i.e. solution of

$$R^p(z) = z.$$

Intuitively, one would expect that as we consider the solutions for all p that we must get an infinite number of solutions, hence periodic points. But the proof of the intuitive fact is certainly non-trivial. We first prove the existence of infinitely many periodic points when $d \geq 2$.

Lemma 5.4. *Let R be a rational map of degree $d \geq 2$. Suppose ξ is a fixed point of R. Moreover, assume that if ξ is a rationally indifferent fixed point its multiplier λ equals 1. Then for all $n \geq 1$, ξ has the same multiplicity in R^n.*

Proof. Since we can use conjugacy, without loss of generality we may assume $\xi = 0$. Thus we may write

$$R(z) = az + bz^k + \dots,$$

with $b \neq 0$, $k > 1$ (since $d > 1$). From the assumption of the lemma either $a = 0$ or $a = 1$, or a nonzero number such that $a^n \neq 1$ for all $n \geq 1$.

Suppose $a = 0$. Then it is straightforward to see that

$$R^n(z) = b^n z^{kn} + \dots.$$

Thus for all $n \geq 1$ $\xi = 0$ is a simple fixed point, i.e. a simple root of $R^n(z) - z$.

Suppose $a = 1$, then it is easy to see

$$R^n(z) = z + nbz^k + \dots.$$

Thus

$$R^n(z) - z = nbz^k + \dots,$$

implying that for all $n \geq 0$, $\xi = 0$ has multiplicity k.

Assume $a \neq 0$ and $a \neq 1$. Since $deg(R^n) = d^n$ we may conclude that

$$R^n(z) = a^n z + b' z^{k'} + \dots,$$

for some $b' \neq 0$ and $k' \geq k$. Thus

$$R^n(z) - z = \alpha z + \beta z^t + \dots,$$

where $\alpha \neq 0$. It follows that under the assumption of the lemma for all $n \geq 1$, $\xi = 0$ is a simple root of $R^n(z) - z$. □

Theorem 5.9. *A rational map R of degree $d \geq 2$ has infinitely many periodic points.*

Proof. First we assume that if ξ is any rationally indifferent fixed point of R, then $R'(\xi) = 1$.

Let $p > 2$ be any prime number. We claim that R^p has a fixed point different from those of R. Let θ be any fixed point of R^p. If it is a fixed point of R, then from the previous lemma it must have the same multiplicity in $R(z) - z$ and $R^p(z) - z$. From Proposition 5.2 R must have $(d+1)$ fixed points. Thus the multiplicity of θ in $R^p(z) - z$ is at most $(d+1)$. If R^p has no other fixed points than those of R, each having multiplicity at most $(d+1)$, then R^p would have at most $(d+1)^2$ fixed points. But by Proposition 5.2 the number of fixed points of R^p is $d^p + 1$ and since $p > 2$ this number exceeds $(d+1)^2$. Hence the existence of a distinct periodic point for a given prime $p > 2$.

Since periodic points corresponding to different prime periods must be distinct, this completes the proof of the infinity of periodic point for the rational maps R having only rationally indifferent fixed points with multiplier equal one.

Next we consider the general case of R. If R has finitely many periodic points it has finitely many rationally indifferent periodic points. If no such periodic point exists, we can use previous argument to conclude the infinity of periodic points. Suppose that R has t periodic points, $\xi_1, \ldots, \xi_t$ of periods $p_1, \ldots, p_t$, multipliers $\lambda_1, \ldots, \lambda_t$, which are respectively, $n_1, \ldots n_t$-th root of unity. Set

$$p = \prod_{i=1}^{t} p_i n_i.$$

Then clearly each ξ is a fixed point of R^p. Moreover from the Chain rule $(R^p)'(\xi) = 1$. R^p has no other rationally indifferent periodic points that $\xi, i = 1, \ldots, t$ since any such point is a rationally indifferent periodic point of R. Thus by the above argument R^p must have an infinite number of periodic points and so does R. □

Remark 5.4. Theorem 5.9 only established the infinity of the periodic points. It turns out that when $deg(R) \geq 2$, for all $p \geq 5$ there is a periodic point of period p (see Beardon (1991), Theorem 6.2.2).

From Theorems 5.6 and 5.9 we have:

Corollary 5.2. *A rational map R of degree $d \geq 2$ has infinitely many repelling periodic points.* □

Lemma 5.5. *If $deg(R) \geq 2$, then R has at most $2d-2$ superattractive periodic points.*

Proof. To prove the lemma to each superattractive periodic point of R we associate a unique critical point of R. Since there are at most $2d-2$ distinct critical points (Theorem 5.7) the proof will follow.

Let $\xi \in \mathbb{C}$ be a superattractive periodic point of R of period p, belonging to a cycle $\xi^{(p)} = \{\xi_j, j = 1, \ldots, p\}$. Since ξ is a superattractive fixed point of R^p, $(R^p)'(\xi) = 0$. But from formula (5.6) we get

$$(R^p)'(\xi) = \prod_{j=1}^{p} R'(\xi_j) = 0,$$

where $\{\xi_j, j = 0, \ldots, p-1\}$ is the corresponding cycle. Thus, $R'(\xi_j) = 0$ for some j. But the cycle containing ξ is unique. Otherwise, ξ is at the intersection of two cycles, hence it must lie on a smaller cycle, contradicting that its period is n. If $\xi = \infty$, we can use a Möbius transformation to map ξ to an element in $\mathbb{C}$ and then repeat the above argument. □

For convenience we give the following definition:

Definition 5.11. A fixed point (periodic point) of a rational map R is said to be an *isolated fixed point* (periodic point) if there exists a neighborhood U of ξ containing no periodic point other than ξ.

Remark 5.5. It turns out that in every neighborhood of a repelling periodic point there is another repelling periodic point. This will be proved later. Clearly there are only countably many periodic points. It is also easy to argue that attractive or superattractive periodic points are isolated.

5.9 Local Behavior of Iterations Near Fixed Points

We wish to analyze the local behavior of fixed point iterations of $R(z)$ near periodic points. It turns out that if we understand the behavior near a fixed point, then we can extend it to a period point of say period p by applying the results to $R^p(z)$. Thus we first consider the case of a fixed point ξ of R.

Theorem 5.10. *Suppose ξ is a fixed point of a rational map R and $\lambda = R'(\xi)$ satisfies $|\lambda| < 1$. Then there exists $r > 0$, $0 < \rho < 1$ such that for all z satisfying $|z - \xi| < r$ the fixed point iterates*

$$|R^n(z) - \xi| \leq \rho^n |z - \xi|.$$

In particular, $R^n(z)$ converges to ξ (uniformly).

Proof. Using conjugacy without loss of generality we may assume $\xi = 0$. Thus for z in disk D around the origin we may write

$$R(z) = R'(0)z + \frac{R''(0)}{2!}z^2 + \frac{R'''(0)}{3!} + \cdots = \lambda z + z^k g(z),$$

where $k \geq 2$, $g(0) \neq 0$. We write

$$R(z) = z(\lambda + z^{k-1}g(z)).$$

There exists $r > 0$ such that

$$|\lambda + z^{k-1}g(z)| < \rho < 1.$$

Thus for all z satisfying $|z| < r$ we have

$$|R(z)| \leq \rho|z|.$$

Iterating we get

$$|R^n(z)| \leq \rho|R^{n-1}(z)| \leq \cdots \leq \rho^n|z|.$$

□

The next theorem characterizes repelling fixed points.

Theorem 5.11. *Suppose ξ is a fixed point of a rational map R and $\lambda = R'(\xi)$ satisfies $|\lambda| > 1$. Then there exists $r > 0$, $\rho > 1$, such that for all $z \neq \xi$ satisfying $|R^j(z) - \xi| < r$, $j = 0, \ldots, n-1$ we have*

$$|R^n(z) - \xi| > \rho^n|z - \xi|.$$

In particular, for each z in the disk $D_r(\xi) = \{z : |z - \xi| < r\}$, there exists $n \geq 1$ such that $R^n(z)$ lies outside of the disk.

Proof. We will first prove the theorem for $n = 1$. As in the previous theorem without loss of generality we may assume $\xi = 0$ and for z in some closed disk D around the origin of radius r_1 write

$$R(z) = \lambda z + z^k g(z),$$

where $|\lambda| > 1$, $k \geq 2$ (since degree of R is at least two), $g(0) \neq 0$. If M is a bound on $|g(z)|$ for $z \in D$, we may write

$$|R(z)| \geq |\lambda||z| - M|z|^k.$$

Now let ρ be a number satisfying $1 < \rho < |\lambda|$. Then for any $z \neq \xi$ satisfying

$$|z| \leq r = \min\{r_1, (\frac{(|\lambda| - \rho)}{M})^{1/(k-1)}\}$$

we have

$$R(z) \geq |\lambda||z| - M|z|^k \geq \rho|z|.$$

Thus we have proved the desired inequality for $n = 1$.

Now given $z \neq \xi$, and $n \geq 1$, suppose $|R^j(z)| < r$, for all $j = 0, \ldots, n-1$. Then by a simple induction we have

$$|R^n(z)| > \rho|R^{n-1}(z)| > \cdots > \rho^n|z|.$$

Clearly, since ρ^n approaches infinity, for each z in $D_r(0)$, there exist $n \geq 1$ such that the norm of $R^n(z)$ will exceed r. □

Remark 5.6. Let us emphasize the contrast between an attractive (or superattractive) fixed point of R and a repelling one. According to Theorem 5.10 if ξ is attractive there is a disk U around ξ so that the image $R(U)$ of the point remains in a contracted disk U' contained in U with a contraction factor $\lambda = R'(\xi)$, see Figure 5.9. Since we can repeatedly apply this, we conclude that $R^n(U)$ converges to ξ as n approaches infinity. In contrast Theorem 5.11 asserts if ξ is a repelling fixed point, there is a disk U around ξ so that under the map R everything in U, except for the fixed point itself, gets thrown out of an extracted disk U' containing U where the extraction factor is $\lambda = R'(\xi)$, see Figure 5.10. In other words the intersection of $R(U)$ and U' is the single point ξ. In fact Theorem 5.11 is a very conservative estimate of how strange and chaotic iterates will behave for points in an open neighborhood of a repelling fixed point. More specifically, as we shall see later, within such a neighborhood, or in fact any open neighborhood of a repelling periodic points, there are forward orbits that will reach all but at most two points on the Riemann sphere, and within a finite number of iterations.

The analysis of the local behavior of the case of indifferent periodic points is much harder. We will consider a rationally indifferent fixed point of a rational map of degree $d \geq 2$, and through conjugacy we may assume without loss of generality the fixed point $\xi = 0$. Furthermore we assume $\lambda = R'(0) = 1$. Thus,

$$R(z) = z - az^{m+1} + O(z^{m+2}), \quad a \neq 0, \quad m \geq 1, \tag{5.3}$$

i.e. 0 is a fixed point of multiplicity $m+1$. We state two results on such fixed point without proof and refer the reader to Beardon (1991) for complete proof. These are stated for the special case when $a = 1$, however with simple modifications they can be stated for general case of a.

Roughly speaking the first theorem implies that a rationally indifferent fixed point on appropriately selected regions behaves as a combination of an attracting and a repelling periodic point.

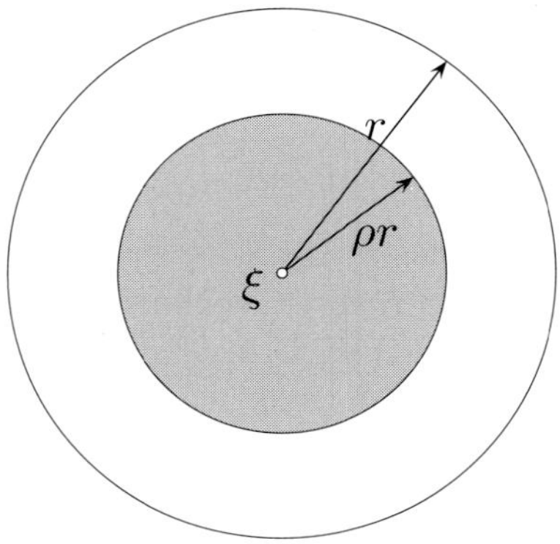

Fig. 5.9 Contraction of a neighborhood around a (super)attractive fixed point.

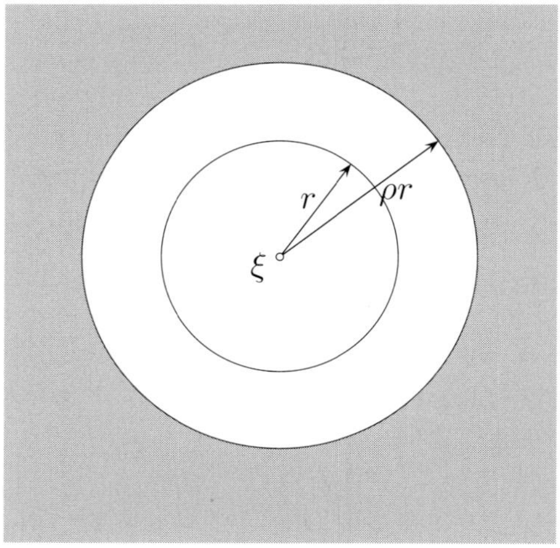

Fig. 5.10 Repulsion of a neighborhood at a repulsive fixed point.

Theorem 5.12. *Let R be a rational satisfying (5.3) where $a = 1$. Let $w_1, \ldots, w_m$ be the m-th root of unity and $\eta_1, \ldots, \eta_m$ be the m-th root of -1. Then there exist radius r_0, and angle θ_0 such that for all $j = 1, \ldots, m$ the sectors S_j and Σ_j defined by*

$$S_j = \{z : 0 < |\frac{z}{w_j}| < r_0, \quad |arg(\frac{z}{w_j})| < r_0\},$$

$$z \in \Sigma_j = \{z : 0 < |\frac{z}{\eta_j}| < r_0, \quad |arg(\frac{z}{\eta_j})| < r_0\}$$

satisfying

$$|R(z)| < |z|, \quad \forall z \in S_j,$$

$$|R(z)| > |z|, \quad \forall z \in \Sigma_j.$$

□

The next theorem is a simple version of what is called the *Petal Theorem*, see Beardon (1991) (Theorem 6.5.4). For another version of this see the *Parabolic Flower Theorem* in Milnor (2006) (Theorem 10.7). These theorems give a detailed description of the dynamics of fixed point iterates near an irrationally indifferent fixed point. Such a fixed point attracts points in regions that like a petal of a flower with the fixed point at its tip. The number of petals happens to be m. Each petal P_k, $k = 1, \ldots, m$ is contained in a simply connected open region L_k, called parabolic Fatou component to be defined later in the chapter. It turns out that the orbit of any point in the parabolic component L_k converges to the parabolic fixed point which is necessarily a boundary point of L_k. Here we have

Theorem 5.13. *Let R satisfy conditions in (5.3). For a given $t > 0$ and each $k = 0, \ldots, m-1$, define the petals*

$$\Pi_k(t) = \{re^{i\theta} : r < t(1 + \cos(m\theta)), \quad |\frac{2k\pi}{m} - \theta| < \frac{\pi}{m}\}.$$

Then, for t small enough R maps each petal into itself. Moreover, for each $z \in \Pi_k(t)$, $O^+(z)$ converges to 0. □

Let ξ be an irrationally indifferent fixed point. Thus the multiplier λ has modulus one, but it is not a root of unity. Consider the special case of $R(z) = \lambda z$. This is a Möbius map and $R^n(z) = \lambda^n z$. The iterates at a given z_0 remain within a circle of radius one at the origin. Here R is a rotation of infinite order on the unit disk centered at ξ. In a sense, through conjugacy a more general case of an indifferent periodic point is reducible to this special case of R, however not through Möbius transformations, but a more general case called analytic conjugacy.

Definition 5.12. Suppose $\xi \in \mathbb{C}$ is a fixed point of a rational map R with multiplier $\lambda \neq 0$. R is said to be *linearizable* at ξ if there exists a neighborhood N of ξ with no poles of R, and an analytic map g such that $g(\xi) = \xi$, it is injective on $N \cup R(N)$, and satisfying

$$gRg^{-1}(z) = \xi + \lambda(z - \xi), \quad \forall z \in g(N).$$

In this case R is said to be analytically conjugate to a rotation of infinite order on the unit disk.

The question of linearizability of a rational map at an irrationally indifferent periodic point has been a deep and delicate question. Cremer proved the existence of rational maps which are not linearizable at some irrationally indifferent fixed points. Such irrationally indifferent fixed point is not an isolated fixed point. Moreover, Siegel (1942) proved the existence of rational map R which are linearizable at some irrationally indifferent fixed points.

Using the terminology of isolated fixed points we have defined, see (5.11), we can state the following theorem which will forgo the technicalities in the characterization of local behavior of fixed points. The complete proof relies on the characterization of rationally indifferent points.

Theorem 5.14. *Suppose $R(z)$ has a fixed point ξ with multiplier $\lambda \neq 0$. Then R is linearizable at ξ if and only if ξ is an isolated fixed point.* □

Definition 5.13. An irrationally indifferent fixed point of a rational map R is called *Cremer point* if R is not linearizable at ξ and a *Siegel point* if R is linearizable at ξ.

Siegel point gives rise to *Siegel disk* as Fatou component, to be discussed later. There are also rational maps having a Cremer point.

Let us now consider the local behavior near a periodic point

$$\xi^{(p)} = \{\xi, R(\xi), R^2(\xi), \ldots, R^{p-1}(\xi)\}.$$

Given a point z_0 we can define

$$z_0^{(p)} = \{z_0, R(z_0), R^2(z_0), \ldots, R^{p-1}(z_0)\},$$

and for each $n \geq 1$ set

$$z_n^{(p)} = R(z_{n-1}^{(p)}) = R^n(z_0^{(p)}) = \{R^n(z_0), R^{n+1}(z_0), \ldots, R^{n+p-1}(z_0)\}.$$

The corresponding orbit is defined as:

$$O^+(z_0^{(p)}) = \{z_n^{(p)} = R^n(z_0^{(p)})\}_{n=0}^{\infty}.$$

Thus the set $z_0^{(p)}$ moves to a new set under successive application of R. The orbit may or may not converge to the cycle $\xi^{(p)}$. The fact that the multiplier $\lambda^{(p)}$ at each element of $z_n^{(p)}$ is the same number guarantees that the behavior of orbit of R^p at $R^i(z_0)$ is invariant under $i = 0, \ldots, p-1$. Figure 5.11 depicts the local behavior at a (super)attractive cycle $\xi^{(p)}$.

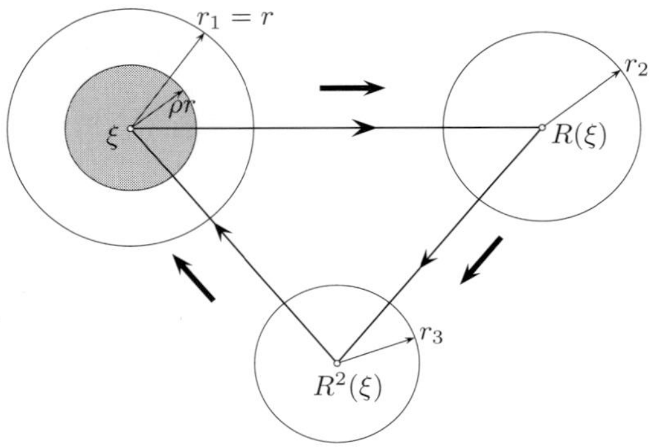

Fig. 5.11 Region of attraction of a (super)attractive periodic point of period $p = 3$.

5.10 Local Behavior of Iterations Near General Points: Equicontinuity and Normality

Consider the sequence of iterates $\{R^n(z)\}_{n=1}^{\infty}$. We wish to analyze the local behavior near general points. We first need to introduce two important concepts: equicontinuity and normality.

First we define the notion of equicontinuity, a notion of continuity for the entire family with the added assumption that given a point and $\epsilon > 0$, the same δ applies to all the members at that point.

Definition 5.14. We will say $\{R^n(z)\}_{n=1}^{\infty}$ is *equicontinuous* at a point $z_0 \in \widehat{\mathbb{C}}$ if given $\epsilon > 0$ there exists $\delta > 0$ such that

$$\sigma(z, z_0) \leq \delta \quad \Longrightarrow \quad \sigma(R^n(z), R^n(z_0)) < \epsilon,$$

for all $n \geq 1$ (as before σ is the chordal metric and is interchangeable with σ_0).

Thus equicontinuity at z_0 implies that when z remains close to z_0, $R^n(z)$ stays close to $R^n(z_0)$, for all n and within the same gap. In other words, when z is close to z_0 the orbits $O^+(z)$ and $O^+(z_0)$ also stay close.

For a given complex number z consider the sequence $z_n = R^{n-1}(z)$, and let z_n^* denote the corresponding points on the Riemann sphere. If we imagine connecting the points z_{n-1}^* to z_n^* via a straight line, then we can interpret the orbit as a path with many broken lines. Thus equicontinuity at z_0 implies that for any point within an ϵ neighborhood of z_0, the paths corresponding to z_0 and z will remain close to within the same ϵ distance (see Figure 5.12). What equicontinuity does not imply is the convergence of $R^n(z_0)$ to a particular limit. In particular, the maps could wander around and have infinitely many accumulation points.

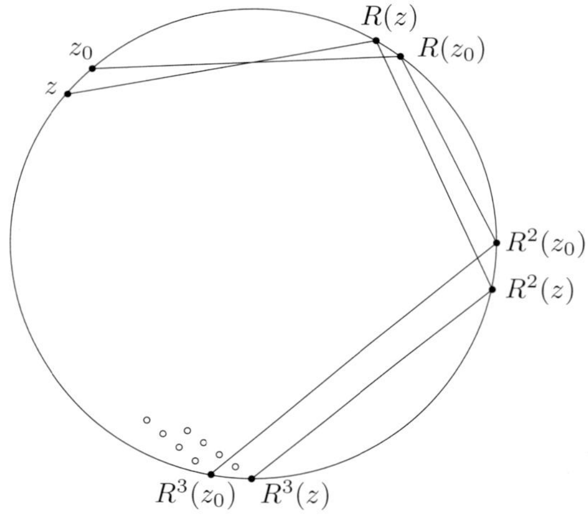

Fig. 5.12 A cross-sectional demonstration of equicontinuity on Riemann sphere.

Definition 5.15. We will say $\{R^n(z)\}_{n=1}^{\infty}$ is *normal* at z_0 if any subsequence $\{R^n(z) : n \in N_1 \subset N\}$, has in turn a subsequence $\{R^n(z) : n \in N_2 \subset N_1\}$ which converges uniformly on some neighborhood U of z_0 to an analytic function f, i.e. given $\epsilon > 0$ there exists n_0 (depending on ϵ) such that

$$\sigma(z, z_0) \leq \delta \quad \Longrightarrow \quad \sigma(R^n(z), f(z)) < \epsilon,$$

for all $n \in N_2$, $n \geq n_0$.

Remark 5.7. By the well-known Weierstrass *Uniform Convergence Theorem* a sequence of analytic functions, if uniformly convergent on a neighborhood, then the limit function is necessarily also analytic. Thus in the above definition the word analytic is superfluous. In fact on compact subsets of the neighborhood, the sequence of derivative also converge uniformly to the derivative of the limit function.

The following result, stated without proof, asserts that in fact equicontinuity and normality are equivalent notions. Thus in the forthcoming proofs we will make use of the two notions interchangeably.

Theorem 5.15 (Arzela-Ascoli). *R^n is equicontinuous at a point $z_0 \in \widehat{\mathbb{C}}$ if and only if it is normal at the point.* □

Theorem 5.16. *If $\xi \in \mathbb{C}$ is an attractive or superattractive periodic point of R, then R^n is normal at ξ.*

Proof. For ξ a fixed point of R the proof is immediate from Theorem 5.10 and the Arzela-Ascoli Theorem. Suppose that ξ has period $p > 1$. Let $\xi_0 = \xi, \dots, \xi_{p-1}$ be the corresponding cycle, with $R(\xi_{i-1}) = \xi_i$, $i = 0, \dots, p-1$. Then for $i = 0, \dots, p-1$, ξ_i is an attractive or superattractive fixed point of $R^{n_p}(z)$, where n_p is n mod p. Thus given $\epsilon > 0$, there exists δ_i such that

$$\sigma(z, \xi_i) \leq \delta_i \quad \Longrightarrow \quad \sigma(R^{n_p}(z), \xi_{n_p}) < \epsilon.$$

To prove equicontinuity it suffices to pick $\delta = \min\{\delta_i, i = 0, \dots, p-1\}$. □

Theorem 5.17. *If $\xi \in \mathbb{C}$ is a repelling periodic point of R, then R^n is not normal at ξ.*

Proof. We give two proofs. Suppose ξ is a fixed point of R. Then the proof that R^n is not normal at ξ is immediate from Theorem 5.11 and Arzela-Ascoli Theorem. If ξ has period p, we apply the previous argument to R^p.

Next we give a proof that does not rely on Arzela-Ascoli Theorem. Suppose that R^n is normal at ξ. Hence by the Weierstrass Uniform Convergence Theorem there exists an open disk D where R^n converges uniformly to an analytic function f. Then $(R^n)'(\xi)$ must converge to $f'(\xi)$. But $f'(\xi)$ is a finite number whereas $(R^n)'(\xi) = (R'(\xi))^n$ approaches infinity. The contradiction implies R^n cannot be normal at ξ. □

Theorem 5.18. *If ξ is a rationally indifferent periodic point of R, $deg(R) \geq 2$, then R^n is not normal at ξ.*

Proof. It suffices to prove the theorem when ξ is an indifferent fixed point. Moveover, without loss of generality we assume that its multiplier $\lambda = 1$. Additionally, through conjugacy we assume $\xi = 0$. Thus

$$R(z) = z + a_r z^r + a_{r+1} z^{r+1} + \cdots$$

where $a_r \neq 0$ and $r \geq 2$. An easy induction gives

$$R^n(z) = z + n a_r z^r + \cdots .$$

Differentiating the above r time reveals

$$(R^n)^{(r)}(0) = n a_r r!.$$

If R^n is normal at 0, then by Weierstrass Theorem in some neighborhood of 0 a subsequence of $R^n(z)$ converges uniformly to some analytic function $f(z)$. Since $R^n(0) = 0$, $f(0)$ must be zero. Thus $f(z)$ must be finite in some neighborhood of the origin, implying that $f^{(r)}(0)$ must be finite. On the other hand, we must have

$$f^{(r)}(0) = \lim_{n \to \infty} (R^n)^{(r)}(0) = \lim_{n \to \infty} n a_r r! = \infty,$$

a contradiction. □

If ξ is an irrationally indifferent periodic point of R, then R^n may or may not be normal there. We will consider this case again in a subsequent section of the chapter. However, here we prove the following.

Theorem 5.19. *Let ξ be an irrationally indifferent periodic point of R. Then R^n is normal at ξ if and only if ξ is an isolated periodic point.*

Proof. Suppose ξ is not an isolated periodic point of R. Then, since all but finitely many periodic points are repelling, in every neighborhood of ξ there is a repelling periodic point. Now it is easy to prove that R^n cannot be normal at ξ.

Assume ξ is an isolated periodic point. Without loss of generality we assume that ξ is a fixed point, and $\xi = 0$. From Theorem 5.14 R is linearizable at 0. Thus, for some analytic function g having the origin as a fixed point, and a neighborhood of the origin we have,

$$gRg^{-1}(z) = \lambda z.$$

This implies

$$gR^n g^{-1}(z) = \lambda z^n.$$

We can assume the neighborhood is such that $z = g(w)$. Thus for w in a neighborhood of 0 we have

$$gR^n(g(w)) = gR^n(w) = \lambda^n g(w).$$

Equivalently,

$$R^n(w) = g^{-1}(\lambda^n g(w)).$$

Since $|\lambda^n| = 1$, given $\epsilon > 0$, we can bound the right-hand side of the above when w is appropriately close to the origin. Hence, R^n is normal at 0. □

5.11 Fatou and Julia Sets and Their Basic Properties

Definition 5.16. The *Julia set* of a given rational function R, denoted by $J(R)$, or just J, is the set of points $z \in \widehat{\mathbb{C}}$ where R^n is not normal at the point. The complement $J(R)$ is called the Fatou set and is denoted by $F(R)$.

The above definition of Fatou and Julia sets is the coarsest way to define these sets. It turns out that these sets can be defined in several different ways. Generally speaking the characterization of these sets is not straightforward. The above definition classifies all the points in $\widehat{\mathbb{C}}$ into two categories. For polynomial root-finding and the Basic Family B_m, these categorizations are not sufficiently satisfactory. We will speak of this issue later in the chapter.

By definition, the Fatou set $F(R)$ is an open set. Hence $J(R)$ is closed. The properties that usually, or at least in interesting cases, are attributed to $J(R)$ are as follows:

- $J(R)$ is fractal, a term coined by Mandelbrot to mean a rough or fragmented geometric shape that reveals repeated self-similarity as one zooms in finer and finer scales.
- $J(R)$ no interior.
- $J(R)$ is a *perfect set*, i.e. closed with no isolated points.
- $J(R)$ has fractional Hausdorff dimension.

We begin by proving a result on normality at a point, but we use equicontinuity.

Lemma 5.6. *Suppose* $\{R^n\}_{n=1}^{\infty}$ *is equicontinuous at* z_0. *Then* $\{R^n\}_{n=1}^{\infty}$ *is equicontinuous at* $R(z_0)$, *as well as any* $w \in R^{-1}(z_0)$.

Proof. Suppose $\{R^n\}_{n=1}^{\infty}$ is equicontinuous at z_0. Then given $\epsilon > 0$, there exists $\delta > 0$ such that

$$\sigma(z, z_0) < \delta, \qquad \Longrightarrow \qquad \sigma(R^n(z), R^n(z_0)) < \epsilon, \quad \forall \ n \geq 1.$$

Let

$$U = \{z : \sigma(z, z_0) < \delta\}, \quad V = \{z : \sigma(z, R(z_0)) < \epsilon\}.$$

Since R is a rational map, it is an open map, i.e. an open set gets mapped to an open set. Thus $R(U)$ is open and so is $R(U) \cap V$. Hence $R(U) \cap V$ contains a disk

$$W = \{z : \sigma(z, R(z_0)) < \delta'\},$$

for some $0 < \delta' \leq \min\{\delta, \epsilon\}$. Since each $z \in W$ must equal $R(w)$ for some $w \in U$, for all $n \geq 1$ we have

$$\sigma[R^n(R(w)), R^n(R(z_0))] = \sigma[R^{n+1}(z) - R^{n+1}(z_0)] < \epsilon.$$

Thus R^n is normal at $R(z_0)$.

Next we prove that R^n is normal at any point $w_0 = R^{-1}(z_0)$. Since R is continuous at z_0, $R^{-1}(U)$ is an open set containing w_0. Hence $R^{-1}(U)$ contains a disk

$$U' = \{z : \sigma(z, w_0) < \delta''\},$$

for some $\delta'' > 0$. If $z \in U'$, then $\sigma(R(z), z_0) < \delta$. Assume without loss of generality that $\delta < \epsilon$. From this assumption and since R^n is normal at z_0, we get

$$\sigma(R^{n-1}(R(z)), R^{n-1}(z_0)) = \sigma(R^n(z), R^n(w_0)) < \epsilon.$$

Hence R^{n-1} is normal at w_0. □

Theorem 5.20 (Invariance of Fatou and Julia Sets). *$F = R(F) = R^{-1}(F)$ and $J = R(J) = R^{-1}(J)$.*

Proof. From Lemma 5.6 we get

$$R(F) \subset F, \quad R^{-1}(F) \subset F.$$

But this implies

$$R^{-1}(R(F)) = F \subset R^{-1}(F), \quad R(R^{-1}(F)) = F \subset R(F).$$

Hence the proof of the first two equalities. Since J is the complement of F the second equalities follow. □

The following is an immediate corollary.

Corollary 5.3. *For all $n \geq 1$, $F = R^n(F) = R^{-n}(F)$ and $J = R^n(J) = R^{-n}(J)$.* □

Theorem 5.21. *For any natural number p we have*

$$F(R) = F(R^p), \quad J(R^p) = J(R) = J.$$

Proof. Suppose $z_0 \in F(R)$. Since the orbit of z_0 with respect to R^p is a subset of its orbit with respect to R, equicontinuity of R at z_0 implies its equicontinuity with respect to R^p. Hence, $z_0 \in F(R^p)$.

Suppose that $z_0 \in F(R^p)$. Since R as a rational function, it is uniformly continuous on C_∞, given $\epsilon > 0$, there exists $\delta_0 > 0$ such that

$$\sigma(u, v) < \delta_0 \quad \Longrightarrow \quad \sigma(R(u), R(v)) < \epsilon.$$

Now given δ_0 above and that $z_0 \in F(R^p)$, there exists $\delta_1 > 0$ such that for all $n \geq 1$

$$\sigma(z, z_0) < \delta_1 \quad \Longrightarrow \quad \sigma(R^{pn}(z), R^{pn}(z_0)) < \delta_0.$$

Thus, letting $u = R^{pn}(z)$ and $v = R^{pn}(z_0)$ from the above two implications we conclude: given $\epsilon > 0$ there exists $\delta_1 > 0$ such that for all $n \geq 1$

$$\sigma(z, z_0) < \delta_1 \quad \Longrightarrow \quad \sigma(R^{pn+1}(z), R^{pn+1}(z_0)) < \epsilon.$$

Using the above inductively we may conclude given $\epsilon > 0$, for each $j = 0, \ldots, p-1$ there exists $\delta_j > 0$ such that for all $n \geq 1$

$$\sigma(z, z_0) < \delta_j \quad \Longrightarrow \quad \sigma(R^{pn+j}(z), R^{pn+j}(z_0)) < \epsilon.$$

We may also select $\delta > 0$ such that for each $i = 1, \ldots, p-1$ we have

$$\sigma(z, z_0) < \delta \quad \Longrightarrow \quad \sigma(R^{p-i}(z), R^{p-i}(z_0)) < \epsilon.$$

Setting $\delta_{\min}$ to be the minimum of δ and δ_i, $i = 1, \ldots, n$ we have thus proved

$$\sigma(z, z_0) < \delta_{\min} \quad \Longrightarrow \quad \forall\ n \geq 1 \quad \sigma(R^n(z), R^n(z_0)) < \epsilon,$$

i.e. R is equicontinuous at z_0. Finally, the fact that $J(R^p) = J(R)$ is obvious since $F(R^p)$ is the complement of $J(R^p)$. □

Remark 5.8. The above theorem has an interesting implication in the context of polynomial root-finding. Suppose that we consider $R(z)$ to be Newton's iteration function with respect to a polynomial $p(z)$, having a simple root ξ. We know that in a neighborhood of ξ Newton iterates will converge quadratically. If for instance we replace R with R^2 and apply the fixed point iterates, the local rate of convergence to ξ will improve by a factor of two. However, as we see the theorem implies that the Fatou sets will remain invariant. Moreover, the set of points whose orbits will converge to ξ will also remain invariant (see basins of attraction in subsequent sections). Thus, other than with respect to the overall computations of the n-th iterate, we will not witness any changes.

5.12 Montel Theorem and Characterization of Fatou and Julia Sets

Montel Theorem reveals an amazing property of rational and more generally analytic function. Here we state the theorem without proof, but give some fundamental consequences to be used in the characterization of Julia and Fatou sets.

Theorem 5.22 (Montel Theorem, see Ahlfors (1979)). *Suppose that $\{f_n(z)\}_{n=1}^{\infty}$ is a sequence of functions, from an open set U of $\widehat{\mathbb{C}}$ into $\widehat{\mathbb{C}}$, and analytic on U. If there exist three distinct points a, b, c in $\widehat{\mathbb{C}}$ which are not included in the range of any of functions, i.e. $f_n(U) \subset \widehat{\mathbb{C}} - \{a, b, c\}$, then f_n is a normal family on U.* □

The notion of normality is analogous to the case of rational functions. The following theorem is an immediate equivalent and more practical statement:

Theorem 5.23 (Montel Theorem). *Suppose that $\{f_n(z)\}_{n=1}^{\infty}$ is a sequence of analytic function from an open neighborhood N of $\widehat{\mathbb{C}}$ into itself. If it is not a normal family, then the set*

$$U = \cup_{n=1}^{n} f_n(N)$$

must contain all of $\widehat{\mathbb{C}}$ with the exception of at most two points. □

In the context of rational functions and their iterations we are often interested in the case where $f_n(z) = R^n(z)$, with $R(z)$ a rational map of degree at least 2.

The following is a more general version of Montel Theorem 5.22 which allows the subset $\{a, b, c\}$ to depend on each of the functions, as long as they stay apart. More precisely

Theorem 5.24 (Montel Theorem). *Suppose that* $\{f_n(z)\}_{n=1}^{\infty}$ *is a sequence of functions, from an open set* U *of* $\widehat{\mathbb{C}}$ *into* $\widehat{\mathbb{C}}$, *and analytic on* U. *If for each* n *there exist three distinct points* a_n, b_n, c_n *in* $\widehat{\mathbb{C}}$ *which are not included in the range of* f_n, *i.e.* $f_n(U) \subset \widehat{\mathbb{C}} - \{a_n, b_n, c_n\}$, *and if* $\sigma(a_n, b_n)$, $\sigma(a_n, c_n)$, *and* $\sigma(b_n, c_n)$ *is bounded below by a positive constant* ϵ *for all* $n \geq 1$, *then* f_n *is a normal family on* U. □

Definition 5.17. An *exceptional point* of a rational map R is a point whose forward and backward orbits are finite sets.

Lemma 5.7. *A point* z *is exceptional if and only if its backward orbit* $O^-(z) = \cup_{n=0}^{\infty} R^{-n}(z)$ *is finite.*

Proof. We only need to show if $O^-(z) = \cup_{n=0}^{\infty} R^{-n}(z)$ is finite, then z is exceptional. Let $O^-(z) = \{w_1, \ldots, w_t\}$. Then we must have $R^{-1}(w_i) \in O^-(z)$ for all i. This says for each i, $R(w_i) = w_j$ for some j. Since $z = w_i$ for some i, we see that for each n, $R^n(z) = w_j$ for some j. Hence z is exceptional. □

We may now state some consequences of Montel Theorem. The first proves an expanding power of any neighborhoods at any Julia point.

Theorem 5.25 (Expanding Neighborhoods Property). *Let* R *be a rational map of degree* $d \geq 2$. *Let* $z \in J$ *and* W *be any open set that contains* z. *Then the set*

$$U = \cup_{n=1}^{\infty} R^n(W)$$

contains all points of $\widehat{\mathbb{C}}$, *except possibly two points which must be exceptional.*

Proof. We know that J is nonempty. From the second version of Montel Theorem, Theorem 5.23 we know that U could miss at most two points of $\widehat{\mathbb{C}}$. But any such missed point z must necessarily be exceptional. Otherwise, $O^-(z)$ is infinite (previous lemma) implying that $R^{-n}(z)$ is contained in $R^m(W)$, for some n, m. But this implies $R^{n+m}(W)$ contains z, a contradiction. □

A consequence of Theorem 5.25 is the following:

Corollary 5.4. *A rational map* R *of degree* $d \geq 2$ *could have at most two exceptional points.*

Proof. Since $deg(R) \geq 2$, $J \neq \emptyset$. Thus previous theorem applies. □

Remark 5.9. A proof of the above corollary can be given which does not rely on Montel Theorem but still relies on other non-trivial results. Indeed for the Basic Family except possibly trivial cases, there are no exceptional points.

As a second corollary of Theorem 5.25 is the following:

Corollary 5.5. *If rational map R of degree $d \geq 2$ has non-empty Fatou set, then J has no interior.*

Proof. Otherwise, if J° is the interior of J then $R^n(J^\circ) \subset R^n(J) = J$ must intersect F, a contradiction. □

Remark 5.10. The corollary implies that the Julia set of the Basic Family for polynomials does not have any interior.

Theorem 5.26. *Suppose $z_0 \in J$ corresponding to a rational map R. Let z_1 be any non-exceptional point of $\widehat{\mathbb{C}}$, not necessarily distinct from z_0. Then any open neighborhood U of z_0 must intersect $O^-(z_1)$ at a point $u \neq z_0$. If z_1 lies in J, then u must lie in J.*

Proof. We consider two cases

(i) $z_0 = z_1$. Assume that z_0 is not a periodic point of R. Let U be any neighborhood of z_0. From Montel Theorem 5.23, there is some point $u \in U$ and k such that $R^k(u) = z_0$. Since z_0 is not a periodic point, $u \neq z_0$. Moreover,

$$u \in R^{-k}(z_0) \subset R^{-k}(J) = J.$$

Now suppose that z_0 is periodic. Then $O^+(z_0)$ is a finite set and since z_0 is not exceptional, there exists $z' \in O^-(z_0) - O^+(z_0)$. Thus there exists $u \in U$, different from z_0 such that $R^m(u) = z' = R^{-t}(z_0)$. Equivalently, $u = R^{m+t}(z_0)$. This completes the proof when $z_0 = z_1$.

(ii) $z_0 \neq z_1$ and z_1 is not exceptional. If z_1 is a periodic point, again the finiteness of the size of $Q^+(z_1)$ and the infinity of the size of $O^-(z_1)$ implies the desired result.

Assume z_1 is not periodic. By Montel Theorem 5.23 there exists $u_0 \in U$ such that $R^m(u_0) = z_1$. If $u_0 \neq z_0$, then we are done. Suppose $u_0 = z_0$, Then z_0 cannot be exceptional. In this case we apply the proof of case (i) to z_0 to conclude that there exists $u_1 \in U \cap O^-(z_0)$, $u_1 \neq z_0$ (see Figure 5.13).

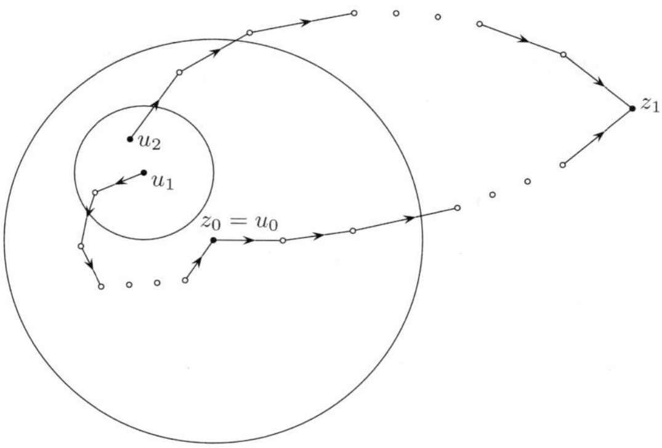

Fig. 5.13 A typical Julia point and backward orbit of a non-exceptional point.

Let U' be a neighborhood of u_1 contained in U. Then by Montel Theorem there exists $u_2 \in U'$ such that $R^m(u_2) = z_1$. Thus $u_2 \in O^-(z_1)$, and $u_2 \neq z_0$. Since $R^{-k}(J) = J = R^m(J)$, if z_1 lies in J so does u_2. □

The following is an important corollary of the above theorem.

Corollary 5.6. *Let z be a point in $\widehat{\mathbb{C}}$ that is not exceptional. Then the closure of $O^-(z)$ strictly contains J if z lies in F, and is equal to J if $z \in J$. In particular, J is a perfect set, i.e. a closed set having no isolated points. Thus J is an uncountable set.*

Proof. Since any neighborhood U of a point $z_0 \in J$ must intersect $O^-(z)$ at a point $u \neq z_0$, the closure of $O^-(z)$ contains J. Since F and J are invariant under R^{-1}, u lies in J if and only if z lies in J. If $z \in J$, then since $O^-(z)$ is contained in J and J is closed, the closure of $O^-(z)$ is also contained in J. But this closure must equal J since any point of J is a limit point of a sequence of points in $O^-(z)$. It also follows that J is a perfect set. By the Baire Category Theorem J is necessarily uncountable. □

Remark 5.11. Sometimes in the literature of dynamical systems the fact that $O^-(z)$ contains J is taken as a source of fast or efficient visualization of fractal images. This assertion cannot be correct unless it does quite a lot of pruning. More specifically, for all but a finite set of values for z, the set $R^{-n}(z)$ contains d^n elements. Thus except for small values of n the computation of $O^-(z)$ is impractical or impossible, both in terms of time or space.

We have seen that R^n is not normal at the repelling period points of R. The repelling periodic points are thus *bad points*. It is also easy to note that R^n cannot be normal at any accumulation point of the repelling periodic points. So that from the point of view of iterations of R these are equally bad.

Definition 5.18. For a rational map R we denote the closure of the set of repelling periodic points by $B(R)$ (the set of bad points). We call this set *The Bad.*

In summary, we have

Proposition 5.4. *For any rational map R we have $B(R) \subset J(R)$.* □

The following non-trivial theorem gives a much deeper view into the Julia set.

Theorem 5.27. *Let R be a rational map of degree $d \geq 2$, then $J(R) = B(R)$.*

Proof. From the above proposition we only need to show that given an element z_0 of J, and any neighborhood N of w. There is a repelling periodic point of R.

To prove this, assume otherwise. Thus there exist $z_0 \in J - B(R)$ and a neighborhood N of z_0 that contains no periodic point of R. Clearly, z_0 is not a critical point of R and since R has a finite number of critical points and a finite number of non-repelling periodic points, we may assume N contains no critical points or periodic points. This implies (see Remark 5.2) there are d branches of R at z_0, say S_j, $j = 1, \ldots, d$, analytic in neighborhood of z_0, say $N_0 \subset N$, mapping N_0 to distinct neighborhoods $N_j, j = 1, \ldots, d$ while satisfying $R(S_j(z)) = z$. Note that for any $z \in N_0$, the set $\{z, S_1(z), S_2(z)\}$ consists of three distinct elements.

The cross-ratio of four distinct elements z_1, z_2, z_3, z_4 is well known to satisfy

$$\frac{(z_3 - z_1)(z_4 - z_2)}{(z_2 - z_1)(z_4 - z_3)} \in C - \{0, 1\}.$$

Thus given $n \geq 1$, if for any $n \in N_0$, $R^n(z) \neq S_j(z)$, $j = 1, 2$, then the domain of the function

$$f_n(z) = \frac{(S_2(z) - z)(R^n(z) - S_1(z))}{(S_1(z) - z)(R^n - S_2(z))}$$

will not contain $\{0, 1, \infty\}$. Clearly the family $\{f_n(z), n \geq 1\}$ is analytic on N_0. Thus by Montel Theorem 5.22 f_n is a normal family on N_0. This in turn easily implies that R^n is normal at z_0, a contradiction. Thus, there exist $n \geq 1$, and $z \in N_0$ such that $R^n(z) = S_j(z)$, $j = 1, 2$. Thus

$$R^{n+1}(z) = R(S_j(z)) = z,$$

contradicting that N_0 contains a periodic point. Thus each neighborhood of a point in J contains a repelling periodic point. □

We conclude this section by proving yet another consequence of Montel Theorem which shows the expanding property of neighborhoods of any Julia point in a uniform sense, and a stronger version of Theorem 5.25.

Theorem 5.28 (Uniform Expanding Neighborhoods Property). *Let R be a rational map of degree $d \geq 2$. Suppose $z \in J$. Let W be any open set that contains z. Then there exists n_0 such that for all $n \geq n_0$, $R^n(W)$ contains J.*

Proof. Since $deg(R) \geq 2$, J is a perfect set, z is not an isolated point. So we can pick three different points z_1, z_2, z_3 of J in W and hence three disjoint neighborhoods W_1, W_2, W_3 centered at z_1, z_2, z_3, and a chordal distance of $\epsilon > 0$ away from each other. We claim there exists m such that for some $i = 1, 2, 3$, $W_i \subset R^m(W_i)$. If the sets W_1, W_2, and W_3 are never contained in $R^n(W_1)$ for any n, then by Montel Theorem 5.24 R^n forms a normal family on W_1 which is impossible since $z_1 \in W_1$. Hence $R^n(W_1)$ contains one of the sets W_1, W_2, W_3. If it is not W_1, but say W_2, then we apply the same argument to W_2 etc. Hence the proof of the claim.

Let $S = R^m$, then the Julia set of S is $J(R^m) = J$. Applying Theorem 5.25 to S and W_i, we conclude that the union of the sets $\{S^n(W_i) : n \geq 1\}$ covers J. As S is a rational map, it is an open map and thus $S^n(W_i)$ is an open set. This implies that the sets $\{S^n(W_i) : n \geq 1\}$ form an open cover of J and since J is compact, it must have a finite cover. Moreover, since $W_i \subset S(W_i)$, it follows that there exists n_1 such that $S^{n_1}(W_i) = R^{m+n_1}(W_i)$ contains J. Let $n_0 = m + n_1$, and since $W_i \subset W$ we have shown

$$J \subset R^{n_0}(W).$$

Since $R(J) = J$ we conclude that for all $n \geq 0$

$$J \subset R^n(W).$$

□

5.13 Fatou and Julia Sets as: The Good, The Bad, and The Undesirable

In the realm of iteration functions for polynomial root-finding, e.g. Newton's method, a naive point of view is to consider the Fatou set as the set of *good points* and the Julia set as the set of bad points. This view would be consistent with the fact that Fatou and Julia sets of a rational map R were defined in terms of normality (equivalently equicontinuity) and the lack of normality, respectively. However, with respect to polynomial root-finding and algorithmic point of view what may be considered as a bad point could turn out to be a good point, and what may be considered as a good point could be dangerous. We shall clarify these points of views.

While the Julia set $J(R)$ has an easy characterization in the sense that it was shown to be equivalent to $B(R)$, the closure of the repelling periodic points, the Fatou set $F(R)$ is not at all easy to characterize. First we give a definition.

Definition 5.19. For a given $R(z)$ of degree $d \geq 2$ we define the set of good points, denoted by $G(R)$, as the set of all points $z \in \widehat{\mathbb{C}}$ such that the orbit $O^+(z)$ converges to an attractive or superattractive periodic cycle. The set of *undesirable points* of R, denoted by $U(R)$ is defined to be $\widehat{\mathbb{C}} - G(R) - J(R)$, i.e. all points of $\widehat{\mathbb{C}}$ that are neither good nor bad.

The characterization of the points into the good, the bad, and the undesirable makes sense when we are considering iteration functions for polynomial root finding. Indeed historically the main task of iteration functions is to find the attractive and superattractive fixed points of a rational map.

More specifically, consider finding the roots of a polynomial $p(z)$ via Newton's iteration function $N(z) = z - p(z)/p'(z)$. We know that the roots of $p(z)$ are attractive or suprattractive fixed points of $N(z)$. We may ask: For what points $z \in \mathbb{C}$ does the orbit $O^+(z)$ converge to a root of $p(z)$? These points are certainly good points. All points in $B(R) = J$ are clearly bad points because the orbits remain within J. But from Montel Theorems we have seen that any neighborhood of a point in J must contain a point whose orbit converges to a root of $p(z)$. Thus in the neighborhood of each bad point there is a good point. However, there could be undesirable points having a whole neighborhood of such points around them. Thus the iterates may get trapped with no chance to escape. Such situations could arise and are characterizable in terms of Siegel disks and Herman rings to be defined in the next section. These are known as rotation domains.

Even if a point converges to an attractive or superattractive periodic point of say Newton's function $N(z)$, there is no guarantee that it will be a root of $p(z)$. This situation too could indeed arise and will be discussed next. Thus even a good point could lead to the wrong fixed point of R.

We remark here that our notion of good and bad points differ with those described in Haesler and Peitgen (1989).

We will now attempt to characterize the points of F that lie in $G(R)$, the good points of R. First, we give a notation for convenience. For a given $z_0 \in \mathbb{C}$, and given natural number $k \geq 1$ define

$$A_k(z_0) = \{z : \lim_{n \to \infty} R^{kn}(z) = z_0\}.$$

From the continuity of R, $A_k(z_0)$ is nonempty if and only if z_0 is a fixed point of R^k.

Definition 5.20. Given an attractive or superattractive fixed point ξ of R, its basin of attraction is defined as

$$A(\xi) \equiv A_1(\xi) = \{z : \lim_{n \to \infty} R^n(z) = \xi\}.$$

More generally, given an attractive or superattractive periodic point ξ of period p with cycle

$$\xi^{(p)} = \{\xi_0 = \xi, \xi_1 = R(\xi), \xi_2 = R^2(\xi), \ldots, \xi_{p-1} = R^{p-1}(\xi)\},$$

its basin of attraction is defined as

$$A(\xi) \equiv A(\xi^{(p)}) = \cup_{j=0}^{p-1} A_p(\xi_j),$$

where for $j = 0, \ldots, p-1$

$$A_p(\xi_j) = \{z : \lim_{n \to \infty} R^{pn}(z) = \xi_j\},$$

i.e. the basin of attraction of the fixed point ξ_j with respect to R^p.

It should be clear that

$$A_p(\xi_j) = R(A_p(\xi_{j-1})), \quad j = 0, \ldots, p-1,$$

where $\xi_{-1} \equiv \xi_{p-1}$.

Definition 5.21. Given a point $z \in \mathbb{C}$ we say $O^+(z)$ converges to a periodic cycle $\xi^{(p)} = \{\xi, R(\xi), R^2(\xi), \ldots, R^{p-1}(\xi)\}$ if $z \in A(\xi^{(p)})$.

Lemma 5.8. *$A(\xi)$ is an open set and is contained in the Fatou set $F(R)$.*

Proof. Without loss of generality assume that $p = 1$, i.e. ξ is a fixed point of R. Thus if $\lambda = R'(\xi)$, $|\lambda| < 1$. Thus there exists an open disk D centered at ξ such that $R(D) \subset D$ so that R^n uniformly converges to the constant function ξ on D (Theorem 5.10). Thus D is contained in $F(R)$. Given $z \in A(\xi)$, there exist $w \in D$, and n such that $z \in R^{-n}(w)$. This implies $z \in F(R)$. To prove $A(\xi)$ is open we note that since R^n is a continuous map, inverse image of an open set is open. Hence $R^{-n}(D)$ is an open set containing z any of whose points converges to ξ. □

From the lemma we conclude that $G(R)$ consists of the finite union of open sets corresponding to basin of attraction of attractive and superattractive periodic points.

Proposition 5.5. *If ξ is a periodic point of R of period p, then for any $k \geq 1$*

$$A(\xi) = A_{p \cdot k}(\xi).$$

In other words the basin of attraction of a periodic fixed point of R remains invariant under the iterations of R^k. □

The following result shows that the basin of attractions shares a common boundary, the Julia set.

Theorem 5.29. *Let ξ be an attractive or superattractive periodic point of R of period p. Then $J = \partial A_p(\xi)$.*

Proof. First assume $p = 1$. Suppose $z \in J$. From Montel Theorem 5.23 it follows that in every neighborhood U of z there is a point u whose forward orbit lies in $A(\xi)$. Thus u itself lies in $A(\xi)$. This implies that z lies in the closure of $A(\xi)$. Since J and $A(\xi)$ do not intersect, z must lie in $\partial A(\xi)$. Thus $J \subset \partial A(\xi)$.

Next we show that if $z \in \partial A(\xi)$, then z must lie in J. Otherwise, z lies in the Fatou set and hence R is equicontinuous there. We use this to prove that the forward orbit of z converges to ξ. This in turn would imply that z lies in $A(\xi)$, an open set, contradicting $z \in \partial A(\xi)$.

Let $\epsilon > 0$ be given. There exists $\delta > 0$ such that $\sigma(z, w) < \delta$ implies $\sigma(R^n(z), R^n(w)) < \epsilon/2$ for all $n \geq 1$. Since $z \in \partial A(\xi)$ there exists $w \in A(\xi)$ for which the above is true. Since the $O^+(w)$ converges to w, there exists $n_0(\epsilon)$ such that $n \geq n_0$ implies $\sigma(R^n(w), \xi) \leq \epsilon/2$. Thus we have shown that given $\epsilon > 0$ there exists $n_0(\epsilon)$ such that for all $n \geq 0$ we have

$$\sigma(R^n(z), \xi) \leq \sigma(R^n(z), R^n(w)) + \sigma(R^n(w), \xi) \leq \epsilon/2 + \epsilon/2 = \epsilon.$$

Thus $z \in A(\xi)$, a contradiction. Hence, $J = \partial A(\xi)$.

Now consider the general case where ξ has period $p > 1$. From the case of $p = 1$ it follows that

$$J(R^p) = \partial A_j(\xi).$$

But from Theorem 5.21, $J(R) = J(R^p)$. □

The basin of attractions is the most straightforward portion of the Fatou set to conceive. In the next section we consider this set in more detail.

5.14 Fatou Components and Their Dynamical Properties

The Fatou set of a rational map R is an open set and thus can be written as a countable union of connected components called Fatou components. In this section we consider the dynamics of Fatou components.

Definition 5.22. Consider an open domain D in $\widehat{\mathbb{C}}$, the Riemann sphere. We say that D is connected if it cannot be written as the union of two disjoint open sets O_1 and O_2 in $\widehat{\mathbb{C}}$. We say D is simply connected if every closed curve in D can be continuously shrunk, while remaining within D, into a single point. If D does not contain ∞, we can define connectivity with respect to the complex plane. If D is connected, it is pathwise connected, i.e. for any two points $u_1, u_2 \in D$, there exists a continuous function f from the unit interval $[0, 1]$ into D such that $f(0) = u_1$, $f(1) = u_2$.

The following lemma is a simple but useful property of a rational map.

Lemma 5.9. *Let $U' = R(U)$ where U is a Fatou component, then U' is a Fatou component.*

Proof. Suppose that two points z_1 and z_2 under R map into u_1 and u_2 in two different Fatou components U_1 and U_2, see Figure 5.14. Since U is connected there is a simple path from z_1 to z_2 remaining entirely in U. Since R is continuous, the image of the path will necessarily intersect the boundary point of U_1 at a point, say u' which must lie in J. But this implies u' lies in F and J, a contradiction. Hence $R(U)$ is contained in a Fatou component U'.

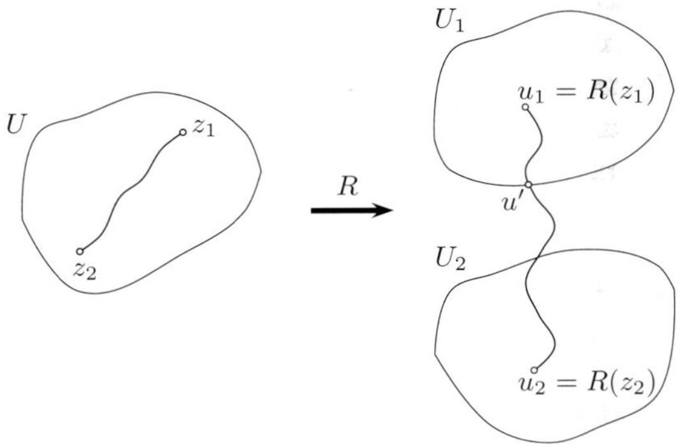

Fig. 5.14 Image of Fatou component under R cannot become disconnected.

Suppose $R(U)$ is properly contained in a Fatou component U'. Let v be a point in U' and the boundary of $R(U)$, see Figure 5.15. Consider $R^{-1}(v)$. On the one hand, this set cannot intersect U. On the other hand, in every neighborhood of v there exists a point in $R(U)$ whose preimage intersects U. Let $\{v_k : k = 1, \dots\}$ be a sequence of points in $R(U)$ converging to v. Let $\{w_k : k = 1, \dots\}$ be the sequence of corresponding preimages in U. Let w be any accumulation point of w_k's. Then we have $R(w) = v$. Thus, $w \in \partial U$ which is a subset of J. But $v \in U'$ implies $v \in F$, a contradiction. Hence $U' = R(U)$.

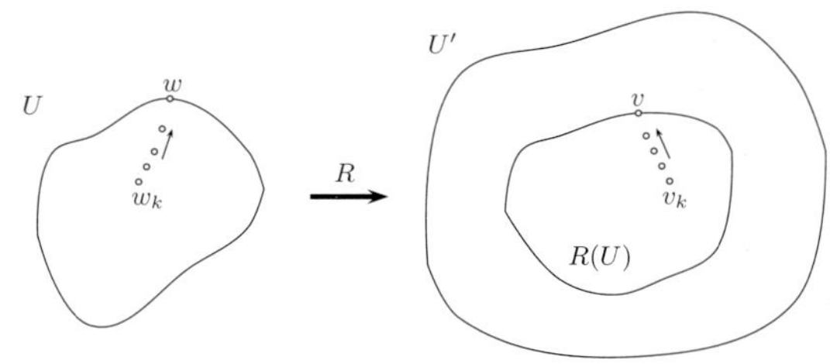

Fig. 5.15 R cannot map a Fatou component to a proper subset of another Fatou component.

□

Remark 5.12. The above result implies that if U is a Fatou component such that for some $u \in U$, $R(u) \in U$, then U is forward invariant, i.e.

$R(U) = U$. It follows that if for some $u \in U$, the set $R^{-1}(u)$ intersects U, then $R(U) = U$.

According to the lemma under the iterations of R a Fatou component U gets mapped to another Fatou component. More generally it follows that if we set $U_n = R^n(U)$, then U_n is a Fatou component. Moreover, for $i \neq j$, either $U_i = U_j$, or $U_i \cap U_j = \emptyset$.

Definition 5.23. We call a Fatou component U to be a *fixed Fatou component* if $R(U) = U$. More generally, a Fatou component U is a periodic Fatou component of period $p \geq 1$ if $R^p(U) = U$ but $R^j(U) \neq U$ for $j = 1, \ldots, p-1$. In this case we denote the cycle of periodic Fatou components as

$$U^{(p)} = \{U, R(U), R^2(U), \ldots, R^{p-1}(U)\}.$$

Remark 5.13. In the literature a fixed Fatou component, is referred as forward invariant. The fixed Fatou component is in analogy with fixed points. In a sense, looking at the dynamics of the Fatou components is a macro view of the dynamics of fixed point iterations. In the course of polynomial root-finding via iteration functions, e.g Newton's method, in a sense it is sufficient to find the immediate basin of attraction of a root, i.e. the Fatou components that contains the root. This then guarantees the convergence of subsequent iterates to the root. Indeed in a subsequent section we will offer graph theoretic views of dynamical properties of Fatou components and analogies that help visualize the overall view of iterations of a rational map.

We will define some very significant Fatou components.

Definition 5.24. Let U be a Fatou component of R. U is a *superattractive component* if it contains a superattractive fixed point ξ.

U is an *attractive component* if it contains an attractive fixed point ξ.

U is a *parabolic component* (also called Leau domain) if there is a parabolic fixed point ξ on its boundary.

U is a *Siegel disk* if it contains an irrationally indifferent periodic point ξ that is a Siegel point.

U is a *Herman ring* if R restricted to U is analytically conjugate to a Euclidean rotation of some annulus onto itself.

The latter two components are called *rotation domains*. It is easy to prove:

Proposition 5.6. *A Fatou component can contain at most one periodic point of R.*

Definition 5.25. The immediate basin of attraction of an attractive or superattractive fixed point ξ of a rational map R is the attractive or superattractive component that contains it. If $\xi^{(p)}$ is an attractive or superattractive cycle, then its immediate basin of attraction is the union of the (disjoint) immediate basin of attraction of each member of $\xi^{(p)}$ with respect to R^p.

The next theorem characterizes the fixed Fatou components. It is a precursor to Sullivan's main theorem, the famous No Wandering domain of his, see e.g. Beardon (1991) for proof. It could be extended to periodic components as well.

Theorem 5.30 (Characterization of Fixed Fatou Components). *Given a rational map R, let U be a fixed Fatou component $((R(U)=U))$. Then only one of the following types can occur:*

(i) U is a superattractive component and $R^n(U)$ converges to a superattractive fixed point $\xi \in U$.

(ii) U is an attractive component and $R^n(U)$ converges to an attractive fixed point $\xi \in U$.

(iii) U is a parabolic component and $R^n(U)$ converges to a parabolic fixed point $\xi \in \partial U$.

(iv) U is a Siegel disk and each $n \geq 1$, $R^n(U)$ is analytically conjugate to a Euclidean rotation of infinite order, $\xi + \lambda(z - \xi)$, on the unit disk, where $\xi \in U$ is a Siegel point.

(v) U is a Herman ring and for each $n \geq 1$, $R^n(U)$ restricted U is analytically conjugate to a Euclidean rotation of some annulus onto itself.

Remark 5.14. The boundary of a Siegel disk has one connected component while that of a Herman ring has two connected components, see Figure 5.16.

Definition 5.26. The *orbit* (forward orbit) of a Fatou component U, denoted as $O^+(U)$, is defined as the sequence of Fatou components $\{U_m = R^m(U), m \geq 1\}$.

Definition 5.27. A Fatou component U is said to be *eventually periodic* if $U_m = R^m(U)$ is periodic for some $m \geq 1$.

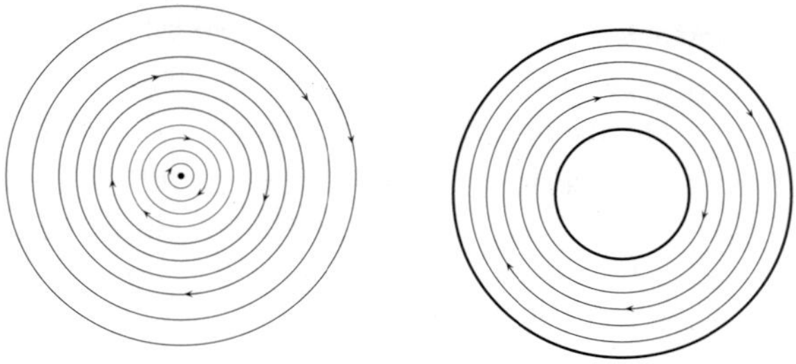

Fig. 5.16 Idealized Siegel disk and Herman ring.

The following theorem is considered to be one of the most significant theorem in the theory of dynamics of rational functions.

Theorem 5.31 (No Wandering Domains, Sullivan (1985)). *Every Fatou component U of a rational map is eventually periodic.* □

5.15 Critical Points and Connection with Periodic Fatou Components

The dynamics of the critical set C of a rational map R play an important role in the dynamics of the Fatou components as well. Moreover, the number of critical points has a delicate connection with the number of periodic Fatou components. In what follows we make these connections clear.

Clearly, a superattractive component contains a critical point, the fixed point itself. Also, periodic superattractive components, by chain rule, must contain a critical point in the immediate basin of attraction of cycle, in particular in a Fatou component containing one of the elements in the cycle.

In the next theorem we will prove it for attractive components. Actually we prove a slight variation. Analogous proof extends to parabolic components.

Theorem 5.32. *Let R be a rational map of degree $d \geq 2$. Suppose ξ is an attractive periodic point of R with multiplier λ. Then the basin of attraction of ξ must contain a critical point of R.*

Proof. Without loss of generality we assume ξ is a fixed point. Suppose the basin of attraction $A(\xi)$ does not contain a critical point of R. Then no neighborhood of ξ can contain a point of $C^+(R)$. There exists a disk $D \subset A(\xi)$ centered at ξ where $R^n(D)$ converges to ξ. Since U does not contain a critical point, D does not contain any point in $C^+(R)$. This by Theorem 5.8 implies that for all $n \geq 1$, $(R^n)'(z)$ does not vanish on D. Hence for each $n \geq 1$ R^n has d^n distinct branches in D. Let $\{S_n : n \geq 1\}$ denote the sequence of all the branches.

We claim $\{S_n : n \geq 1\}$ is normal on D. To prove it, note that D cannot contain any periodic point other than ξ. On the other hand, since $d \geq 2$, there exists a repelling periodic cycle $\{\theta_1, \dots, \theta_p\}$ of R of period $p \geq 3$. The range of S_n cannot contain any θ_i. Otherwise, $S_n(z) = \theta_i$ for some $z \in D$. But then for some $m \geq 1$, where R^m is the inverse of S_n, we get $R^m(\theta_i) = z$. But this implies z lies in F and in J, a contradiction.

Thus if D does not contain any point in $C^+(R)$, S_n is normal on D. Suppose that S_n is a branch of R^m for some $m \geq 1$. Then by the chain rule we have

$$S_n'(\xi) = \lambda^{-m}.$$

But this implies that the sequence $\{S_n'(\xi) : n \geq 1\}$ approaches infinity, contradicting the normality of S_n at ξ. Hence the proof. □

It can actually be argued that the immediate basin of attraction contains a critical point. We will state the general theorem on the connection between periodic Fatou components and critical points. First we make a remark.

Remark 5.15. If c is a critical point whose orbit $O^+(c)$ converges to a fixed point ξ which is not superattractive, then $O^+(c)$ has infinite cardinality. This is immediate because if $R^n(c) = \xi = R(\xi)$ then from chain rule $R'(\xi) = 0$, a contradiction.

The next result describes the connection between a parabolic fixed point and the Fatou components with respect to which the fixed point is a boundary point (see e.g. Beardon (1991) or Milnor (2006)).

Theorem 5.33. *If ξ is a parabolic fixed point of R of multiplicity $m+1$, then the immediate basin of attraction of ξ is the union of m disjoint Fatou components that have ξ on their boundary.* □

In Figure 5.17 we see the actual visualization of this for $p(z) = z - z^{m+1}$ where the origin is a parabolic fixed point. As we see m components come together to touch at the origin.

Fig. 5.17 Parabolic fixed point of $p(z) = z - z^{m+1}$, $m = 1, 2, 3, 4$.

The following theorem describes the connection between critical points and fixed Fatou components of various kinds. More generally, the statements of the theorem are applicable to periodic Fatou components.

Theorem 5.34. *Let U be a fixed Fatou component of a rational map R of degree $d \geq 2$ and let C be its critical set, and set*

$$C^{+}(R) = \cup_{i=1}^{\infty} R^{n}(C).$$

If U is a superattractive, attractive, or parabolic Fatou component of R, it must contain a critical point.

If U is a Siegel disk or Herman ring then the boundary of U is contained in the closure of $C^{+}(R)$. □

The above theorem accounts for all the periodic points except for irrationally indifferent points that lie in $J(R)$, the Cremer points. The following describes their connection to critical points.

Theorem 5.35. *Any irrationally indifferent periodic point of R that lies in $J(R)$ is a limit point of $C^+(R)$.* □

The more intimate connection between critical points of a rational map and the periodic cycle of Fatou components is summarizable in the following significant theorem:

Theorem 5.36 (Bound on Components, Shishikura (1987)). *Let R be a rational map of degree $d \geq 2$. The total number of cycles of periodic Fatou components, i.e. cycles of periodic attractive components, superattractive components, parabolic components, Siegel disks, and Herman rings is bounded above by $2d - 2$.* □

The theorem is proved by associating a distinct critical point to each cycle that is attractive, superattractive, parabolic, or irrationally indifferent, and two distinct critical points to each cycle of Herman rings. The bound of $2d - 2$ then follows from the fact that this number bounds the critical points. It should be mentioned that the cycles could end up having a large number of Fatou components, independent of d.

We end this section by stating some results on the number of Fatou components and on the nature of dynamics of these components. These are helpful in the understanding and visualization of the dynamics of rational maps.

Theorem 5.37. *The Fatou set of a rational map R has either* $0, 1, 2$ *or infinitely many components. In particular, $B_m(z)$ has infinitely many components as applied to a polynomial with at least* 3 *distinct roots.* □

The fact that more than 2 components implies infinity of them can be argued from Montel Theorem, the expanding property of neighborhoods of any Julia point proved as Theorem 5.25.

The next is essentially a consequence of basic properties of a rational map and Fatou components.

Theorem 5.38. *Suppose that U is a Fatou component of a rational map R of degree d not containing any critical values of R. Then $R^{-1}(U)$ consists of d Fatou components $U_1, \ldots, U_d$ such that R maps U_i onto U homeomorphically. Moreover, U_i is simply connected.* □

Given an arbitrary Fatou components of a rational map R of degree 2, $R^{-1}(U)$ consists of $t \leq d$ distinct components $U_1, \ldots, U_t$ which are pairwise disjoint (if $t \geq 2$). The restriction of R to U_i is a rational map of degree d_i with the property that $\sum_{i=1}^{t} d_i = d$. Thus the inverse image of a point in $z \in \mathbb{C}$ has d_i copies in the component U_i. Equivalently, the restriction of R on U_i is a d_i-fold map.

As an example consider the map

$$R(z) = \frac{z}{z^2+1} = z(1 - z^2 + z^4 - \cdots) = z - z^3 + O(z^5).$$

Its only fixed point is 0 of multiplicity 3. It is a parabolic fixed point, discussed in detail earlier, and hence two Fatou components, say U_1 and U_2 are incident to them. It has two critical points which are solutions of $R'(z) = (z^2 + 1 - 2z^2)/(z^2 + 1) = (-z^2 + 1)/(z^2 + 1) = 0$, i.e. ± 1. Note that their number coincides with $2d - 2 = 4 - 2 = 2$. Each of U_1 and U_2 must contain a critical point. Thus the orbit of $+1$ and -1 must converge to 0. It can be shown that $R^{-1}(U_i) = U_i$, $i = 1, 2$, i.e. the components are completely invariant. It would follow that there are no other Fatou components. Indeed $J(R)$ is the y-axis. Figure 5.18, top-left image, shows the corresponding image.

As a more general version consider $R_a(z) = az/(1 + z^2)$, where a is a nonzero real number. With the same critical point as $R(z)$, the fixed points are $\xi_0 = 0$, $\xi_1 = \sqrt{a-1}$, $\xi_2 = -\sqrt{a-1}$. The corresponding multipliers satisfy $|\lambda_0| = |a|$, $|\lambda_1| = |\lambda_2| = |2 - a|/|a|$. We can see that when $a > 1$ both ξ_1 and ξ_2 are attractive while 0 is repulsive. When a lies in the interval $(-1, 1)$ both ξ_1 and ξ_2 are repulsive while 0 is attractive. When $a < -1$ all three fixed points become repulsive. In Figure 5.18 we show example images for the three different cases, bottom images. For $a < -1$ we have used R^2 to be able to give a better image of Fatou components and the Julia set which are of course the same as those of R.

Let us consider Newton's method for $z^3 - 1$, i.e. $N(z) = (2z^3 + 1)/3z^2$, considered before. Consider the immediate basin of attraction of $\xi = 1$, say U_1. Then $R^{-1}(U_1)$ consists of two components, U_1 itself and one more component U_2. The corresponding degrees are $d_1 = 2$, $d_2 = 1$. In other words the inverse image of each point in U_1 consists of two points in the basin U_1 and one point in U_2. The only point that has a duplicate copy of inverse image is the fixed point $\xi = 1$ itself.

Note that if U is a rotation domain, i.e. Siegel disk or Herman ring, then if $d \geq 2$ it follows that $R^{-1}(U)$ consists of at least of components different from U itself. This is because the degree of R restricted to such rotation domain is one.

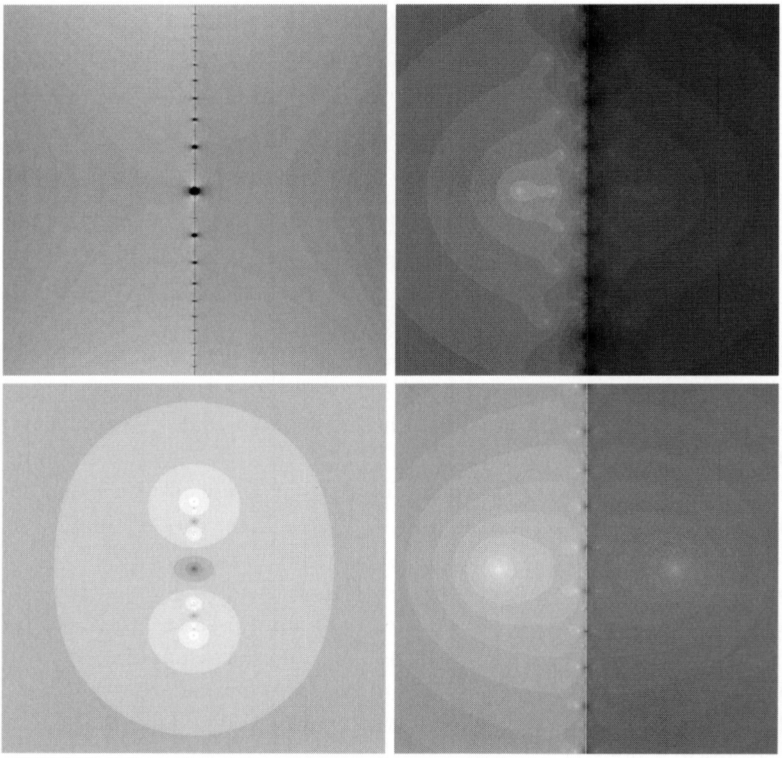

Fig. 5.18 Image of $R(z) = z/(1+z^2)$ (top-left image). Images corresponding to $R_{1.5}(z)$ (top-right), $R_{0.5}(z)$ (bottom-left), and $R^2_{-2}(z)$ (bottom-right).

We end this section with a summary of the significant classifications.

$$\xi : \begin{cases} \textit{superattractive}, & \lambda = 0; \\ \textit{attractive}, & 0 < |\lambda| < 1; \\ \textit{repelling}, & |\lambda| > 1; \\ \textit{indifferent}, & |\lambda| = 1. \end{cases}$$

$$\textit{indifferent} : \begin{cases} \textit{rationally indifferent}, & \lambda^n = 1 \text{ for some } n \geq 1; \\ \textit{irrationally indifferent}, & \lambda^n \neq 1, \text{ for all } n \geq 1. \end{cases}$$

$$\textit{irrationally indifferent}: \begin{cases} \textit{Siegel point}, & \xi \in F(R); \\ \textit{Cremer point}, & \xi \in J(R). \end{cases}$$

$$\textit{fixed Fatou} - \textit{component}:$$

$$\begin{cases} \textit{superattractive}, & \text{contains superattractive fixed point;} \\ \textit{attractive}, & \text{contains attractive fixed point;} \\ \textit{parabolic}, & \text{boundary contains parabolic fixed point;} \\ \textit{Siegel disk}, & \text{analytically equivalent to unit disk;} \\ \textit{Herman ring}, & \text{analytically equivalent to annulus.} \end{cases}$$

From a pedagogical point of view, the relationship between a non-repelling period point and the Fatou component(s) that get associated to them can be seen as a sort of "partnership." Each (super)attractive fixed point ξ gets wedded to the Fatou component containing it. This wedding is in the sense that the orbit of each point in the Fatou component converges to ξ. In that sense a parabolic ξ is also attractive, in fact it attracts at least two Fatou components. Thus the label "indifferent" may not be very inappropriate. That label perhaps is more appropriate for describing the relationship between a Siegel point and the Fatou component containing it. In summary, a (super)attractive fixed point has a monogamous relation with its Fatou component, while a parabolic fixed point has a polygamist relationship with its Fatou components, and a Siegel point is indifferent to its Fatou partner. In the more general case where ξ is periodic of period, under each iteration of R the point ξ and its Fatou partner(s) get relocated to an analogous periodic point and its Fatou partner(s).

5.16 Fatou-Julia and Topological Fatou-Julia Graphs: Analogies for Visualization and Conceptualization of Dynamics

Definition 5.28. Given a rational function $R(z)$ of degree $d \geq 2$, we define *Fatou-Julia graph*, $G_R(V, E)$, to be the following directed graph.

(i): There is a one-to-one correspondence between the Fatou components of R and the vertices in V.

(ii): Given two vertices $v_i, v_j \in V$, there is a directed edge (v_i, v_j) from v_i to v_j if and only if the corresponding Fatou components U_i, U_j satisfy $R(U_i) = U_j$.

The graph $G_R(V, E)$ inherits some of the properties from iterations of R. The indegree of each vertex is at most d. The *outdegree* of each vertex is 1, except for those that correspond to fixed Fatou components, i.e. corresponding to fixed attractive, superattractive, parabolic, Siegel and Herman components. The outdegree for the latter vertices is zero. A predecessor of a given vertex v is any vertex $u \neq v$ from which there is a directed path from u to v.

The graph $G_R(V, E)$ however does not reflect the topological properties of the Fatou components in relationship with each other. For this reason we next define the topological Fatou-Julia graphs, defined independently of a rational map $R(z)$.

Definition 5.29. Consider a directed graph $G(V, E)$ whose vertex set V consists of a set of points on the Riemann sphere S, and E is a collection of directed arcs that connect two given vertices along the geodesic arc (determined by the great circle passing through the points on the Riemann sphere). We say the directed graph $G(V, E)$ is a *topological Fatou-Julia graph* if the following properties are satisfied.

(i) V is countable.

(ii) The indegree of all vertices is bounded by a number $d \geq 2$.

(iii) The outdegree of each is either zero or one.

(iv) Given any vertex $v \in V$ and an edge $e = (u, v) \in E$, there exists a point z on e with the property that for every neighborhood N of z, and for every predecessor u' of u, there exists a vertex $w \neq v$ which lies in N, where there is a directed path from w to u'. Figure 5.19 shows this property.

Definition 5.30. A vertex in a topological Fatou-Julia graph $G(V, E)$ is a *fixed vertex* if it has no children. More generally, a *periodic vertex* of period $p \geq 1$ is a vertex that lies on a cycle of size p.

Definition 5.31. We say that a topological Fatou-Julia graph has *no wandering domain property* if every vertex $u \in V$ is either periodic or there is a directed path from u to a periodic vertex.

Definition 5.32. We say that a topological Fatou-Julia graph has the *finite-cycle property* if it has a finite number of periodic cycles.

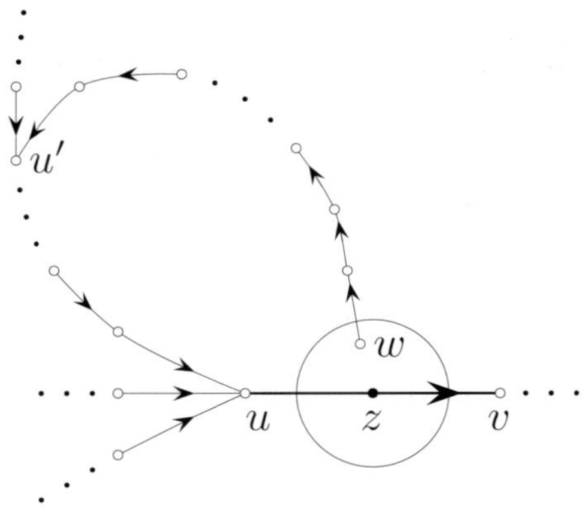

Fig. 5.19 Depiction of topological Fatou-Julia graph property around any neighborhood of the special point z on an edge (u, v).

Some obvious but interesting questions are: Can we associate a topological Fatou-Julia graph to a given rational map R? Do these always satisfy the no wandering property? (A property analogous to Sullivan's no wandering domain property?). If not, under what additional condition do they satisfy this property? When do graphs satisfy the finite-cycle property? (a property analogous to Shishikura's bound on the number of attractive, superattractive, parabolic, Siegel domains, as well as Herman rings). We show that the answer to the first question is in the affirmative.

Given a rational map R of degree $d \geq 2$, consider the Fatou-Julia graph $G_R(V, E)$ with the added assumption that for each Fatou component $U \subset \widehat{\mathbb{C}}$ we select the corresponding vertex to be an arbitrary point $u \in U$. Now given two vertices $u, u' \in V$ with the corresponding Fatou components U and U' satisfying $R(U) = U'$, we connect u to u' along their corresponding geodesic arc. We may refer to this as *embedded Fatou-Julia graph.*

Theorem 5.39. *Given a rational map R any embedded Fatou-Julia graph $G_R(V, E)$ is a topological Fatuo-Julia graph.*

Proof. We only need to prove that condition (iv) of previous definition holds. Each edge $e = (u, v) \in E$ connects two different components U and V. Hence it must intersect the boundary of the components at a point, say z that necessarily lies in J. Let N be any neighborhood of

z. Now consider any Fatou component U' such that $R^m(U') = U$ for some $m \geq 1$. If such component does not exist, the condition is vacuously satisfied. Otherwise, by Montel Theorem 5.23 there exists a point w in F, but not in U such that $R^n(w) \in U'$. Let W be the Fatou component containing w. Then $R^n(W) = U'$. Without loss of generality we may assume w is the representative of W in the graph, i.e. w is a vertex in V. $\square$

To appreciate the topological Fatou-Julia graph let us give another view into the Fatou-Julia graphs in terms of a directed d-ary tree. Such tree has a distinguished root with at most d children. Each of the children in turn has at most d children and so on. Binary trees and ternary trees are widely used. A connected component of the Fatou-Julia graph of a rational function of degree d is a directed d-ary tree. The direction is from a child to a parent. We know that there are at least $2d-2$ connected components. As an example if we consider Newton's for z^3-1 the Fatou-Julia graph will consist of union of three complete ternary trees, as shown in Figure 5.20. In fact in this case the tree, ignoring the root, is a complete ternary tree, meaning that each vertex has exactly three children.

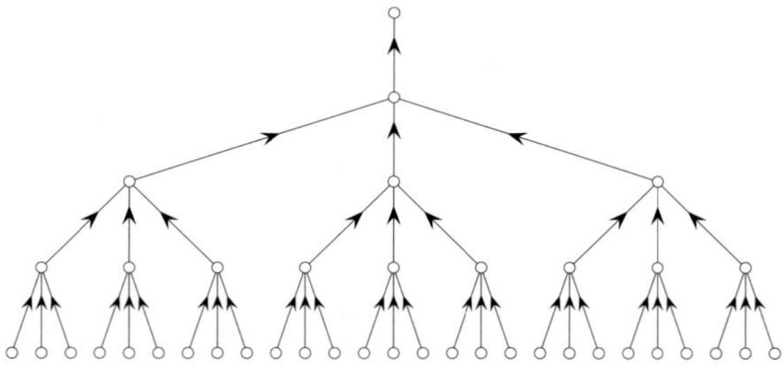

Fig. 5.20 Complete ternary tree corresponding to a component of Fatou-Julia graph of z^3-1.

In Figure 5.21 we show the segment of the connected component of the Fatou-Julia graphs on the corresponding polynomiograph of $z^3 - 1$ with respect to the root $\xi = 1$. We have omitted the directions.

In Figure 5.22 we try to show the growth of this tree and the intricate way in which it evolves. The first image (left) shows the three inverse images of a typical Fatou component that map into that component. It shows that the three inverse images of a component are scattered around. The next two images show the growth for the component containing one.

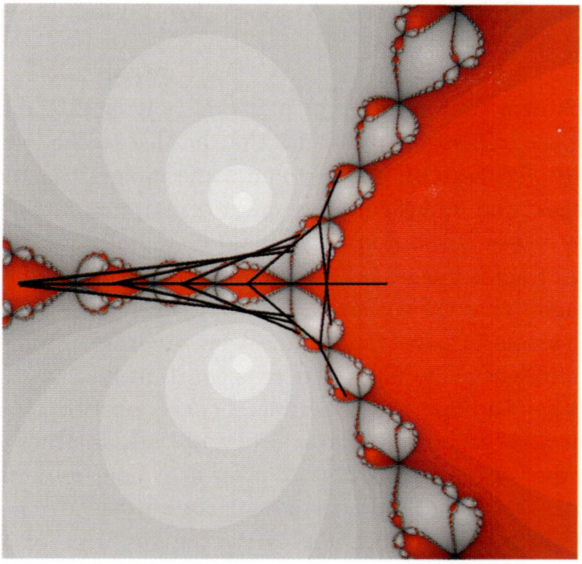

Fig. 5.21 Depiction of Fatou-Julia graph of Newton's for $z^3 - 1$.

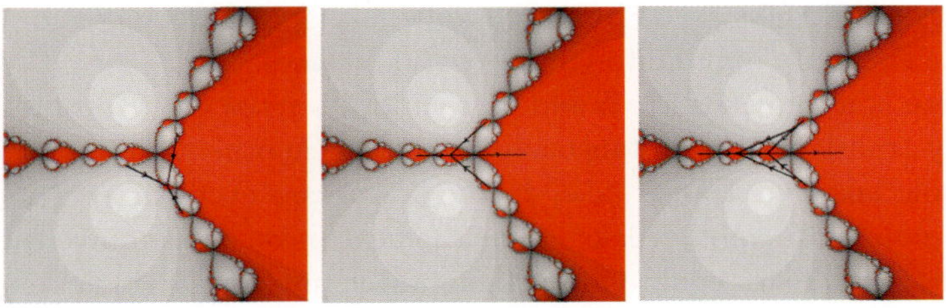

Fig. 5.22 Evolving flow of Fatou-Julia graph.

If we were to compactly draw a ternary tree as is normally done (Figure 5.20), if there are many vertices we would soon run out of space. In contrast considering the polynomiograph of $z^3 - 1$ under Newton's and the corresponding Fatou-Julia graph, we can see that not only we get ideas on how to place an infinite ternary tree in a finite space but to interweave three of them at the same time. In the next section we give a three-dimensional visualization of the flow of Fatou components.

5.17 Lakes and Waterfalls: Analogy for Dynamics of Rational Maps

When we look at a fractal image or a fractal polynomiograph, we do not get a sense of the underlying dynamics that has led to the image. For that reason we will offer here an analogy and some images that hopefully will reveal or suggest the underlying dynamics of Fatou components. Indeed this gives rise to new kind of appreciation for fractal polynomiographs, as well as general fractal images corresponding to a rational map.

Let us imagine that the surface of the earth, ideally the Riemann sphere, consists of a system of lakes that connect one another through waterfalls, often times infinitely many lakes. There is constant flow of water from one lake into another through the waterfalls. There are two general types of lakes: *periodic lakes* and *transit lakes*. A periodic lake is either an individual lake or a member of a finite number of lakes that pour into one another in a cyclic fashion. A cycle of more than one lake whose water pours into one another is perhaps best visualizable as the famous work of M.C. Escher, "Waterfall," Figure 5.23, a cyclic flow that defies the force of gravity.

To describe the lake-waterfall relations, there are two significant properties:

(i): Every lake is either a periodic lake or eventually through a finite number of waterfalls pours into a periodic lake.

(ii): There are only finitely many periodic lakes.

There is a *universal clock* and at each of its *pulse* every drop of water in every transit lake simultaneously pours into another lake that is either a periodic lake or a transit lake. In the latter case every drop of its water in turn pours into another lake, all at the next pulse of the universal clock. Several transit lakes could empty their water into one periodic lake. But there is a fixed number d so that for every lake there are at most d transit lakes that pour into it.

Each transit lake has an elevation that is higher than the one it pours into. The periodic lakes have zero elevation. The ones that pour into it have elevation one, the ones that pour into the elevation one lakes have elevation two, and so on. Since there are only finitely many periodic lakes, there are finitely many lakes of a particular elevation, and thus if there are infinitely many lakes, then there are transit lakes of arbitrary elevation. So there is really no source to where all the water originates from!

Fig. 5.23 M.C. Escher's "Waterfall." © 2008 The M.C. Escher Company-Holland. All rights reserved. www.mcescher.com

There is one other peculiar property:

(iii): Pick any lake L, then pick any lake L' whose water eventually pours into L. Then in any close proximity of L we can find another lake L'', possibly very tiny, whose water pours into L'.

The periodic lakes, despite their finiteness, have different characteristic. In some such lakes all the water at successive pulses of the clock move closer toward a *sink* or a cycle of *sinks*, but with the exception of a few *drops* of water (at most d drops), no drop ever reaches the sink or cycle of sinks.

There are other kinds of periodic lakes: *rotation lakes*. In the simplest kind of these where it consists of a single lake its water perpetually rotates about a particular sink. If we trace the path of a single drop of water in a rotation lake, as determined by successive pulses of the universal clock, it follows a closed loop, but never visits the same spot on this loop. Rotation lakes are like slow moving whirlpools and there could be a cyclic system of these lakes. Actually they could look like fast moving whirlpools, depending upon the speed of pulse, but the speed is not significant to the concept.

There is yet another kind of rotation lakes, similar to them, but as if a central part containing the sink or cycle of sinks are completely removed.

Another category of lakes consists of several individual periodic lakes which share a common sink, but at their boundary. These too could occur as a cyclic system of such lakes.

Fig. 5.24 A 3D depiction of topological Fatou-Julia Graph of $z^3 - 1$ based on Newton's.

In Figure 5.24 we give an image corresponding to these lake-waterfall system, based on the polynomiograph of Newton method as applied to $z^3 - 1$. The figure should help to visualize the topological Fatou-Julia graph. We would have to start from level zero and show the levels rising. This 3D impression gives more insight on the dynamics of Fatou components. For an animation of this see www.polynomiography.com.

5.18 General Convergence: Algorithmic Limitation of Iterations

Consider again Newton's iteration function $N(z) = z - p(z)/p'(z)$ for a polynomial $p(z)$. The iteration function has the significant property that every root of $p(z)$ is an attractive or superattractive fixed point. Moreover, $N(z)$ is a rational function of z defined in terms of z, $p(z)$, and $p'(z)$.

We are interested in the performance of Newton's iteration function from the point of view of its application to an individual polynomial $p(z)$, as well as a map

$$N : p(z) \longrightarrow N_p(z) = z - \frac{p(z)}{p'(z)}$$

assigning to each general polynomial $p(z)$ of degree d a rational function $N_p(z)$, which happens to be also of degree d.

The same consideration applies to other iteration functions for polynomial root-finding such as any member of the basic family, except that the degree of the rational function is some function of d, say $k = k(d)$. For clarity we give some definitions.

Definition 5.33. Given a polynomial $p(z)$ of degree d, a rational function $T_p(z)$ of degree $k = k(d)$ defined in terms of z, $p(z)$, and its first t derivatives is said to be an *iteration function* for that polynomial if each root of $p(z)$ is an attractive or superattractive fixed point of $T_p(z)$. If this property holds for all $p(z)$ of degree d we say the rational map T defined as $T(p(z)) = T_p(z)$ is a *purely iterative algorithm.*

Clearly Newton's method is a purely iterative map. Some authors like to give a more relaxed definition for an iteration function:

Definition 5.34. Given a polynomial $p(z)$ of degree d, a rational function $T_p(z)$ whose coefficients are rational functions of the coefficients of $p(z)$ is said to be an iteration function for the polynomial if each root of $p(z)$ is an attractive or superattractive fixed point of $T_p(z)$. If this property holds for all $p(z)$ of degree d we shall say the rational map $T : p(z) \longrightarrow T_p(z)$ is a *purely iterative algorithm.*

Consider for a polynomial $p(z)$ the rational function $T_p(z) = p(z)/(1 + p(z)^2)$. The only time roots of $p(z)$ would be fixed points is when $p(z) = z^d$. However, even for $p(z) = z$, $T_p'(0) = 1$. Therefore $T_p(z)$ is not an iteration function for $p(z) = z$, despite the fact that any point not lying on the y-axis converges to zero, see Figure 5.18, top-left image.

We may ask how good is Newton's method in finding roots of a polynomial? More specifically, does Newton's converge for almost every initial guess? We give a general definition.

Definition 5.35. Given a polynomial $p(z)$, an iteration function $T_p(z)$ is said to be *generally convergent* if for almost all $z \in \mathbb{C}$ the orbit $O^+(z)$ converges to a root of $p(z)$.

Later we will use our knowledge about the dynamics of rational maps to refine the above intuitive notion. All numerical analysis textbooks, and even some calculus textbooks, do cover the most basic property of the method, i.e. when z is "close" to a root ξ of $p(z)$, the corresponding orbit converges to ξ. The rate of convergence is quadratic when ξ is a simple root of $p(z)$ and linear if it is a multiple root. But the question is what does "close" mean? If we pick a random point z in $\mathbb{C}$ what happens to its orbit $O^+(z)$?

For a quadratic polynomial with distinct roots, the fact that Newton's method is generally convergent is well known and goes back to Cayley (1897) and Schröder (1870) (see also Peitgen *et al.* (1984)). In fact an arbitrary point converges to the root which happens to be closer to the point. The failure in convergence happens only on the line that is the perpendicular bisector of the line joining the two roots. In other words the basin of attraction of each root is its Voronoi region.

This property for quadratics has also been proved for some other iteration functions, e.g. for Halley's method, equivalent to $B_3(z)$, see e.g. Kneisl (2001). In fact this property is true for the entire Basic Family member $B_m(z)$ and is a consequence of the fact that the orbit of any point z_0 under $B_m(z)$ is a subsequence of the sequence $\{B_m(z_0) : m = 0, 1, \dots\}$, i.e. the Basic Sequence. This fact is a unique property of quadratics.

The fact that Newton's method is not generally convergent for polynomials was investigated by Barna (1956). For instance he considered $p(z) = 3z^5 - 10z^3 + 23z$. Newton's has $\{-1, 1\}$ as a superattracting cycle. Its polynomiography was shown in Figure 5.1 shows the white area near the cycle.

Proposition 5.7. *For all $n \geq 3$ Newton's method is not generally convergent for the polynomial*

$$p(z) = z^n - (n-1)z + (n-1).$$

To justify the above we follow an argument of Smale (1985) who considered this example for $n = 3$, i.e. $p(z) = z^3 - 2z + 2$, while raising the question of convergence of Newton's method. Their polynomiographies for some values of n are given below.

To arrive at the example consider the polynomial $p(z) = z^n + az + b$. We now attempt to select the coefficients so that Newton's method will have a superattractive periodic point of period 2 at the origin. More specifically, we want

$$N(0) = 1, \quad N(1) = 0, \quad N'(0) = 0, \quad N'(1) \neq \infty.$$

This and Chain rule would give

$$N^2(0) = 0, \quad (N^2)'(0) = N'(N(0))N'(0) = N'(1)N'(0) = 0.$$

In other words, 0 would be a superattractive fixed point of N^2. Hence there exists an open neighborhood of the origin U such that the fixed point iterates do not converge to any of the roots of $p(z)$. Consequently Newton's method is not generally convergent for this polynomial. The condition $N(0) = 1$ implies $a = -b$. The condition $N(1) = 0$ implies $a = -n+1$. Since $N'(z) = p(z)p''(z)/p'(z)^2$ all the other conditions can be trivially checked. Thus we have shown that the polynomial $p(z) = z^n - (n-1)z + (n-1)$ makes Newton's method fail to converge to a root of $p(z)$ over a set of positive (Lebesgue) measure.

We may next ask if Newton's method could give rise to other periodic cycles, i.e. a periodic cycle that is attractive, repulsive, parabolic, Siegel, or Cremer. We may in fact even try the same cycle $\{0, 1\}$. Indeed we find that all these are possible for a cubic polynomial. More precisely we have:

Proposition 5.8. *Let λ be any complex number not equal to 4. Then Newton's method for the polynomial*

$$p(z) = z^3 + (a-2)z^2 - az + a, \qquad a = \frac{1}{2}\left(1 \pm \sqrt{1 + \frac{32}{4-\lambda}}\right)$$

has the cycle $\{0, 1\}$ with its multiplier equal to λ. In particular Newton's method for cubic polynomials could give rise to Fatou components that are superattractive, attractive, parabolic, or Siegel. Moreover, Cremer points are also possible.

Proof. The recipe for constructing the polynomial is simple. We set

$$p(z) = z^3 + a_2 z^2 + a_1 z + a_0.$$

Then we try to solve for the coefficients so that

$$N(0) = 1, \quad N(1) = 0, \quad N'(0)N'(1) = \lambda.$$

Equivalently the above equations give:

$$\frac{p(0)}{p'(0)} = -1, \qquad \frac{p(1)}{p'(1)} = 1, \qquad \frac{p''(0)}{p'(0)}\frac{p''(1)}{p'(1)} = \lambda. \tag{5.4}$$

The first equation above implies $a_1 = a_0$ while the second equation implies $a_2 = a_0 - 2$. Thus setting $a_0 = a$ we will test the feasibility of the polynomial

$$p(z) = z^3 + (a-2)z^2 - az + a.$$

Substituting into the third equation in (5.4) we get

$$\frac{2(a-2)}{a}\frac{(2a+2)}{(a-1)} = \lambda.$$

This gives rise to the quadratic equation

$$(4-\lambda)a^2 - (4-\lambda)a - 8 = 0.$$

The solution then is as claimed. Note that no value of λ which is different from 4 can make $a = 0$ or $a = 1$ because in either case they imply $32/(4-\lambda) = 0$. In other words the third equation in (5.4) is well-defined. □

Note that when $\lambda = 0$ we get two solutions for $a = -1$ and $a = 2$. The latter gives the previously considered polynomial, $p(z) = z^3 - 2z + 2$.

Having proved that Newton's method is not generally convergent, we will consider the general member of the Basic Family with respect to general convergence. We essentially consider the same polynomials for obtaining superattractive cycle for Newton's.

Theorem 5.40. *For a given $m \geq 2$, let $n \geq m+1$, $p(z) = z^n + az - a$, where a is a parameter, and consider the corresponding basic family member*

$$B_m(z) = z - p(z)\frac{D_{m-2}(z)}{D_{m-1}(z)},$$

$$D_m(z) = \begin{vmatrix} p'(z) & \frac{p''(z)}{2!} & \dots & \frac{p^{(m-1)}(z)}{(m-1)!} & \frac{p^{(m)}(z)}{(m)!} \\ p(z) & p'(z) & \ddots & \ddots & \frac{p^{(m-1)}(z)}{(m-1)!} \\ 0 & p(z) & \ddots & \ddots & \vdots \\ \vdots & \vdots & \ddots & \ddots & \frac{p''(z)}{2!} \\ 0 & 0 & \dots & p(z) & p'(z) \end{vmatrix}.$$

For each $m \geq 1$ set

$$q_m(a) = D_m(1) = \begin{vmatrix} (n+a) & \binom{n}{2} & \dots & \binom{n}{m-1} & \binom{n}{m} \\ 1 & (n+a) & \ddots & \ddots & \binom{n}{m-1} \\ 0 & 1 & \ddots & \ddots & \vdots \\ \vdots & \vdots & \ddots & \ddots & \binom{n}{2} \\ 0 & 0 & \dots & 1 & (n+a) \end{vmatrix},$$

a polynomial of degree m in a. If a is any solution satisfying $q_{m-1}(a) - q_{m-2}(a) = 0$, also satisfying $q_{m-2}(a) \neq 0$, then we have

$$B_m(0) = 1, \quad B_m(1) = 0, \quad B'_m(0) = 0, \quad B'_m(1) \neq 0.$$

In other words, the origin is a superattractive periodic point of period 2. In particular $B_m(z)$ is not generally convergent for this polynomial.

Proof. For any $a \neq 0$ the polynomial $p(z) = z^n + az - a$ implies

$$p'(z) = nz^{n-1} + a, \quad \frac{p^{(i)}(z)}{i!} = \binom{n}{i} z^{n-i}.$$

Evaluating these at $z = 0$ and substituting for $k = m - 2$ and $k = m - 1$ into $D_k(z)$ we get $D_k(0) = a^k$. Substituting into the equation $B_m(0) = 1$ we get

$$1 = 0 - p_n(0)\frac{D_{m-2}(0)}{D_{m-1}(0)} = -a\frac{a^{m-2}}{a^{m-1}} = -1.$$

To argue that $B_m(1) = 0$ when a in addition is an appropriate root of $q_{m-1}(a) - q_{m-2}(a) = 0$, we note that since $p(1) = 1 + a - a = 1$, the choice of a implies $D_{m-2}(1)/D_{m-1}(1) = 1$. Thus we have justified the existence of a for which the polynomial $p(z) = z^n + az - a$ results in 0 being a periodic point of $B_m(z)$ of period 2.

To complete the proof it remains to show $B'_m(0) = 0$, $B'_m(1) \neq \infty$ so that 0 will be superattracting. We have

$$B'_m(z)$$
$$= 1 - p'_n(z)\frac{D_{m-2}(z)}{D_{m-1}(z)} - p_n(z)\frac{D'_{m-2}(z)D_{m-1}(z) - D_{m-2}(z)D'_{m-1}(z)}{D_{m-1}(z)^2}.$$

We claim $D'_k(0) = 0$ for $k = m - 2, m - 1$. To justify this we need to observe that $D'_k(z)$ is the sum of product terms each of which must contain at least one term of the form $p_n^{(i)}(z)/i!$, $i \geq 2$. Hence proving the claim. This together with the fact that the first two sums in $B'_m(0)$ are $1 - a \cdot a^{m-2}/a^{m-1} = 0$ shows $B_m(0) = 0$.

The proof that $B'_m(1) \neq \infty$, note that this is clear because neither $q_{m-2}(a)$ nor $q_{m-1}(a)$ are zero. □

Remark 5.16. The statement of the theorem is most likely more conservative that necessary. Firstly, it appears that any solution a to the equation $q_{m-1}(a) - q_{m-2}(a) = 0$ satisfies the condition that $q_{m-2}(a) \neq 0$. A condition that could ensure this is that $q_m(a)$ has distinct roots (see Problem 6). Secondly, it should be possible to compute explicit polynomials of degree $d \geq 4$ for which the corresponding $B_m(z)$ would result in attractive cycles of period 2 (see Problem 7).

Suppose $T_p(z)$ is an iteration function for a given polynomial $p(z)$. From the general theory on dynamical systems developed in this chapter we conclude:

Theorem 5.41. *An iteration function $T_p(z)$ for a polynomial $p(z)$ is generally convergent if and only if the measure of Julia set J is zero and for all $z \in F$ the corresponding orbit $O^+(z)$ converges to a root of $p(z)$.*

A further refinement of general convergence is the following:

Theorem 5.42. *An iteration function $T_p(z)$ for a polynomial $p(z)$ is generally convergent if and only if the measure of Julia set J is zero, moreover there are no cycle of attractive periodic points of period greater than one, no cycle of parabolic domains, no cycle of Siegel disks, and no cycle of Herman rings.*

In the case of Newton's method we have seen its failure for cubic polynomials in the sense that Newton's for $p(z) = z^3 - 2z + 2$ gave rise to attractive periodic cycle of period greater than one. We may ask could Newton's method fail in different ways? Newton's method could not give rise to Herman rings for any polynomial because the Julia set is always connected, see Theorem 5.47. However, one could try to find examples of parabolic periodic points, or a Siegel point, by considering a monic polynomial that would have such a cycle at 0 and 1.

Neither of the conditions in the above theorem is easy to check or to guarantee. A condition that could give rise to a means for checking general convergence relies on the behavior of critical points. First we need a definition.

Definition 5.36. A critical point c of a rational map R is said to be *pre-periodic* if $R^n(c)$ is periodic for some $n \geq 1$.

Theorem 5.43 (See McMullen (1994), McMullen (2004)).
Let $R(z)$ be a rational function which has an attracting cycle. If all critical points are either pre-periodic or converge to attractive cycles, then the measure of $J(R)$ is zero. Moreover, for any $z \notin J(R)$ the orbit converges to an attractive cycle.

If all the critical points of R converge to an attractive cycle, then R is said to be *hyperbolic*. The above theorem which is proved using the theory of orbifolds says that any point not in the Julia set of such rational function follows the path of a critical point. As an example consider the function $R(z) = az/(z^2+1)$ we examined earlier, see Figure 5.18.

From the above theorem we have

Theorem 5.44. *An iteration function $T_p(z)$ for a polynomial $p(z)$ is generally convergent if any of its critical points is either pre-periodic or its orbit converge to a root of $p(z)$.*

The above theorem is useful in the sense that it allows examining critical points. For example since the critical points of $N(z)$ are either the roots of $p(z)$ or $p''(z)$ we immediately get:

Corollary 5.7. *Consider Newton's method $N(z) = z - p(z)/p'(z)$ for a polynomial. If the roots of $p''(z)$ are either pre-periodic or converge to a root $p(z)$, then Newton's method is generally convergent for $p(z)$.*

Considering now the example $p(z) = z^3 - 2z + 2$, $p''(0) = 0$, but as we have seen 0 is an attractive periodic point and not a root of $p(z)$. Hence we could have anticipated the failure of the method. However, $N(z)$ is hyperbolic for this polynomial because there are 4 attractive cycle consisting of the three roots of $p(z)$ and the cycle $\{0, 1\}$, and we know that the number of critical points is $2d - 2 = 6 - 2 = 4$. In particular, the Julia set has zero measure. Thus the white areas we see in the polynomiograph in Figure 5.3 must consist of the union of two disjoint basins of attractions under $N^2(z)$: those of 0 and 1. If we were to look at the polynomiograph corresponding to $N^2(z)$ we would see 5 regions. This argument verifies the correctness of the polynomiograph in Figure 5.3.

Consider now $p(z) = z^3 - 1$. Then the only critical point in Newton's method is $z = 0$ which is pre-periodic since it converges to ∞, a fixed point. In particular the Julia set for this polynomial has measure zero. More generally, it is easy to show that Newton'method is generally convergent for $p_n(z) = z^n - 1$, for all $n \geq 1$. However even in the case of Basic Family members $B_m(z)$ as applied to this simple polynomial, there seems to be no easy way to prove the general convergence for arbitrary m and arbitrary n. A partial proof is given:

Theorem 5.45 (Jin and Kalantari (2007)). *For $m = 3, 4$ $B_m(z)$ is generally convergent to $p_n(z) = z^n - 1$ for all $n \geq 1$.*

In Jin and Kalantari (2007) some conjectures are given whose proof would imply the convergence for general m and n. Moreover, these conjectures have been verified for given specific values of n and m.

A purely iterative algorithm is a rational map T that assigns to each polynomial $p(z)$ of degree d an iteration function $T_p(z)$. It can be viewed as a map from the space of polynomials of degree d to the space of rational functions $k = k(d)$. Without loss of generality we assume that $p(z)$ is monic. Denoting the space of all monic polynomials of degree d by $Poly_d$, and the space of rational functions of degree $k = k(d)$ by Rat_k we have

$$T : Poly_d \longrightarrow Rat_k.$$

The space $Poly_d$ can be identified with $\mathbb{C}^d$, the space of d-dimensional complex numbers according to the following association

$$p(z) = z^n + a_{d-1}z^{d-1} \cdots + a_1 z + a_0 \longleftrightarrow (a_{d-1}, \ldots, a_0).$$

Definition 5.37. A purely iterative algorithm $T : Poly_d \longrightarrow Rat_k$ is said to be generally convergent if there exists an open, dense, full-measure subset of $Poly_d$ such that for each $p(z)$ in this set $T(p(z))$ is generally convergent for $p(z)$.

A variation of the notion of general convergence initially raised by Smale (1985) is the following:

Definition 5.38. A purely iterative algorithm $T : Poly_d \longrightarrow Rat_k$ is generally convergent if for almost all input z and almost all $p(z)$ in $Poly_d$ the orbit $O^+(z)$ with respect to $T(p(z))$ converges to a root of $p(z)$.

The example $p(z) = z^3 - 2z + 2$ which proves Newton's is not generally convergent for that polynomial, through a continuity argument, also demonstrate the failure of Newton's method as a purely iterative generally convergent algorithm for $Poly_3$.

We have seen that in the case of a cubic polynomial Newton's method could fail in the sense that an attracting cycle could exist. For a polynomial Newton's method cannot result in a Herman ring because in this case the Julia set can be shown to be connected (Theorem 5.47). The existence of a Herman ring would contradict the connectivity of the Julia set. The unavoidable appearance of a Siegel disk for some polynomials in Newton's method, or more general iteration functions is a consequence of a technical rigidity result proved by McMullen, see McMullen (1987), McMullen (1988) which we will discuss next. However, it appears to be an open question as to whether or not the Julia set with respect to iteration functions could have positive measure, another source of trouble for almost everywhere convergence of orbits to roots of the underlying polynomial.

Theorem 5.46 (Algorithmic Limitations, McMullen (1987)). *There does not exist a generally convergent purely iterative algorithm iteration function for solving polynomials of degree* $d \geq 4$.

McMullen's theorem is a deep and technical result. It is based on proving that any generally convergent purely iterative algorithm must necessarily satisfy a certain rigidity property which in turn would imply the existence of a Möbius transformation that would have the ability to permute the roots of any two arbitrary polynomials p and q of degree $d \geq 2$. The lack of such Möbius transformation when $d \geq 4$ then implies the non-existence.

The question one may raise is: How could an iteration function T fail to be generally convergent? We have seen that in case of Newton's method attractive cycles are possible by exhibiting an explicit polynomial and continuity implies that for a whole open domain in $Poly_d$, given any $d \geq 3$. More generally $B_m(z)$ could fail to be generally convergent for $Poly_d$, admitting a polynomial with attracting cycle of period two. Hence by continuity it would not be generally convergent for the corresponding polynomial spaces.

McMullen result on the lack of generally convergent algorithms for a general polynomial of degree $d \geq 4$ is a consequence of more general result on the rigidity of algebraic and analytic rational maps. This result proves a certain stability, see Theorem 4.2 in McMullen (1994), in particular implies that any iteration function $T : Poly_d \longrightarrow Rat_k$ must admit an attractive cycle of arbitrary large period. Furthermore, through parametrization and continuity, the existence Siegel disks can be established: If $T_t(z)$ is a family of rational maps with a cycle that changes from repelling to attracting as t varies, then this cycle also becomes the center of a Siegel disk for some parameter t. Hence $B_m(z)$ could admit attractive cycles of arbitrary size and Siegel disks, for any $m \geq 2$.

While Newton's method is not generally convergent for cubic polynomial McMullen (1987) designed a generally convergent algorithm for a general cubic polynomial which without loss of generality may be assumed to be $p(z) = z^3 + az + b$. McMullen's generally convergent algorithm is

$$T_p(z) = z - \frac{p(z)(3az^2 + 9bz - a^2)}{(3az^4 + 18bz^3 - 6a^2z^2 - 6abz - 9b^2 - a^3)}.$$

This algorithm happens to be *superconvergent*. A superconvergent iteration function for a polynomial $p(z)$ has the property that its critical points coincide with the roots of $p(z)$.

For example for $p(z) = z^3 - 1$ and $p(z) = z^3 - 2z + 2$ we get:

$$T_p(z) = \frac{(z^4 + 2z)}{(2z^3 + 1)}, \quad T_p(z) = \frac{(9z^4 - 16z^3 + 36z^2 - 36z + 4)}{(3z^4 - 18z^3 + 12z^2 - 12z + 14)}.$$

In fact $T(z)$ can be obtained by applying Newton's method to the rational function

$$q(z) = \frac{p(z)}{3az^2 + 9bz - a^2}.$$

In the case of $p(z) = z^3 - 1$, $T(z)$ happens to coincide with $B_3(z)$, also known as Halley's method, applied to this polynomial. In particular its rate of convergence is cubic. For the general cubic $p(z) = z^3 + az + b$ the rate of convergence can also be shown to be cubic, e.g. by showing that T_p' and T_p'' are zero at any root of p. However, for the general cubic polynomial $T(z)$ and $B_3(z)$ do not coincide. In Figure 5.1, the first image in the second row, shows the polynomiography of $p(z) = z^3 - 2z + 2$ with respect to $T(z)$.

The fact that McMullen's iteration for cubic polynomials has order 3 is proved by Hawkins (2002) who gives a characterization of generally convergent iteration functions for cubics and showing that any such iteration function consists of a generally convergent iteration function for $z^3 - 1$ and its conjugation with a certain Möbius map. Using the characterization Hawkins gives fifth-order generally convergent for cubic polynomials. The general convergence of $B_m(z)$ for $z^3 - 1$ (see Jin and Kalantari (2007)) and her recipe could give rise to higher-order generally convergent algorithms for cubic polynomials. This would only be of theoretical interest since solving a cubic equation is among the easiest of root-finding problems.

Although there does not exist any generally convergent purely iterative method, Shub and Smale (1986) have shown that if one allows conjugation generally convergent algorithms do exist. In another line of approach for solving the roots of polynomial equations through iterations of rational functions is the construction of a *tower of algorithm*, i.e. algorithms which consist of inputting the limit of one sequence of iterations from a rational function into another one and so on. Doyle and McMullen (1989) have given algorithms of such for quintic polynomials. Crass and Doyle (1997).

We close this section by some remarks. In a sense the work of McMullen's theorem on the lack of generally convergent algorithms is a profound negative result and somewhat analogous to the unsolvability by radicals the solution of general polynomial equation of degree 5 or higher and connections with Galois theory. These are both historic results on polynomials. However, neither of these is really limiting the actual approximation of roots of polynomials. Firstly, from the convergence point of view the pointwise evaluation of the Basic Family, the basic sequence $\{B_m(a) : m = 1, \cdots\}$, almost surely will converge to a root. Secondly, from the point of view of computation, efficient and practical algorithms for polynomial root-finding demand preprocessing techniques such as bounding the roots of a polynomial combined with binary search type of algorithm, e.g. Weyl's method, and then apply iteration functions such as Newton's or other basic family members once sufficiently close to a root. We will address this again in the book.

5.19 A Summary for the Behavior of Iteration Functions

From the point of view of an iteration function $R(z)$ for finding the root of a polynomial $p(z)$, it is convenient to partition $\widehat{\mathbb{C}}$ not just as Julia and Fatou sets $J(R)$ and $F(R)$, but as the Good, the Bad, and the Undesirable sets and even further partition of these sets. More specifically, the Bad set, denoted by $B(R)$, is simply J where the behavior of the orbits are bad as expected. We partition $F(R)$ as the Good set, denoted by $G(R)$, the set of points $z \in F(R)$ whose orbit $O^+(z)$ converges to an attractive or superattractive cycle of R. Then Undesirable set, $U(R) = F(R) - G(R)$. The set $G(R)$ can further be partitioned as the set of Ideal point, $I(R)$, the points whose orbit converges to a root of $p(z)$. The Non-ideal points of R are $N(R) = G(R) - I(R)$. We may further partition $U(R)$ as the set of Siegel disks, $S(R)$, the parabolic components, $P(R)$, and the Herman rings, $H(R)$. The Venn diagram in Figure 5.25 shows this partitioning.

<table>
<tr><th>$G(R)$</th><th>$B(R)$</th><th>$U(R)$</th></tr>
<tr><td rowspan="3">$I(R)$</td><td rowspan="6">$J(R)$</td><td rowspan="2">$P(R)$</td></tr>
<tr></tr>
<tr><td rowspan="2">$S(R)$</td></tr>
<tr><td rowspan="3">$N(R)$</td></tr>
<tr><td rowspan="2">$H(R)$</td></tr>
<tr></tr>
</table>

Fig. 5.25 Further partition of the Good, the Bad, and the Undesirable.

As far as $B_m(z)$, cycles of periodic attractive components, superattractive components, parabolic components, and Siegel disks can be formed. However, it is not clear if cycle of Herman rings could be formed. Herman rings cannot be formed if the Julia set can be proved to be connected. This is known for Newton's method due to the following theorem:

Theorem 5.47 (Shishikura (1990)). *If a rational map has only one fixed point which is repelling or parabolic with multiplier 1, then its Julia set is connected. In other words, every component of the complement of the Julia set is simply connected. In particular, the Julia set of the Newtons method for a non-constant polynomial is connected (since ∞ is the only repelling fixed point). Consequently, each Fatou component is simply connected.* □

This above theorem however is not applicable to the entire Basic Family (see Problem 9). Let us summarize the behavior of an iteration function $R(z)$ for finding the roots of a polynomial $p(z)$ of degree n. Each root of $p(z)$ will be an attractive or superattractive fixed point so that it will have its own Fatou components, its immediate basin of attraction. There may be other attractive fixed points. This however will not happen for $B_m(z)$. There will generally be a finite number of attractive or superattractive periodic points of period $p > 1$. Each of these will have its own Fatou component. If there is any parabolic periodic point, depending upon its multiplicity $m + 1$, it will attract m Fatou components.

Each periodic Siegel point, if it exists, will have its own Fatou component. The critical points or their forward orbits will either land in a periodic Fatou component or they accumulate around the boundary of Siegel disks or Herman rings.

5.20 Undecidability Issues in Rational Functions

In this section we would like to give a brief and mainly informal introduction to the work of Blum *et al.* (1998) on decidability issues with regard to the Mandelbrot set; the set of points of $\mathbb{C}$ for which Newton's method converges to a root of a given polynomial $p(z)$; and the Julia set of rational maps.

The foundation of the classical notion of decidability is laid by the work of such logicians as Turing and Gödel leading to some of the deepest discoveries in mathematics and computer science. Traditionally the notion of decidability is defined with respect to countable sets, but not for instance over the reals. Intuitively decidability of a set S within a given universe U of elements means if there is a procedure or algorithm that can decide, within a finite number of well-defined steps, if a given $x \in U$ is member of S.

Intuitively, decidability of Newton's method for a given polynomial $p(z)$ means if there is an algorithm that, given a complex number c, can decide if the corresponding orbit would be convergent to a root. This question could be posed in the approximate sense, i.e. if for a given $\epsilon > 0$ the orbit would lead to a point z satisfying $|p(z)| < \epsilon$.

Blum *et al.* (1998) prove that these problems are undecidable with respect to a notion of decidability they define which in turn requires a notion of a machine over the real numbers. Such machine is formally definable with respect to a directed graph with certain required type of nodes corresponding to e.g, input and output. The halting set of a machine is the set of all inputs on which the machine halts.

Definition 5.39. A set S in $\mathbb{R}^n$ is *decidable* if given an x in $\mathbb{R}^n$ there is an algorithm that determines in a finite number of steps whether or not it lies in S.

Definition 5.40. A set S is *semidecidable* if whenever x lies in S there is an algorithm that can validate the membership in a finite number of steps. A semidecidable set is the *halting set* of a machine.

Proposition 5.9. *A set S is decidable if both S and its complement are semidecidable.*

Definition 5.41. A set S is *basic semialgebraic* if it satisfies a finite system of polynomial equalities and inequalities over $\mathbb{R}$. A *semialgebraic* set is a finite union of basic semialgebraic sets.

A main result of Blum *et al.* (1998) theory is the following:

Theorem 5.48. *The halting set of a machine is necessarily a countable disjoint union of basic semialgebraic sets.*

Their result proves another limitation on iteration functions for polynomial root-finding, even Newton's method:

Theorem 5.49. *The set of points under which Newton's method converge for a polynomial is generally undecidable, even for a cubic over the reals.*

To prove Newton's method is generally undecidable one can make use the following theorem due to Barna (1956):

Theorem 5.50. *Let $p(x)$ be real polynomial of degree $d > 3$ which has only real roots. Then the set of points on the real axis for which Newton's method does not converge to a root is a Cantor set with zero measure.*

On the other hand, over the reals a semialgebraic set is necessarily the finite union of intervals. But a Cantor set has no interior point and hence it is not the countable union of semialgebraic sets.

However, the halting set must have integral Hausdorff dimension. Since for most rational functions the Julia set has fractional Hausdorff dimension one can conclude that they are not decidable. Blum *et al.* (1998) also show that, within the confines of their theory, the Mandelbrot set is undecidable despite the fact that its boundary is known to have Hausdorff dimension 2, shown in Shishikura (1994).

One may wonder if the notion of decidability could be extended and or considered based on rational skeleton of the underlying problems. However, since the rational skeleton of a problems over the reals may not provide sufficient information, Blum *et al.* (1998) reject the idea of using rational inputs in extending the notion of decidability. For instance, as they point out the curve defined by $x^3 + y^3 = 1$ has no nontrivial rational solutions in the first quadrant. However for some problems it seems to make perfect sense to restrict the decidability question to some rational or algebraic skeleton, e.g. see Problems 3 and 4 below. We end this chapter with a few problems.

Problem 1. Does the Basic Family $\{B_m(z) : m \geq 2\}$ for a polynomial $p(z)$ form a normal family on the interior of Voronoi regions of the roots of $p(z)$?

Problem 2. Is the Julia set of the Basic Family for polynomials a set of measure zero?

Problem 3. (Wolfram's question (Wolfram (2004))) Given $p(z) = z^3 - 1$ and an input z_0 (think of a pixel) with rational real and imaginary parts, can we determine if under the iterations of Newton's method it converges, if so to which root? (if so we can quickly decide how to color the pixel).

Problem 4. (Generalization of Wolfram's) Given $p(z) = z^n - 1$ and an input z_0 with rational real and imaginary parts, can we determine if under the iterates of $B_m(z)$ it converges, and if so which of the roots does it converge?

Problem 5. Is $B_m(z)$ generally convergent for $p(z) = z^n - 1$? (for partial answer see Jin and Kalantari (2007)).

Problem 6. Does the matrix A_m have distinct characteristic roots for any $m \geq 2$?

$$A_m = \begin{pmatrix} \binom{n}{1} & \binom{n}{2} & \cdots & \binom{n}{m-1} & \binom{n}{m} \\ 1 & \binom{n}{1} & \ddots & \ddots & \binom{n}{m-1} \\ 0 & 1 & \ddots & \ddots & \vdots \\ \vdots & \vdots & \ddots & \ddots & \binom{n}{2} \\ 0 & 0 & \cdots & 1 & \binom{n}{1} \end{pmatrix}.$$

If so, then for each solution a to the equation $\det(A_{m-1} - zI_m) - \det(A_{m-2} - zI_{m-2}) = 0$, and for each $n \geq m+1$ the polynomial $p(z) = z^n + az - a$ will have $\{0, 1\}$ as an attractive cycle of $B_m(z)$, hence not generally convergent for this polynomial (see Theorem 5.40).

Problem 7. Is it possible to compute an explicit polynomial of degree $d \geq 4$ for which the corresponding $B_m(z)$ would result in attractive cycle of period 2? Try this first for $B_3(z)$, Halley's method.

Problem 8. Given an iteration function $R(z)$, defined as a rational function in terms of z, and $p^{(i)}(z), i = 0, \ldots, t$, for some $t \geq 1$, can we always select a polynomial so that we can create an attractive or super attractive cycle of period two at say 0 and 1? (i.e. can we give a systematic method that proves such an iteration function is not generally convergent without needing to resorting to the general theory of McMullen).

Problem 9 Is the Julia set of $B_m(z)$ for polynomials connected? Can $B_m(z)$ admit a Herman ring?

Problem 10 Analyze the critical points of the Basic Family and their location with respect to the roots of the underlying polynomial.

Problem 11. Given any fixed polynomial $p(z)$, does there exist $m \geq 2$

for which $B_m(z)$ is a generally convergent iteration function for $p(z)$?

Problem 12. Given two distinct n digit numbers with digits in $0-9$ set

$$A = a_{n-1} \cdots a_1 a_0, \quad B = b_{n-1} \cdots b_1 b_0,$$

let

$$P(A) = a_{n-1} z^n + \cdots + a_0, \quad P(B) = b_{n-1} z^n + \cdots + b_0$$

be their polynomials having zero set

$$Z_A, \quad Z_B$$

In terms of n, find a lower bound on the Hausdorff distance of the sets Z_A and Z_B.

Problem 13. Given a polynomial $p(z)$ and an input z_0 in the Voronoi region of a root θ of $p(z)$, and an $\epsilon > 0$, determine the smallest m for which $|B_m(z) - \theta| < \epsilon$.

Problem 14. Given an iteration function defined for any polynomial in terms of z, $p(z)$ and its derivatives, can we determine explicit polynomial of degree 3 or higher that would admit a cycle of a given size?

Problem 15. Suppose $R(z)$ is an iteration function for a polynomial $p(z)$. When is every periodic cycle a subset of the boundary of the Voronoi region of the roots of $p(z)$?

Problem 16. Suppose that we have a rational function $R(z)$. How to determine if R is factorizable? i.e. $R(z) = Q^t(z)$ for some rational function Q and integer t? For example can we determine if $R(z) = N^t(z)$ where $N(z)$ is Newton's for some $p(z)$? How about the general case of the question when $R(z)$ a polynomial?

Problem 17. Given a point u in $F(R)$, can we decide if u and $R(u)$ are within the same component?

Problem 18. Given a rational map $R(z)$ can we decide if it has infinitely many Fatou components?

Problem 19. Given a rational map $R(z)$, in terms of its coefficients find bounds on the norm of non-repelling periodic points (see Chapter Chap. 15 on bounds on zeros of polynomials).

Chapter 6

Fixed Points of the Basic Family *

In this chapter we will analyze the fixed points of the individual members of the Basic Family. These include the roots of the underlying polynomial as well as the fixed points that are not roots of the polynomial. The latter are called *extraneous* fixed points. For each particular Basic Family member, all polynomial roots turn out to be either attractive or superattractive while all extraneous fixed point are repulsive. This is an important property of the Basic Family.

6.1 Introduction

Consider a complex polynomial $p(z)$. We recall that for each $m \geq 2$ the Basic Family member is defined as

$$B_m(z) = z - p(z)\frac{D_{m-2}(z)}{D_{m-1}(z)},$$

where $D_0(z) = 1$, and for each $m \geq 1$,

$$D_m(z) = \begin{vmatrix} p'(z) & \frac{p''(z)}{2!} & \dots & \frac{p^{(m-1)}(z)}{(m-1)!} & \frac{p^{(m)}(z)}{(m)!} \\ p(z) & p'(z) & \ddots & \ddots & \frac{p^{(m-1)}(z)}{(m-1)!} \\ 0 & p(z) & \ddots & \ddots & \vdots \\ \vdots & \vdots & \ddots & \ddots & \frac{p''(z)}{2!} \\ 0 & 0 & \dots & p(z) & p'(z) \end{vmatrix}$$

A root θ of $p(z)$, simple or not, is necessarily a fixed point of $B_m(z)$, i.e. $p(\theta) = 0$ implies $B_m(\theta) = \theta$. This follows since $D_m(\theta)$ simplifies into $p'(\theta)^m$.

*Part of this chapter has been reprinted from On Extraneous Fixed-points of the Basic Family of Iteration Functions, *BIT*, Vol. 43 (2003) 435–458, B. Kalantari and Y. Jin. With kind permission of Springer Science and Business Media.

The converse however may not be true. When $B_m(\theta) = \theta$, but $p(\theta) \neq 0$, then θ is said to be an *extraneous fixed point* of B_m. Note that an extraneous fixed point is necessarily a root of $D_{m-2}(z)$.

The Newton function $B_2(z)$, clearly has no extraneous fixed points. However, for $m > 2$ there are many extraneous fixed points. Vrscay and Gilbert (1988) proved that Halley's iteration function, i.e. $B_3(z)$, has the property that each extraneous fixed point is repulsive. We prove here that this property hold for all $m \geq 2$ and give explicit value of the multiplier. This result is important in the sense that it implies a convergent orbit corresponding to any specific member of the Basic Family will necessarily converge to a zero of $p(z)$.

It should however be emphasized that the fact that $B_m(z)$ has no extraneous (super)attractive fixed point does not imply the same assertion with respect to its periodic points. In fact as we have seen in Chapter 5 even in the case of Newton's function there are examples of a cubic polynomials, e.g. $p(z) = z^3 - 2z + 2$, with periodic points that are attractive. The roots of $p(z)$ will however always turn out to be attractive or superattractive fixed point of $B_m(z)$.

6.2 Properties of the Fixed Points of the Basic Family

The following theorem gives a complete characterization of the fixed points of the Basic Family.

Theorem 6.1 (Kalantari and Jin (2003)). *Let θ be a fixed point of $B_m(z)$, $m \geq 2$ corresponding to a polynomial $p(z)$ of degree n. Then*

$$B'_m(\theta) = \begin{cases} 0, & \text{if } p(\theta) = 0,\ p'(\theta) \neq 0; \\ \frac{s-1}{s+m-2}, & \text{if } \theta \text{ a root of multiplicity } s; \\ 1 + \frac{m-1}{k}, & \text{if } p(\theta) \neq 0, \text{ but } \theta \text{ is a root of } D_{m-2}(z) \text{ of multiplicity} \end{cases}$$

Thus, simple roots of $p(z)$ are superattractive; multiple root of $p(z)$ are attractive; and all the extraneous fixed points of $B_m(z)$ are repulsive. In particular, $B_m(z)$ has no indifferent fixed points.

Remark 6.1. We note that at any fixed point θ the multiplier is independent of the actual value of θ, either zero or dependent only on m and multiplicity as a root of $p(z)$ or a root of $D_{m-2}(z)$.

6.3 Proof of Main Theorem

We now prove Theorem 6.1. Differentiating $B_m(z)$ we get

$$B'_m(z) = 1 - p'(z)\frac{D_{m-2}(z)}{D_{m-1}(z)} - p(z)\frac{D'_{m-2}(z)D_{m-1}(z) - D_{m-2}(z)D'_{m-1}(z)}{(D_{m-1}(z))^2}.$$

Suppose that θ is a simple root of $p(z)$. Then from the determinantal form of $D_m(z)$ it follows that

$$D_m(\theta) = p'(\theta)^m.$$

Substituting $z = \theta$ in B'_m we see that

$$B'_m(\theta) = 1 - p'(\theta)\frac{p'(\theta)^{m-2}}{p'(\theta)^{m-1}} = 0.$$

This proves the first part of the formula in the theorem.

Next we prove the third part of the formula in the theorem. From Corollary 4.3 in Chapter 4, written for $m-2$ instead of m and regrouping we have

$$D_{m-1}(z) = p'(z)D_{m-2}(z) - \frac{1}{m-1}p(z)D'_{m-2}(z). \tag{6.1}$$

Suppose that $B_m(\theta) = \theta$, but $p(\theta) \neq 0$. Since

$$B_m(z) = z - p(z)\frac{D_{m-2}(z)}{D_{m-1}(z)},$$

it follows that

$$D_{m-2}(\theta) = 0.$$

Assume that θ is a root of multiplicity k of $D_{m-2}(z)$. Then

$$D_{m-2}(z) = (z-\theta)^k W(z), \tag{6.2}$$

for some polynomial $W(z)$ where $W(\theta) \neq 0$. Thus

$$D'_{m-2}(z) = k(z-\theta)^{k-1}W(z) + (z-\theta)^k W'(z). \tag{6.3}$$

Substituting (6.2) and (6.3) into (6.1), the quantity

$$p(z)\frac{D_{m-2}(z)}{D_{m-1}(z)}$$

after factoring the term $(z-\theta)^{k-1}$ reduces to

$$R(z) = p(z)\frac{U(z)}{V(z)},$$

where

$$U(z) = (z - \theta)W(z),$$

and

$$V(z) = p'(z)(z-\theta)W(z) - \frac{p(z)}{m-1}[kW(z) - (z-\theta)W'(z)].$$

Then,

$$B'_m(z) = 1 - R'(z) = 1 - p'\frac{U(z)}{V(z)} - p(z)\frac{U'(z)V(z) - U(z)V'(z)}{V^2(z)}.$$

Since $U(\theta) = 0$, $U'(\theta) = W(\theta)$, and $V(\theta) = -p(\theta)kW(\theta)/(m-1)$, it follows that

$$B'_m(\theta) = 1 - \frac{p(\theta)W(\theta}{-p(\theta)kW(\theta)/(m-1)} = 1 + \frac{m-1}{k}.$$

Hence we have proved the third part of the formula.

The proof of the second formula in the theorem can be established using the theory of symmetric function from which it follows that:

$$\lim_{z\to\theta} \frac{B_m(z) - \theta}{z - \theta} = \frac{s-1}{s+m-2}.$$

For the proof of this property we refer the reader to Jin and Kalantari (2005b). □

Problem 1. Analyze the extraneous fixed points of the Basic Family and find bounds on their magnitude (see Chapter 15). How are these extraneous fixed points related to the location of the roots of $p(z)$ and their convex hull?

Chapter 7

Algebraic Derivation of the Basic Family and Characterizations

In this chapter we give an algebraic derivation of the Basic Family and as a result derive many characterizations of this fundamental family of iteration functions which reveal their optimality with respect to several criteria, giving further evidence to their significance and uniqueness among iteration functions for root finding. We also offer recipes for constructing many other families of iteration functions, including one credited to Euler as well as Schröder.

7.1 Introduction

Let $p(z)$ be a polynomial of degree $n \geq 2$ with coefficients in a subfield K of the complex numbers. For each natural number $m \geq 2$, let $L_m(z)$ be the $m \times m$ lower triangular matrix whose diagonal entries are $p(z)$ and for each $j = 1, \ldots, m-1$, its j-th subdiagonal entries are $p^{(j)}(z)/j!$, i.e.

$$L_m(z) = \begin{pmatrix} p(z) & 0 & 0 & \ldots & 0 \\ p'(z) & p(z) & 0 & \ldots & 0 \\ \frac{p''(z)}{2!} & p'(z) & p(z) & \ldots & 0 \\ \vdots & \vdots & \vdots & \ddots & \vdots \\ \frac{p^{(m-1)}(z)}{(m-1)!} & \frac{p^{(m-2)}(z)}{(m-2)!} & \frac{p^{(m-3)}(z)}{(m-3)!} & \ldots & p(z) \end{pmatrix}. \tag{7.1}$$

For $i = 1, 2$, let $L_m^{(i)}(z)$ be the $(m-i) \times (m-i)$ matrix obtained from $L_m(z)$ by deleting its first i rows, and its last i columns, i.e.

$$L_m^{(1)}(z) = \begin{pmatrix} p'(z) & p(z) & 0 & \dots & 0 \\ \frac{p''(z)}{2!} & p'(z) & p(z) & \dots & 0 \\ \vdots & \vdots & \vdots & \ddots & \vdots \\ \frac{p^{(m-1)}(z)}{(m-1)!} & \frac{p^{(m-2)}(z)}{(m-2)!} & \frac{p^{(m-3)}(z)}{(m-3)!} & \dots & p'(z) \end{pmatrix},$$

$$L_m^{(2)}(z) = \begin{pmatrix} \frac{p''(z)}{2!} & p'(z) & \dots & 0 \\ \frac{p'''(z)}{3!} & \frac{p''(z)}{2!} & \dots & 0 \\ \vdots & \vdots & \ddots & \vdots \\ \frac{p^{(m-1)}(z)}{(m-1)!} & \frac{p^{(m-2)}(z)}{(m-2)!} & \dots & \frac{p''(z)}{2!} \end{pmatrix},$$

with $L_1^{(1)}(z) \equiv 1$. Setting

$$D_k(z) = \det(L_k^{(1)}(z))$$

for each k the Basic Family member defined in previous chapters is

$$B_m(z) = z - p(z)\ \frac{D_{m-2}(z)}{D_{m-1}(z)}.$$

Definition 7.1. For each pair of nonnegative integers m and M, $M \geq m$, let $S(m, M)$ be the set of all $g(z) \in K(z)$ (rational functions with coefficient in K) so that for all roots θ of $p(z)$ we have

$$g(z) = \theta + \sum_{i=m}^{M} \gamma_i(z)(\theta - z)^i, \tag{7.2}$$

where $\gamma_i(z) \in K(z)$, for $i = m, \dots, M$, and well-defined at any simple root θ, i.e. $\lim_{x\to\theta} \gamma_i(z) \equiv \gamma_i(\theta)$ exists. Moreover, $\gamma_m(z)$ and $\gamma_M(z)$ are not identically zero.

Remark 7.1. Given $g \in S(m, M)$ with $m \geq 2$, from the continuity of $\gamma_i(z)$'s at a simple root θ, and since $m \geq 2$ it follows that there exists a neighborhood of θ for which $|g(z) - \theta| \leq C|x - \theta|^m$, for some constant C. This implies the existence of a neighborhood I_θ of θ for which for any $z_0 \in I_\theta$ the fixed-point iteration

$$z_{k+1} = g(z_k), \quad k \geq 0, \tag{7.3}$$

converges to θ, and clearly

$$\lim_{k\to\infty} \frac{(\theta - z_{k+1})}{(\theta - z_k)^m} = -\gamma_m(\theta), \tag{7.4}$$

i.e. the rate of convergence of the sequence $\{z_k\}_{k=0}^{\infty}$ is of order m. Also note that if z_0 is in K, then so are all the iterates in (7.3).

Denoting $\gamma_m(z)$ corresponding to $B_m(z)$ by $\gamma_m^{(m)}(z)$, we show that for a simple root θ, the asymptotic constant of convergence for $B_m(z)$ is

$$\gamma_m^{(m)}(\theta) = (-1)^m \frac{\det(L_{m+1}^{(2)}(\theta))}{\det(L_m^{(1)}(\theta))} = \frac{(-1)^m}{p'(\theta)^{m-1}} \det(L_{m+1}^{(2)}(\theta)). \tag{7.5}$$

In particular for $m = 2$ and $m = 3$ we get

$$\gamma_2^{(2)}(\theta) = \frac{p''(\theta)}{2p'(\theta)}, \qquad \gamma_3^{(3)}(\theta) = -\frac{2p'(\theta)p'''(\theta) - 3p''(\theta)^2}{12p'(\theta)^2}.$$

We show that if all roots of p are simple, $B_m(z)$ is the unique member of $S(m, m+n-2)$.

The following definition provides means by which members from $S(m, M)$ can be compared.

Definition 7.2. Given $g \in S(m, M)$, let the *order* of g to be defined as m; the *coefficient vector* of g to be the vector $\Gamma(z) = (\gamma_m(z), \ldots, \gamma_M(z))$; the *leading coefficient* of g to be γ_m; and the *width* of g to be $M - m + 1$. The *depth* of g is defined to be d if the formula for g depends on $p^{(j)}$, $j = 0, \ldots, d$. Similarly the *depth* of the leading coefficient of g is defined. The *simple-root-depth* of g is defined to be ρ, if for any simple root θ of $p(z)$, $\gamma_m(\theta)$ depends on $p^{(j)}(\theta)$, $j = 0, \ldots, \rho$.

Remark 7.2. If g and h are in $S(m, M)$, so is any affine combination, $\alpha g + \beta h$, where $\alpha + \beta = 1$. As $B_m(z)$ will be shown to belong to $S(m, m+n-2)$, it follows that for any $m \geq 2$, and $M \geq m+n-2$, the set $S(m, M)$ is nonempty. If $p(z)$ has simple roots, as $B_m(z)$ will be shown to be the unique element of $S(m, m+n-2)$, it follows that $S(m, M)$ is empty if $M < m+n-2$. Thus, $B_m(z)$ is also *minimal* with respect to width. Traub (1964), has shown that any m-th order one-point iteration function must have depth at least equal to $m-1$. Thus, in the sense of depth too, $B_m(z)$ is minimal.

Remark 7.3. Given $g \in S(m, M)$, $m \geq 2$, from Taylor's Theorem and the continuity of γ_i's at a simple root θ, it is easy to conclude that $g^{(i)}(\theta) = 0$, for all $i = 1, \ldots, m-1$, and

$$\gamma_m(\theta) = (-1)^m \frac{g^{(m)}(\theta)}{m!}. \tag{7.6}$$

It should be noted that as functions $\gamma_m(z)$ and $(-1)^m g^{(m)}(z)/m!$ are not identical. For example take $g(z) = z - p(z)/p'(z)$. In other words, $g^{(m)}(z)$ and $\gamma_m(z)$ may have different depths, but they have the same simple-root-depth.

Definition 7.3. Let $\overline{S}(m, M)$ be the set of all $g(z) \in K(z)$ which can be represented as (7.2) but with the relaxation of the condition that for $i = m, \ldots, M$, $\gamma_i(\theta)$ is well-defined.

Definition 7.4. Let $S^\circ(m, M)$ be the set of all $g(z) \in K(z)$ such that $p(y)$ divides

$$\gamma_0(y) = -g(z) + y + \sum_{i=m}^{M} \gamma_i(z)(y - z)^i, \tag{7.7}$$

in $K(z)[y]$ (i.e. polynomials in y with coefficients in $K(z)$) where $\gamma_i(z) \in K(z)$, for $i = m, \ldots, M$, and $\gamma_m(z)$ and $\gamma_M(z)$ are not identically zero.

Remark 7.4. If θ is a root of p, then $\gamma_0(\theta) \equiv 0$. Thus, it follows that $S^\circ(m, M)$ is a subset of $\overline{S}(m, M)$. Also, if all the roots of $p(z)$ are simple (in particular if it is irreducible), then $S^\circ(m, M) = \overline{S}(m, M)$. To see this, pick $g \in \overline{S}(m, M)$, and define a corresponding $\gamma_0(y)$. Then there exists $q(y), r(y) \in K(z)[y]$ such that

$$\gamma_0(y) = p(y)q(y) + r(y), \tag{7.8}$$

where the degree of $r(y)$ is less than n. Since $\gamma_0(\theta) \equiv 0$, for all roots of p, it follows that $r(y) \equiv 0$ (see e.g. van der Waerden (1970)). Note that in particular we must have

$$g(z) = z - \gamma_0(z). \tag{7.9}$$

In general however, $\overline{S}(m, M)$ and $S^\circ(m, M)$ are not the same. For example if $p(z) = z^2$, then both $g_1(z) = x/2$ and $g_2(z) = z^2$ lie in $\overline{S}(2, 2)$, but $g_2(z)$ is not in $S^\circ(2, 2)$.

We will give two formulas for generating new members of $S(m, M)$ with higher orders. Both formulas make use of the following relationship which is a consequence of Taylor's Theorem, derived purely by algebraic-combinatorial means in Chapter 2:

$$0 = \sum_{i=0}^{n} \frac{p^{(i)}(z)}{i!}(\theta - z)^i. \tag{7.10}$$

From (7.10) we have

$$-p(z) = \sum_{i=1}^{n} \frac{p^{(i)}(z)}{i!}(\theta - z)^i. \tag{7.11}$$

Writing

$$z = \theta - (\theta - z), \tag{7.12}$$

and adding corresponding sides to (7.11) we get

$$B_1(z) \equiv z - p(z) = \theta + (p'(z)-1)(\theta - z) + \sum_{i=2}^{n} \frac{p^{(i)}(z)}{i!}(\theta - z)^i \in S(1,n). \quad (7.13)$$

From (7.10) we also immediately get Newton's function as a member of $S(2,n)$

$$B_2(z) = z - \frac{p(z)}{p'(z)} = \theta + \sum_{i=2}^{n} \frac{p^{(i)}(z)}{i!p'(z)}(\theta - z)^i. \quad (7.14)$$

As we shall see from $B_1(z)$ and $B_2(z)$ together with the application of the two formulas one can construct a large class of iterative solutions which include the Basic Family, and the Euler-Schröder family. Finally, we show that the iteration functions within $S(m,M)$ can be extended to any arbitrary smooth function f, with the uniform replacement of $p^{(j)}$ with $f^{(j)}$ in g, and in the asymptotic constant of convergence $\gamma_m(\theta)$.

In the following sections, first we give an algebraic proof of the existence of $B_m(z)$. Next, we derive a closed formula for $B_m(z)$. Then, we describe two formulas for the generation of new iteration functions. Finally, we consider the extension of iteration function within $S(m,M)$ to arbitrary smooth functions.

7.2 Algebraic Proof of Existence of the Basic Family

In this section we prove algebraically the existence of an iteration function in $\overline{S}(m, m+n-2)$, and its uniqueness under the assumption that $p(z)$ has simple roots. We shall only be concerned with the existence of $B_m(z)$ as opposed to its closed form. The theorem of this section will be used in the subsequent sections in deriving the closed form, as well as in proving the equivalence of $\overline{S}(m, m+n-2)$, $S^\circ(m, m+n-2)$, and $S(m, m+n-2)$. The proof of the theorem also motivates the definition of $S^\circ(m, m+n-2)$ which in turn gives rise to the closed formula for $B_m(z)$. Moreover, the theorem will be used to prove that the Basic Family can also be obtained from a recursive formula.

Theorem 7.1. *For any natural number $m \geq 2$, $\overline{S}(m, m+n-2)$ is nonempty, and has a unique element if $p(z)$ has simple roots.*

To prove Theorem 7.1 we first need an auxiliary lemma. Given two natural numbers s and t let

$$\binom{s}{t} \equiv \frac{s(s-1)\dots(s-t+1)}{t!}, \quad (7.15)$$

the binomial coefficient for $s \geq t$, and zero, for $s < t$. For a given pair of natural numbers m and r define the following $r \times r$ matrix

$$U_{m,r} = \begin{pmatrix} \binom{m}{1} & \binom{m+1}{1} & \dots & \binom{m+r-1}{1} \\ \binom{m}{2} & \binom{m+1}{2} & \dots & \binom{m+r-1}{2} \\ \vdots & \vdots & \ddots & \vdots \\ \binom{m}{r} & \binom{m+1}{r} & \dots & \binom{m+r-1}{r} \end{pmatrix}.$$

Lemma 7.1. *For any natural numbers m and r, $U_{m,r}$ is invertible (in fact $\det(U_{m,r}) = \binom{m+r-1}{r}$).*

Proof. To prove the lemma we will show that by performing elementary operations on $U_{m,r}$, we can reduce it to the following matrix

$$W_{m,r} = \begin{pmatrix} m & (m+1) & \dots & (m+r-1) \\ m^2 & (m+1)^2 & \dots & (m+r-1)^2 \\ \vdots & \vdots & \ddots & \vdots \\ m^r & (m+1)^r & \dots & (m+r-1)^r \end{pmatrix}.$$

Since $W_{m,r} = V^T D$, where V is an invertible Vandermonde matrix, and $D = diag(m, m+1, \dots, m+r-1)$, it follows that $W_{m,r}$ is invertible. To obtain this reduction, we first multiply the j-th row of $U_{m,r}$ by $j!$. Let U^1 denote the new matrix. The first column of U^1 has entries which can be written as the following polynomials in m: $f_1(m) = m$, $f_2(m) = m^2 - m, \dots, f_r(m) = m^r + \alpha_{r-1} m^{r-1} + \alpha_{r-2} m^{r-2} + \dots + \alpha_1 m$, for some coefficients α_i, $i = 1, \dots, r-1$. The second column of U^1 can be written as the same polynomials evaluated at $(m+1)$, and so on. Thus, U^1 is the $r \times r$ matrix whose i-th row vector is given by

$$U^1 = \begin{pmatrix} f_1(m) & f_1(m+1) & \dots & f_1(m+r-1) \\ f_2(m) & f_2(m+1) & \dots & f_2(m+r-1) \\ \vdots & \vdots & \ddots & \vdots \\ f_r(m) & f_r(m+1) & \dots & f_r(m+r-1) \end{pmatrix}.$$

By adding scalar multiples of the first row of U^1 to other rows, we obtain a new matrix U^2 whose first and second rows are the corresponding rows of $W_{m,r}$, and whose remaining rows are polynomials free of the linear terms. Next, by adding scalar multiples of the second row of U^2 to its i-th rows, $i = 3, 4, \dots, r$, we obtain U^3 whose first three rows are those of $W_{m,r}$ and whose remaining rows are polynomials free of the linear and quadratic terms. Clearly, repeating this process we arrive at $W_{m,r}$. □

Proof of Theorem 7.1. Let θ be any root of $p(z)$. Recalling (7.2) and replacing $\gamma_k(z)$ by $(-1)^k\gamma_k(z)$ for convenience, we will prove the existence of rational functions $g(z)$ and $\gamma_k(z)$, $k = m, \ldots, (m+n-2)$ with coefficient over K so that

$$-g(z) + \theta + \sum_{k=m}^{m+n-2} \gamma_k(z)(z-\theta)^k = 0. \tag{7.16}$$

Using that $p(\theta) = 0$, for any $k \geq n$ we can write

$$\theta^k = \sum_{l=0}^{n-1} \alpha_l \theta^l, \tag{7.17}$$

where $\alpha_l \in K$. From (7.17) it is easy to see that for each $j = 0, \ldots, (n-2)$,

$$(z-\theta)^{m+j} = \sum_{i=0}^{n-1} \theta^i P_i^{m+j}(z), \tag{7.18}$$

where $P_i^{m+j}(z)$'s are polynomials with coefficients in K. Furthermore, for $i \leq m+j$,

$$P_i^{m+j}(z) = \binom{m+j}{i} x^{m+j-i} + \textit{lower order terms}, \tag{7.19}$$

and for $i > m+j$, $P_i^{m+j}(z) \equiv 0$, which from (7.15) can be written as

$$P_i^{m+j}(z) = 0 = \binom{m+j}{i} x^{m+j-i}. \tag{7.20}$$

Using (7.18), we can rewrite (7.16) as

$$-g(z) + \sum_{i=0}^{n-1} \theta^i h_i(z) = 0, \tag{7.21}$$

where

$$h_1(z) = 1 + \sum_{k=m}^{m+n-2} \gamma_k(z) P_1^k(z), \tag{7.22}$$

$$h_i(z) = \sum_{k=m}^{m+n-2} \gamma_k(z) P_i^k(z), \quad i = 0,\ i = 2, \ldots, n. \tag{7.23}$$

Thus, $\overline{S}(m, m+n-2)$ is nonempty if the system of equations $h_i(z) \equiv 0$, $i = 1, \ldots, (n-1)$, is solvable, in which case

$$g(z) = h_0(z) = \sum_{k=m}^{m+n-2} \gamma_k(z) P_0^k(z) \in \overline{S}(m, m+n-2). \tag{7.24}$$

Equivalently, $\overline{S}(m, m+n-2)$ is nonempty if the system of linear equations

$$\Pi_m(z)\Gamma_m(z) = e, \tag{7.25}$$

is solvable where

$$\Gamma_m(z) = [\gamma_m(z), \gamma_{m+1}(z), \ldots, \gamma_{m+n-2}(z)]^T, \quad e = [-1, 0, \ldots, 0]^T, \tag{7.26}$$

and where $\Pi_m(z)$ is the following $(n-1) \times (n-1)$ matrix

$$\Pi_m(z) = \begin{pmatrix} P_1^m(z) & P_1^{m+1}(z) & \ldots & P_1^{m+n-2}(z) \\ P_2^m(z) & P_2^{m+1}(z) & \ldots & P_2^{m+n-2}(z) \\ \vdots & \vdots & \ddots & \vdots \\ P_{n-1}^m(z) & P_{n-1}^{m+1}(z) & \ldots & P_{n-1}^{m+n-2}(z) \end{pmatrix}.$$

All that remains to be done is to show that the determinant of $\Pi_m(z)$ which is a polynomial in the indeterminate x, is not identically zero, in which case

$$\Gamma_m(z) = \Pi_m^{-1}(z)e. \tag{7.27}$$

To show that $\det(\Pi_m(z))$ is not identically zero, it suffices to prove that its highest term is not zero. From (7.19) and (7.20), the highest term of $\det(\Pi_m(z))$ is the determinant of $H_m(z)$, the $(n-1) \times (n-1)$ matrix given by

$$H_m(z) = \begin{pmatrix} \binom{m}{1} x^{m-1} & \binom{m+1}{1} x^m & \ldots & \binom{m+n-2}{1} x^{m+n-3} \\ \binom{m}{2} x^{m-2} & \binom{m+1}{2} x^{m-1} & \ldots & \binom{m+n-2}{2} x^{m+n-4} \\ \vdots & \vdots & \ddots & \vdots \\ \binom{m}{n-1} x^{m-n+1} & \binom{m+1}{n-1} x^{m-n+2} & \ldots & \binom{m+n-2}{n-1} x^{m-1} \end{pmatrix}.$$

It is easy to check that

$$\det(H_m(z)) = x^{(m-1)(n-1)} \det(U_{m,n-1}), \tag{7.28}$$

hence not identically zero by Lemma 7.1.

We must also show that $\gamma_m(z)$ and $\gamma_{m+n-2}(z)$ are not identically zero. To prove this first suppose that $p(z)$ has simple roots and consider the expression $-g(z) + \sum_{i=0}^{n-1} \theta^i h_i(z)$ as a polynomial over $K(z)$ in the indeterminate θ. Every root of $p(z)$ is a root of this polynomial. On the other hand its degree is $n-1$. It follows that $-g(z) + h_0(z) \equiv 0$, and $h_i(z) \equiv 0$, for $i = 1, \ldots, (n-1)$. In other words, (7.27) must be satisfied implying the uniqueness of $g(z)$. Let us denote this unique solution by

$$B_m(z) = \theta + \sum_{i=m}^{m+n-2} \gamma_i^{(m)}(z)(\theta - z)^i. \tag{7.29}$$

Next we prove that $B_m(z)$ and $B_{m+1}(z)$ are distinct. Otherwise, we must have $\Gamma_m(z) = \Gamma_{m+1}(z)$. From (7.25) the following must hold

$$\Pi_m(z)\Gamma_m(z) = \Pi_{m+1}(z)\Gamma_m(z). \tag{7.30}$$

Equivalently,

$$(\Pi_{m+1}(z) - \Pi_m(z))\Gamma_m(z) = 0. \tag{7.31}$$

But by the same reasoning as before, the leading coefficient of the determinant of $(\Pi_{m+1}(z) - \Pi_m(z))$ is the determinant of $H_{m+1}(z)$, hence nonzero. But this implies $\Gamma_m(z) \equiv 0$, a contradiction. Hence the proof of distinctness.

Next we prove that for all $m \geq 2$, $\gamma_m^{(m)}(z)$, $\gamma_{m+n-2}^{(m)}(z)$ are not identically zero. As we have seen for $m = 2$, the function $z - p(z)/p'(z)$ (see (7.14)) lies in $\overline{S}(2, n)$. From uniqueness we must have $B_2(z) = z - p(z)/p'(z)$. Now assume $m > 2$ and consider $B_{m-1}(z)$, $B_m(z)$, and $B_{m+1}(z)$. If $\gamma_m^{(m)}(z)$ is identically zero, then

$$\frac{1}{2}B_m(z) + \frac{1}{2}B_{m+1}(z) \in \overline{S}(m+1, m+n-1),$$

contradicting uniqueness. If $\gamma_{m+n-2}^{(m)}(z)$ is identically zero, then

$$\frac{1}{2}B_m(z) + \frac{1}{2}B_{m-1}(z) \in \overline{S}(m-1, m+n-3),$$

again a contradiction. From the above and the fact that $B_2(z) \in \overline{S}(2, n)$, we can inductively arrive at the desired conclusion.

To complete the proof we must show that $\gamma_m^{(m)}(z)$ and $\gamma_{m+n-2}^{(m)}(z)$ remain to be nonzero even if $p(z)$ has multiple roots. But in this case too we can arrive at the same set of polynomials $P_i^{m+j}(z)$ and hence the same set of equations (7.24) and (7.25). □

7.3 Derivation of Closed Form of the Basic Family

In this section we shall derive the closed form for members of the Basic Family. In doing so, it is more convenient to consider functions in $S^\circ(m, M)$, with $m \geq 2$, and $M \geq m+n-2$. Recall that $g \in S^\circ(m, M)$ implies there exist $\gamma_i(z) \in K(z)$, $i = m, \ldots, M$, such that $p(y)$ divides

$$\gamma_0(y) = -g(z) + y + \sum_{l=m}^{M} \gamma_l(z)(y-z)^l, \tag{7.32}$$

in $K(z)[y]$. Equivalently, if we let $w = y - z$, then $f(w) = p(w+z)$ must divide $h(w) = \gamma_0(w+z)$. From Taylor's Theorem as applied to the polynomial $p(y)$ we have

$$f(w) = \sum_{i=0}^{n} \frac{p^i(w)}{i!} w^i. \tag{7.33}$$

Since $\gamma_0(z) = z - g(z)$, we have

$$h(w) = \gamma_0(z) + w + \sum_{l=m}^{M} \gamma_l(z) w^l. \tag{7.34}$$

From the divisibility condition, there exists a polynomial

$$q(w) = \sum_{j=0}^{M-n} v_j(z) w^j, \tag{7.35}$$

with $v_j(z) \in K(z)$ such that

$$h(w) = f(w)q(w). \tag{7.36}$$

Substituting (7.33)-(7.35) in (7.36) and equating coefficients of the corresponding powers of w we get the system of equations

$$p(z)v_0(z) = \gamma_0(z), \tag{7.37}$$

$$p'(z)v_0(z) + p(z)v_1(z) = 1, \tag{7.38}$$

$$\sum_{i+j=l} \frac{p^{(i)}(z)}{i!} v_j(z) = 0, \quad l = 2, \ldots, m-1, \tag{7.39}$$

$$\sum_{i+j=l} \frac{p^{(i)}(z)}{i!} v_j(z) = \gamma_l(z), \quad l = m, \ldots, M, \tag{7.40}$$

where $v_{M-n+1}(z) = \cdots = v_M(z) \equiv 0$. The matrix corresponding to the first m equations, i.e. (7.37), (7.38), and (7.39) is precisely the matrix $L_m(z)$ (see (7.1)). Letting $b(z)$ to be the m-vector $(\gamma_0(z), 1, 0, \ldots, 0)^T$ and $v(z)$ the m-vector $(v_0(z), \ldots, v_{m-1}(z))^T$, we get

$$L_m(z)v(z) = b(z). \tag{7.41}$$

Lemma 7.2. *For any natural number k the matrix*

$$Z_{n,k} = \begin{pmatrix} \binom{n}{1} & \binom{n}{0} & 0 & \dots & 0 \\ \binom{n}{2} & \binom{n}{1} & \binom{n}{0} & \dots & 0 \\ \vdots & \vdots & \vdots & \ddots & \vdots \\ \binom{n}{k} & \binom{n}{k-1} & \binom{n}{k-2} & \dots & \binom{n}{1} \end{pmatrix},$$

is invertible.

Proof. All that is needed is to reduce $Z_{n,k}$ to a matrix of the type $U_{m,r}$ which is invertible by Lemma 2.1. Using the identity $\binom{n}{i} + \binom{n}{i-1} = \binom{n+1}{i}$, starting with the last column of $Z^1 = Z_{n,k}$ we add to each column its previous column to get a new matrix Z^2. We repeat this process to columns of Z^2 except for its second column and get a new matrix Z^3. Next we repeat the process to columns of Z^3 except for its third column. This process will eventually result in a matrix of the type $U_{m,r}$. □

From Lemma 7.2, analogous to the proof of invertibility of $H_m(z)$, we immediately conclude

Corollary 7.1. *For each natural number $m \geq 2$, the matrix $L_m^{(1)}(z)$ is invertible.*

Theorem 7.2 (Kalantari *et al.* (1997)). *Let $B_m(z)$ be the function in $\overline{S}(m, m+n-2)$ whose existence was proved in Theorem 7.1. Then*

$$B_m(z) = z - p(z)\,\frac{\det(L_{m-1}^{(1)}(z))}{\det(L_m^{(1)}(z))}.$$

Moreover, $B_m(z)$ lies in $S(m, m+n-2)$ and if θ is a simple root of $p(z)$ then

$$\gamma_m^{(m)}(\theta) = (-1)^m \frac{\det(L_{m+1}^{(2)}(\theta))}{\det(L_m^{(1)}(\theta))}.$$

Proof. In order to obtain the claimed closed form for $B_m(z)$ it suffices to consider the case where $p(z)$ has only simple roots. Recall that in this case $\overline{S}(m, m+n-2)$ and $S^\circ(m, m+n-2)$ are identical (Remark 7.4). Letting $\Delta_j(z)$ to be the determinant of $L_m(z)$ with the j-th column replaced with $b(z)$, from Cramer's rule as applied to the system (7.41) we get

$$v_j(z) = \frac{\Delta_j(z)}{\det(L_m(z))}, \quad j = 0, \dots, m-1. \tag{7.42}$$

Since $p(z)$ is nonzero, $\det(L_m(z))$ is not identically zero, thus each $v_j(z)$ is well-defined and unique. Since for $B_m(z)$, $M = m+n-2$, we must have $v_{m-1}(z) \equiv 0$. By expanding the determinant of $\Delta_{m-1}(z)$ along its last column, and then expanding the second resulting determinant along its first row, we get

$$\Delta_{m-1}(z) = \gamma_0(z)\det(L_m^{(1)}(z)) - p(z)\det(L_{m-1}^{(1)}(z)) \equiv 0. \tag{7.43}$$

From Corollary 7.1, $L_m^{(1)}(z)$ is invertible so that (7.43) can be solved for $\gamma_0(z)$ which results in the claimed closed formula for $B_m(z)$.

Assume that θ is a simple root of $p(z)$. To derive the formula for $\gamma_m^{(m)}(\theta)$, consider the $m+1$ equations corresponding to (7.37), (7.38), (7.39), and the first equation of (7.40) at $x=\theta$. Since the first of these equations is zero, if we let $b^{(m)}(\theta)$ denote the m-vector $(1,0,\ldots,0,\gamma_m^{(m)}(\theta))^T$, we get the following $m \times m$ system

$$L_{m+1}^{(1)}(\theta)v(\theta) = b^{(m)}(\theta). \tag{7.44}$$

The above system has a unique solution since the coefficient matrix is a lower triangular with diagonal entries $p'(\theta) \neq 0$. On the one hand, $v_{m-1}(\theta) = 0$. On the other hand, by Cramer's rule $v_{m-1}(\theta)$ is the determinant of the $L_{m+1}^{(1)}(\theta)$ with its last column replaced by $b^{(m)}(\theta)$. By expansion of this determinant along the last column we get

$$\gamma_m^{(m)}(\theta)\det(L_m^{(1)}(\theta)) - (-1)^{m-1}\det(L_{m+1}^{(2)}(\theta)) = 0. \tag{7.45}$$

Solving the above for $\gamma_m^{(m)}(\theta)$ gives the desired formula.

Since $\gamma_m^{(m)}(\theta)$ is well-defined, from (7.42) $v_0(\theta),\ldots,v_{m-2}(\theta)$ are well-defined. From this and (7.40) it follows that $\gamma_l^m(\theta)$ is well-defined for $l = m,\ldots,m+n-2$. Thus, $B_m(z)$ lies in $S(m, m+n-2)$. □

Let us treat $\gamma_0(z)$ in equation (7.37) also as a variable, and consider the m equations in $m+1$ unknowns corresponding to equations (7.37)-(7.39), i.e.

$$L_m(z)v(z) - \gamma_0(z)e_1 = e_2, \tag{7.46}$$

where e_i is the column m-vector with the i-th component equal to 1, and all other components equal to 0. In matrix form this is

$$\begin{pmatrix} p(z) & 0 & 0 & \ldots & 0 & -1 \\ p'(z) & p(z) & 0 & \ldots & 0 & 0 \\ \frac{p''(z)}{2!} & p'(z) & p(z) & \ldots & 0 & 0 \\ \vdots & \vdots & \vdots & \ddots & \vdots & 0 \\ \frac{p^{(m-1)}(z)}{(m-1)!} & \frac{p^{(m-2)}(z)}{(m-2)!} & \frac{p^{(m-3)}(z)}{(m-3)!} & \ldots & p(z) & 0 \end{pmatrix} \begin{pmatrix} v_0(z) \\ v_1(z) \\ v_2(z) \\ \vdots \\ v_{m-1}(z) \\ \gamma_0(z) \end{pmatrix} = \begin{pmatrix} 0 \\ 1 \\ 0 \\ \vdots \\ 0 \\ 0 \end{pmatrix}.$$

The coefficient matrix of the system above is an $m \times m+1$ matrix. In general given a matrix equation $Ax = b$ where A is $m \times n$, $n \geq m$, a solution is said to be a *Basic Solution* if it is a solution with at most m nonzero components. It turns out that every other solution is a linear combination of the basic solutions. The following is evident from previous analysis. It is the main reason behind the naming of the Basic Family.

Theorem 7.3. *The basic solutions of the system of equations in (7.46) corresponding to $v_i(z) = 0$, $i = 0, \ldots, m-1$ are $B_{i+1}(z)$, where $B_1(z) = x - p(z)$, and setting $\gamma_0(z) = 0$ gives $B_0(z) = x$.* □

Remark 7.5. If $g(z) \in S^{\circ}(m, M)$ has the property that its corresponding equation in (7.41) satisfies $v_{m-1}(z) = 0$, and $v_m(z) = 0$ (a condition that is necessarily satisfied for $M = (m + n - 2))$, then $\gamma_m(z)$ depends only on $p^{(i)}(z)$, $i = 0, \ldots, m$, i.e. the simple-root-depth of g will be m.

Remark 7.6. Since Taylor's Theorem holds in any field of characteristic zero, the existence of $B_m(z)$ extends to the case where K is such a field.

7.4 Two Formulas for Generation of Iteration Functions

In this section we derive two formulas that will result in the generation of very large class of iteration functions. Both will rely on (7.11).

To derive the first formula, suppose that $g(z) \in S(m, M)$, with $m \geq 1$, i.e.

$$g(z) = \theta + \sum_{i=m}^{M} \gamma_i(z)(\theta - z)^i. \tag{7.47}$$

From (7.11) we have

$$(-p(z))^m = \sum_{i=m}^{M'} \mu_i^{(m)}(z)(\theta - z)^i, \tag{7.48}$$

where

$$\mu_i^{(m)}(z) = \sum_{i_1+\cdots+i_m=i, i_j \geq 1} \frac{p^{(i_1)}(z) \cdots p^{(i_m)}(z)}{i_1! \cdots i_m!}, \qquad i = m, \ldots, mn. \tag{7.49}$$

Multiplying (7.48) by $-\gamma_m(z)/\mu_m^{(m)}(z) = -\gamma_m(z)/p'(z)^m$ and adding the result to (7.47) we get

$$h(z) = g(z) - (-1)^m \gamma_m(z)(\frac{p(z)}{p'(z)})^m = \theta + \sum_{i=m+1}^{M'} \eta_i(z)(\theta - z)^i, \tag{7.50}$$

where

$$M' = \max\{M, mn\},$$

$$\eta_i(z) = \gamma_i(z) - \gamma_m(z)\frac{\mu_i^{(m)}(z)}{p'(z)^m}, \quad i = m+1, \dots, M', \tag{7.51}$$

with $\gamma_i(z) \equiv 0$ for $i > M$, and $\mu_i^{(m)}(z) \equiv 0$ for $i > mn$.

Theorem 7.4. *Assume that the leading coefficient of $h(z)$ is not identically zero. Then, $h(z)$ lies in $S(\overline{m}, \overline{M})$, for some $\overline{m} \geq m+1$, $\overline{M} \leq M'$. If $\overline{m} = m+1$ and θ is a simple root of $p(z)$, then*

$$\eta_{m+1}(\theta) = \gamma_{m+1}(\theta) - \gamma_m(\theta)\frac{m}{2}\frac{p''(\theta)}{p'(\theta)}. \tag{7.52}$$

Proof. Since $g(z) \in S(m, M)$, for $i = m, \dots, M$, $\gamma_i(\theta)$ is well-defined. Clearly, for $i = m+1, \dots, mn$, $\eta_i(\theta)$ is well-defined. The equation (7.52) follows from (7.51) and that $\mu_{m+1}^{(m)}(z) = (m/2)(p''(z)p'(z)^m)$. □

Next we derive another formula. Assume that we are given $g(z) \in S(m-1, M_{m-1})$ and $h(z) \in S(m, M_m)$, where $m \geq 1$ and M_{m_1} and M_m are naturals greater than $m-1$ and m, respectively. Let $M_{m+1} = \max\{M_m, M_{m-1}, m+n-1\}$. We write

$$g(z) = \theta + \sum_{i=m-1}^{M_{m+1}} \gamma_i(z)(\theta - z)^i, \tag{7.53}$$

where

$$\gamma_i(z) \equiv 0, \quad \forall i > M_{m-1},$$

$$h(z) = \theta + \sum_{i=m}^{M_{m+1}} \eta_i(z)(\theta - z)^i, \tag{7.54}$$

where

$$\eta_i(z) \equiv 0, \quad \forall i > M_m.$$

From (7.53) and (7.54) we get

$$g(z) - h(z) = \gamma_{m-1}(z)(\theta - z)^{m-1} + \sum_{i=m}^{M_{m+1}} (\gamma_i(z) - \eta_i(z))(\theta - z)^i. \tag{7.55}$$

Also, multiplying (7.11) by $(\theta - z)^{m-1}$, and since $p^{(j)}(z) \equiv 0$, for $j > n$, we get

$$0 = p(z)(\theta - z)^{m-1} + \sum_{i=m}^{M_{m+1}} \frac{p^{(i+1-m)}(z)}{(i+1-m)!}(\theta - z)^i. \tag{7.56}$$

Multiplying (7.55) by $p(z)$, and (7.56) by $-\gamma_{m-1}(z)$ and adding the resulting equations we get

$$p(z)(g(z)-h(z)) = \sum_{i=m}^{M_{m+1}} \left(p(z)(\gamma_i(z)-\eta_i(z))-\gamma_{m-1}(z)\frac{p^{(i+1-m)}(z)}{(i+1-m)!}\right)(\theta-z)^i. \tag{7.57}$$

Multiplying (7.57) by $-\eta_m(z)/D(z)$, where

$$D(z) = [p(z)(\gamma_m(z) - \eta_m(z)) - p'(z)\gamma_{m-1}(z)], \tag{7.58}$$

and adding the result to (7.54) we get

$$f(z) = h(z) - \eta_m(z)\frac{p(z)(g(z) - h(z))}{D(z)} = \theta + \sum_{i=m+1}^{M_{m+1}} \phi_i(z)(\theta - z)^i, \tag{7.59}$$

where

$$\phi_i(z) = \eta_i(z) - \frac{\eta_m(z)}{D(z)}\left(p(z)(\gamma_i(z)-\eta_i(z))-\gamma_{m-1}(z)\frac{p^{(i+1-m)}(z)}{(i+1-m)!}\right). \tag{7.60}$$

Theorem 7.5. *Suppose that $D(z)$ is not identically zero. Then $f(z) \in S(\overline{m}, \overline{M})$, where $\overline{m} \geq m+1$, $\overline{M} \leq M_{m+1}$. If $\overline{m} = m+1$ and θ is a simple root of $p(z)$, then*

$$\phi_{m+1}(\theta) = \eta_{m+1}(\theta) - \eta_m(\theta)\frac{p''(\theta)}{2p'(\theta)}. \tag{7.61}$$

Proof. Clearly $f(z)$ lies in $\overline{S}(\overline{m}, \overline{M})$ for some $m+1 \leq \overline{m} \leq \overline{M} \leq M_{m+1}$. To show $f(z)$ lies in $S(\overline{m}, \overline{M})$ we must show that for $i = m+1, \ldots, M_{m+1}$ $\phi_i(\theta)$ is well-defined. From (7.60) we have

$$\lim_{x \to \theta} \phi_i(z) = \eta_i(\theta) - \eta_m(\theta)\frac{p^{(i+1-m)}(\theta)}{(i+1-m)!p'(\theta)},$$

hence well-defined since $p'(\theta) \neq 0$. $\square$

Next we consider two applications of these formulas.

Theorem 7.6. *Suppose that in the formula (7.59), $g(z) = B_{m-1}(z)$ and $h(z) = B_m(z)$, then $f(z) = B_{m+1}(z)$, for all $m \geq 2$.*

Proof. It suffices to prove the theorem under the assumption that $p(z)$ has only simple roots. We first need to show that the function $D(z)$ is not identically zero. From (7.57) we have

$$p(z)(g(z) - h(z))$$

$$= D(z) + \sum_{i=m+1}^{m+n-1} \left(p(z)(\gamma_i(z) - \eta_i(z)) - \gamma_{m-1}(z) \frac{p^{(i+1-m)}(z)}{(i+1-m)!} \right) (\theta - z)^i. \quad (7.62)$$

If $D(z)$ is identically zero, then

$$B_{m+1}(z) + p(z)(g(z) - h(z)) \in S(m+1, m+n-1). \quad (7.63)$$

But this contradicts the uniqueness result proved in Theorem 7.1. Thus, $D(z)$ is not identically zero, implying that $f(z) \in S(m+1, m+n-1)$. Again by Theorem 7.1, we must have $f(z) = B_{m+1}(z)$. □

7.5 Deriving the Euler-Schröder Family

Theorem 7.7. *Consider the family $E_1(z), E_2(z), E_3(z), \ldots,$ of iteration functions obtained from the repeated application of (7.50) starting with $E_1(z) = B_1(z) = z - p(z)$. For $m \geq 3$, $E_m(z) \in S(m, mn)$. Moreover, the depth of $E_m(z)$ is $m - 1$.*

Proof. The fact that $E_m(z)$ lies in $S(m, mn)$ is a trivial consequence of Theorem 4.1. The claim on the depth can be proved inductively using that theorem and (7.51). □

In the reaming of the section we will generate $E_2(z), \ldots, E_5(z)$. Suppressing x for brevity, for a given i let us denote the γ_i's corresponding to E_m by $\gamma_i^{(m)}$'s. We now obtain $\gamma_m^{(m)}$ for $m = 1, 2, 3, 4,$ and 5. From (7.50) this will give $E_2, E_3, E_4,$ and E_5 since we have

$$E_{m+1}(z) = E_m(z) - (-1)^m \gamma_m^{(m)}(z) (\frac{p(z)}{p'(z)})^m.$$

From (7.51) we may write

$$\gamma_i^{(m+1)} = \gamma_i^{(m)} - \gamma_m^{(m)} \frac{\mu_i^{(m)}}{(p')^m}, \quad i \geq m+1. \tag{7.64}$$

From (7.13) we have $\gamma_1^{(1)} = (p' - 1)$. This gives

$$E_2(z) = z - \frac{p}{p'}.$$

It follows that $\gamma_2^{(2)} = p''/2p'$. Thus we have

$$E_3(z) = E_2(z) - (-1)^2\gamma_2^{(2)}(z)(\frac{p}{p'})^2 = z - \frac{p}{p'} - (\frac{p}{p'})^2\frac{p''}{2p'}.$$

To compute $E_4(z)$, from (7.64) we need

$$\gamma_3^{(3)} = \gamma_3^{(2)} - \gamma_2^{(2)}\frac{\mu_3^{(2)}}{p'^2}.$$

From (7.49) and expansion of E_2 we have

$$\mu_3^{(2)} = p'p'', \quad \gamma_3^{(2)} = \frac{p'''}{3!p'}.$$

Thus,

$$\gamma_3^{(3)} = \frac{p'''}{6p'} - \frac{p''}{2p'^2}.$$

Hence

$$E_4 = z - (\frac{p}{p'}) - (\frac{p}{p'})^2\frac{p''}{2p'} + (\frac{p}{p'})^3(\frac{p'''}{6p'} - \frac{p''}{2p'^2}).$$

Next we need

$$\gamma_4^{(4)} = \gamma_4^{(3)} - \gamma_3^{(3)}\frac{\mu_3^{(2)}}{p'^3}.$$

This in turn requires the computation of $\mu_3^{(2)}$ and

$$\gamma_4^{(3)} = \gamma_4^{(2)} - \gamma_2^{(2)}\frac{\mu_4^{(2)}}{p'^2}.$$

Note that $\mu_3^{(2)} = p'p''$, $\gamma_4^{(2)} = p^{(4)}/4!p'$, and $\mu_4^{(2)} = 2p'p'''/3! + p''^2/4$. Substituting these and simplifying we get

$$\gamma_4^{(4)} = \frac{p^{(4)}}{4!p'} - \frac{5p''p'''}{12p'^2} + \frac{5p''^3}{8p'^3}.$$

Thus,

$$E_5 = z - (\frac{p}{p'}) - (\frac{p}{p'})^2\frac{p''}{2p'} + (\frac{p}{p'})^3(\frac{p'''}{6p'} - \frac{p''}{2p'^2}) - (\frac{p}{p'})^4(\frac{p^{(4)}}{4!p'} - \frac{5p''p'''}{12p'^2} + \frac{5p''^3}{8p'^3}).$$

The iteration functions $E_2 - E_5$ coincide with the first few members of a family credited both to Euler as well as Schröder. For historical comments as well as other approaches for deriving this family see Smale (1985).

7.6 Extension to Non-Polynomial Root Finding

In the previous sections we have shown the existence of many fixed point iteration functions for finding roots of a given polynomial $p(z)$, namely functions within $S(m, M)$, any of whose members, say $g(z)$, has m-th order rate of convergence to any simple root θ of $p(z)$. In this section we will prove that these high order methods derived for polynomials extend to high order methods for arbitrary smooth functions, having the same order of convergence to simple roots. Essentially given a sufficiently smooth function f, defined over the real or complex numbers, the analysis amounts to replacing the polynomial p and its higher derivatives with f and its higher derivatives, including in the asymptotic constant of convergence. In particular, both depth and simple-root-depth will be unchanged. For simplicity we consider real functions.

Theorem 7.8. *Assume that there exists $h : \mathbb{R}^m \to \mathbb{R}$, $m \geq 2$, so that for any real polynomial $p(x)$ of degree at most $(2m-1)$ the following properties hold:*

If $\theta \in \mathbb{R}$ is a simple root of $p(x)$, then h is in C^m (i.e. continuously differentiable m times) in a neighborhood of the point

$$x_p(\theta) = (\theta, p(\theta), p'(\theta), \ldots, p^{(m-1)}(\theta)) \in \mathbb{R}^m,$$

and the function

$$g_p(x) = h(x, p(x), p'(x), \ldots, p^{(m-1)}(x))$$

satisfies

$$g_p(\theta) = \theta, \qquad (i)$$

$$g_p^{(i)}(\theta) = 0, \quad i = 1, \ldots, m-1. \qquad (ii)$$

Then, the above properties can be extended to any function $f(x)$ which is in C^{2m-1} in a neighborhood of a simple real root θ in the sense that $p^{(j)}$ can be replaced with $f^{(j)}$. Moreover

$$g_f^{(m)}(\theta) = g_{P_n}^{(m)}(\theta), \qquad (iii)$$

where

$$P_n(x) = \sum_{i=0}^{n} \frac{f^{(i)}(\theta)}{i!}(x-\theta)^i,$$

the Taylor polynomial of degree $n = 2m - 1$ at θ.

Proof. Since f is in C^n,

$$f^{(j)}(x) = P_n^{(j)}(x) + R_n^{(j)}(x), \quad j = 0, \ldots, n,$$

where $R_n^{(0)}(x)$ is the remainder term. Since $f^{(j)}(\theta) = R_n^{(j)}(\theta)$ for all $j = 0, \ldots, n$ we have

$$R_n^{(j)}(\theta) = 0, \quad j = 1, \ldots, n. \tag{7.65}$$

In particular, θ is a simple root of $P_n(x)$, and properties (i) and (ii) apply to $P_n(x)$. Also, the fact that h is in C^n in a neighborhood of the point $x_f(\theta)$ follows immediately. From the chain rule we have

$$g'_f = \frac{\partial h}{\partial x} + \frac{\partial h}{\partial f} f' + \frac{\partial h}{\partial f'} f'' + \cdots + \frac{\partial h}{\partial f^{(m-1)}} f^{(m)}. \tag{7.66}$$

If we now substitute for $f(x)$ and $f^{(j)}(x)$ in (7.66), using (7.65) and the continuity of the partial derivatives of h we conclude that $g'_f(x)$ is continuous in an interval containing $x_f(\theta)$. From the repeated application of the chain rule, the assumption that h is in C^m, that f is in C^{2m-1}, and the use of (7.65), it follows by induction that for all $j = 0, \ldots, m$, $g_f^{(j)}(x)$ is continuous in an interval containing $z_f(\theta)$. Moreover $g_f^{(j)}(\theta) = g_{P_n}^{(j)}(\theta)$, for all $j = 0, \ldots, m$. □

Corollary 7.2. *The depth of $B_m(x)$ with respect to $f \in C^{2m-1}$ remains to be $m-1$. Moreover, the simple-root-depth is m.* □

As an application of the above consider the replacement of f in Newton's method with $f(x)/f'(x)$ which results in a new iteration function $g(x)$ credited to Schröder (e.g. see Scavo and Thoo (1995)). Then, from the corollary $g''(\theta)/2$ can easily be computed in terms of f and its higher derivatives.

Remark 7.7. The assumption $f \in C^{2m-1}$ in the above corollary can be relaxed for $B_m(x)$. It suffice to have $f \in C^m$. For example for $B_2(x) = x - p(x)/p'(x)$, it is enough to assume that $f \in C^2$, despite the fact that $B_2''(x)$ relies on the third derivative of f.

7.7 Conclusions

In this chapter we have derived the Basic Family of iteration functions from an algebraic point of view, leading to certain characterizations that are not in view when we consider the closed form formulas. In other words, we have derived an expansion formula for B_m which enables us to compute its rate of convergence to be m, its precise asymptotic constant of convergence, and several minimality conditions that allow us to compare the Basic Family members to other iteration function and to see why they are optimal with respect to certain criteria. The expansion formula can be viewed as a generalization of the special case of Taylor's Theorem, also considered in Chapter 2. This will be treated in much more generality in Chapter 10.

The formulas and the properties proved here hold in the more general case where the underlying field K is any field of characteristic zero, i.e. repeated addition of a nonzero number will never equal zero. We have shown that the members of this family are optimal with respect to the notions of depth, width, simple-root-depth (see Definition 7.2, and Remark 7.2). Under the assumption of the simplicity of the roots, the m-th order member, $B_m(z)$, was shown to be the unique member of the class $S(m, m+n-2)$. This in particular tells us that Newton's method is optimal among all iteration functions having quadratic rate of convergence for simple roots.

In the chapter we also described two simple recursive formulas for generation of new iteration functions which in particular resulted in the generation of the Euler-Schröder family whose m-th order member belongs to $S(m, mn)$, $m > 2$. Smale (1985) refers to an "incremental" version of Euler-Schröder family as "the most appropriate for practically computing zeros of complex polynomials" (see Example 5_k, p. 110). In view of the development of the Basic Family, the closed forms, and properties established in this chapter as well as previous chapters, the Basic Family is not only more efficient, but it is fair to venture that the incremental version of the Basic Family too would be at least as good as the Euler-Schröder family.

Problem 1. Given a polynomial $p(z)$, prove that every rational iteration function for the polynomial can be represented as a certain linear combination of the Basic Family members. Here the coefficients in the linear combination are meant to be over an appropriate subfield. As an example, for the Euler-Schröder family we have $E_2 = B_2$. Represent E_3 in terms of B_2 and B_3. More generally, represent E_m in terms of $B_2, \ldots, B_m$. Solution to this problem would imply that the Basic Family can be viewed as a basis for all rational iteration functions.

Chapter 8

The Truncated Basic Family and the Case of Halley Family

In this chapter we introduce the *Truncated Basic Family* and in particular analyze the special case of *Halley Family*. The Halley Family and their derivative-free variants offer alternatives to the traditional root-finding methods, such as secant, Newton, and Muller methods, as well as Halley's method itself. More generally, the Truncated Basic Family offers alternatives to the Basic Family itself and gives rise to new strategies and algorithms for root-finding. In this chapter we also give some polynomiography corresponding to the Truncated Basic Family and contrast these with those of the Basic Family. We will see the emergence of Mandelbrot-like set in some example polynomiographs of the Truncated Basic Family.

8.1 The Halley Family

In this section we consider the root-finding problem for smooth functions of a single real variable, $f(x)$. For each natural number $m \geq 3$, we give an iteration function $H_m(x)$, cubically convergent for simple roots. For quadratic polynomials however, the order of convergence of $H_m(x)$ is m. We will refer to this family as *Halley Family* since its first member is the well-known method of Halley. As in Halley's method, each $H_m(x)$ depends on the input x, $f(x)$, $f'(x)$, and $f''(x)$. Each $H_m(x)$ is described in terms of matrix determinants that are also computable recursively.

To describe the family, for each $m \geq 1$, let $A_m(x)$ be the $m \times m$ matrix having the following properties: all its diagonal entries are $f'(x)$, all its subdiagonal entries are $f(x)$, all its superdiagonal entries are $f''(x)/2$, and all other entries are zero. Thus,

$$A_m(x) = \begin{pmatrix} f'(x) & \frac{f''(x)}{2} & 0 & \dots & 0 \\ f(x) & f'(x) & \frac{f''(x)}{2} & \ddots & 0 \\ 0 & f(x) & f'(x) & \ddots & \vdots \\ \vdots & \vdots & \ddots & \ddots & \frac{f''(x)}{2} \\ 0 & 0 & 0 & \dots & f'(x) \end{pmatrix}. \tag{8.1}$$

Let

$$h_m(x) = \det(A_m(x)).$$

As we shall see $h_m(x)$ satisfies the recursion:

$$h_m(x) = f'(x)h_{m-1}(x) - \frac{1}{2}f(x)f''(x)h_{m-2}(x). \tag{8.2}$$

For each $m \geq 3$, define

$$H_m(x) = x - f(x)\ \frac{h_{m-2}(x)}{h_{m-1}(x)}.$$

The function

$$H_3(x) = x - f(x)\frac{f'(x)}{f'(x)^2 - \frac{1}{2}f(x)f''(x)}$$

is the iteration function of Halley's method. The method was first considered by the astronomer Halley (1694). Halley's method and its asymptotic error constant has been derived by many authors, see Chapter 3 for many references on this famous iteration function.

The Halley Family has the property that for all $m \geq 4$, the asymptotic error constant of H_m is invariant. We will show that once we have evaluated H_m at an input x_0, $H_{m+1}(x_0)$ can be computed in constant number of arithmetic operations, independent of m. These iteration functions and their derivative-free variants offer alternatives to the traditional root-finding methods, such as secant method and Newton's methods which are based on linear approximation, Muller's method which is a derivative-free method based on quadratic approximation (see McNamee (2007)), as well as Halley's method itself.

The Halley Family is closely related to the Basic Family discussed before. i.e.

$$B_m(x) = x - f(x)\frac{D_{m-2}(x)}{D_{m-1}(x)}, \quad m \geq 2.$$

Each $B_m(x)$ depends on the input x, $f(x)$, and its first $m-1$ derivatives. For each natural number t it is possible to define a family of iteration function $\{B_{m,t}(x)\}_{m=t+1}^{\infty}$, where in $B_m(x)$ we set all derivatives higher than the t-th derivative equal to zero. We shall refer to $B_{m,t}(x)$ as the *Truncated Basic Family of order* t.

In the present section we consider the special case of the Halley Family, $\{H_m(x) \equiv B_{m,2}(x)\}_{m=3}^{\infty}$, i.e. the Truncated Basic Family of order 2. We prove that each member has cubic order. More generally, each member of the Truncated Basic Family $\{B_{m,t}(x)\}_{m=t+1}^{\infty}$ has order $t+1$. An additional interesting property of the Truncated Basic Family is that for polynomials of degree at most t, the order of $B_{m,t}(x)$ is m. This property is a direct consequence of the fact that for such polynomials $B_{m,t}(x)$ coincides with the corresponding $B_m(x)$.

The family $\{B_{m,1}(x)\}_{m=2}^{\infty}$ consists only of a single element, namely Newton's $B_2(x) = x - f(x)/f'(x)$. The Halley Family is thus the first non-trivial Truncated Basic Family of order t. The Halley Family gives rise to the following root-finding algorithm:

Step 1. Given an input x_0, let

$$P_2(x) = f(x_0) + f'(x_0)(x - x_0) + \frac{f''(x_0)}{2}(x - x_0)^2.$$

Step 2. Approximate a root of $P_2(x)$ using the sequence $\{H_m(x_0)\}_{m=3}^{\infty}$ according to:

Strategy I. Let $x_1 = H_3(x_0)$.

Strategy II. Fix $\epsilon > 0$. Let $x_1 = H_m(x_0)$, where m is the least $m \geq 4$ such that

$$|H_m(x_0) - H_{m-1}(x_0)| \leq \epsilon.$$

Step 3. Replace x_0 with x_1. Go to Step 1.

The above algorithm, if it only invokes Strategy I, is simply Halley's method. The justification in using Strategy II lies in the following fact:

If f is a polynomial with real or complex coefficients, and θ a root, there exists a neighborhood of θ so that for any input x_0 in this neighborhood, we have

$$\theta = \lim_{m \to \infty} B_m(x_0).$$

In particular, if x_0 is a reasonably good approximation to a root θ of the quadratic approximation to $f(x)$, say $P_2(x)$, then

$$\theta = \lim_{m \to \infty} H_m(x_0).$$

Thus, the above algorithm can be viewed as one that attempts to obtain an approximation to a root of the quadratic Taylor polynomial at the current iterate, x_0, and where this approximation is achieved via the Halley Family, all evaluated at x_0.

It is also possible to define multipoint versions of Halley's method. These are special cases of the multipoint versions of the Basic Family described in Kalantari (2000a), and analyzed in Kalantari (1999). A version of Halley's method can be defined using only the first derivative, thus comparable with Newton's method. However, its order of convergence is 2.41. Also, a derivative-free version of Halley's method can be defined having order of convergence equal to 1.84. The derivative-free version of Halley's method is comparable with the well-known Muller's method, having identical order of convergence with that method. It is also possible to define a multipoint Halley Family.

8.2 The Order and Asymptotic Error of Halley Family

We restrict ourselves to the problem of approximating real roots of functions defined over an interval.

Theorem 8.1 (Kalantari (1998b)). *Assume that $f(x)$ is three times continuously differentiable in an open interval containing a simple root θ. For each $m \geq 3$, there exists $r > 0$ such that given any $x_0 \in N_r(\theta) = \{x : |x - \theta| < r\}$, the fixed-point iteration*

$$x_{k+1} = H_m(x_k), \quad k = 1, 2, \ldots,$$

is well-defined, it converges to θ having order 3, satisfying

$$\lim_{k \to \infty} \frac{(\theta - x_{k+1})}{(\theta - x_k)^3} = \begin{cases} -\frac{1}{6}\frac{f'''(\theta)}{f'(\theta)}, & \text{if } m > 3; \\ -\frac{1}{6}\frac{f'''(\theta)}{f'(\theta)} + \frac{3}{12}\frac{f''(\theta)^2}{f'(\theta)^2}, & \text{if } m = 3. \end{cases}$$

Note that for $m > 3$, the asymptotic error constant remains unchanged, it does not depend on the second derivative, and it is zero for the case where $f(x)$ is a quadratic polynomial. In fact in this case the order of convergence of H_m is m. The latter result follows from the m-th order convergence rate of $B_m(x)$.

In order to prove Theorem 8.1 we first state and prove several auxiliary lemmas.

Lemma 8.1. *Let $h_m(x) = det(A_m(x))$. We have,*

$$h_m = f'h_{m-1} - \frac{1}{2}ff''h_{m-2}. \qquad (i)$$

$$h'_m = f''h_{m-1} + f'h'_{m-1} - \frac{1}{2}\left(f'f'' + ff'''\right)h_{m-2} - \frac{1}{2}ff''h'_{m-2}. \qquad (ii)$$

$$h''_m = f'''h_{m-1} + 2f''h'_{m-1} + f'h''_{m-1} - \frac{1}{2}\left((f'')^2 + 2f'f''' + ff^{(iv)}\right)h_{m-2}$$

$$-(f'f'' + ff''')h'_{m-2} - \frac{1}{2}ff''h''_{m-2}. \qquad (iii)$$

Proof. We will only prove the first equation. The second and third equations are derived by straightforward differentiation. By expanding $h_m(x)$ along its first column we get,

$$h_m(x) = f'(x)h_{m-1}(x) - f(x)det(\widehat{A}_{m-1}(x)),$$

where $\widehat{A}_{m-1}(x)$ is the $(m-1) \times (m-1)$ matrix that replaces the first row of $A_{m-1}(x)$ by the vector $(f''(x)/2, 0, \dots, 0)$. By expanding $det(\widehat{A}_{m-1}(x))$ along its first row we get the desired formula. □

Lemma 8.2. *For all $m \geq 1$, we have*

$$h_m(\theta) = f'(\theta)^m. \qquad (I)$$

$$h'_m(\theta) = \frac{(m+1)}{2}f'(\theta)^{m-1}f''(\theta). \qquad (II)$$

For $m = 1$, $h''_1(\theta) = f'''(\theta)$, and for all $m > 1$, we have

$$h''_m(\theta) = \frac{m(m+1)}{4}f'(\theta)^{m-2}f''(\theta)^2 + f'(\theta)^{m-1}f'''(\theta). \qquad (III)$$

Proof. The proof of (I) follows directly from the fact that $A_m(\theta)$ is upper triangular. Equivalently, it follows from the recursive equation implied by Lemma 2.1:

$$h_m(\theta) = f'(\theta)h_{m-1}(\theta).$$

To prove (II), from (ii) we get the recursive formula

$$h'_m(\theta) = \frac{1}{2}f'(\theta)^{m-1}f''(\theta) + f'(\theta)h'_{m-1}(\theta).$$

This implies that for each natural number $j \leq m-1$, we have

$$h'_m(\theta) = \frac{j}{2}f'(\theta)^{m-1}f''(\theta) + f'(\theta)^j h'_{m-j}(\theta).$$

In particular, setting $j = m-1$, and using that $h'(\theta) = f''(\theta)$, the proof of (II) follows.

For $m = 1$, the proof of (III) is trivial. For $m > 1$, from (iii), (I), (II), and that $f(\theta) = 0$, we get the following recursive equation

$$h''_m(\theta) = \frac{m}{2} f'(\theta)^{m-2} f''(\theta)^2 + f'(\theta) h''_{m-1}(\theta).$$

Repeated application of the above recursion gives

$$h''_m(\theta) = \frac{1}{2} f'(\theta)^{m-2} f''(\theta)^2 [m + (m-1) + \cdots + 3] + f'(\theta)^{m-2} h''_2(\theta).$$

It is easy to show that

$$h''_2(\theta) = \frac{3}{2} f''^2(\theta) + f'(\theta) f'''(\theta).$$

Substituting the above into the previous equation and simplifying, we obtain the proof of (III). □

Lemma 8.3. *For each $m \geq 3$, we have*

$$H'_m(\theta) = H''_m(\theta) = 0.$$

$$H'''_3(\theta) = -\frac{f'''(\theta)}{f'(\theta)} + \frac{3}{2} \frac{f''(\theta)^2}{f'(\theta)^2}.$$

For all $m > 3$, we have

$$H'''_m(\theta) = -\frac{f'''(\theta)}{f'(\theta)}.$$

Proof. Let

$$R_m(x) = \frac{h_{m-2}(x)}{h_{m-1}(x)}.$$

Thus, $H_m(x) = x - f(x)R(x)$. Repeated differentiation gives

$$H'_m(\theta) = 1 - (f'(\theta)R(\theta)).$$

$$H''_m(\theta) = -(f''(\theta)R(\theta) + 2f'(\theta)R'(\theta)).$$

$$H'''_m(\theta) = -(f'''(\theta)R(\theta) + 3f''(\theta)R'(\theta) + 3f'R''(\theta)).$$

We have

$$R'_m = \frac{1}{h^2_{m-1}} \Big(h'_{m-2} h_{m-1} - h_{m-2} h'_{m-1} \Big).$$

$$R''_m = \frac{1}{h_{m-1}^2}\left(h''_{m-2}h_{m-1} - h_{m-2}h''_{m-1}\right) - 2\frac{h'_{m-1}}{h_{m-1}}R'_m.$$

From Lemma 8.2, we get

$$R_m(\theta) = \frac{1}{f'(\theta)}.$$

Also,

$$R'_m(\theta) = \frac{1}{f'(\theta)^{2m-2}}\left(\frac{m-1}{2}f'(\theta)^{2m-4}f''(\theta) - \frac{m}{2}f'(\theta)^{2m-4}f''(\theta)\right) = -\frac{1}{2}\frac{f''(\theta)}{f'(\theta)^2}.$$

From the same Lemma, for $m > 3$, we have

$$R''_m(\theta) = \frac{1}{f'(\theta)^{2m-2}}\left(\frac{1}{4}[(m-2)(m-1) - (m-1)m]f'(\theta)^{2m-5}f''(\theta)^2\right) + \frac{m}{2}\frac{f''(\theta)^2}{f'(\theta)^3}.$$

Thus,

$$R''_m(\theta) = \frac{1}{2}\frac{f''(\theta)^2}{f'(\theta)^3}, \quad m > 3.$$

For $m = 3$, we get

$$R''_3(\theta) = \frac{1}{f'(\theta)^4}\left(f'''(\theta)f'(\theta)^2 - f'(\theta)(\frac{3}{2}f''(\theta)^2 + f'(\theta)f'''(\theta)\right) + \frac{3}{2}\frac{f''(\theta)^2}{f'(\theta)^3} = 0.$$

From the values of $R_m(\theta)$, $R'_m(\theta)$, and $R''_m(\theta)$, the lemma follows. □

Proof. (Theorem 8.1) To complete the proof of Theorem 8.1, we only need to apply Taylor's Theorem. Given that x_k is sufficiently close to the simple root θ, we have

$$H_m(x_k) = \theta + H'_m(\theta)(x_k - \theta) + \frac{H''_m(\theta)}{2!}(x_k - \theta)^2 + \frac{H'''_m(\xi_k)}{3!}(x_k - \theta)^3,$$

where ξ_k lies between θ and x_k. This implies

$$\frac{H_m(x_k) - \theta}{(x_k - \theta)^3} = \frac{H'''_m(\xi_k)}{3!}.$$

The above implies that the fixed-point iteration is well-defined, it converges, and it gives the asymptotic error constant as $H'''_m(\theta)/6$, derived explicitly in Lemma 8.3. □

8.3 The Truncated Basic Family

In this section we describe the Truncated Basic Family for a real or complex polynomial $p(x)$ and state a main property. The Basic Family member B_m depends on $p(x)$ and its first $m-1$ derivatives. Traub (1964) showed that for simple roots, the order of convergence of any root-finding iteration function that depends on such derivatives, is at most m. Thus B_m has the highest possible order of convergence. If we restrict the use of derivatives of p up to the t-th derivative in B_m when $m \geq t+2$, the order of convergence of the method in this "truncated" family is at most $t+1$. However, these methods still achieve the optimal order of convergence, namely $t+1$, and furthermore, they share the same asymptotic error constant.

Definition 8.1 (Kalantari (2000a)). *(The Truncated Basic Family of Order t) For each integer $t \geq 1$, the Truncated Basic Family of order t is a family of iteration functions $\{B_{m,t}(x)\}_{m=t+1}^{\infty}$, where $B_{m,t}(x)$ is obtained from $B_m(x)$ by replacing derivatives of $p(x)$ of order higher than t by 0.*

The Truncated Basic Family of order t also has a recursive definition: for each integer $m \geq 1$, define

$$D_{m,t}(x) = \sum_{i=1}^{\min(m,t)} (-1)^{i-1} p(x)^{i-1} \frac{p^{(i)}(x)}{i!} D_{m-i,t}(x), \tag{8.3}$$

where $D_{0,t}(x) = 1$. Then, for each integer $m \geq t+1$, the rational function $B_{m,t}(x)$ is defined as

$$B_{m,t}(x) = x - p(x)\frac{D_{m-2,t}(x)}{D_{m-1,t}(x)}. \tag{8.4}$$

For $t = 1$,

$$D_{m,1}(x) = \begin{vmatrix} p'(x) & 0 & 0 & \dots & 0 \\ p(x) & p'(x) & 0 & \ddots & 0 \\ 0 & p(x) & p'(x) & \ddots & \vdots \\ \vdots & \vdots & \ddots & \ddots & 0 \\ 0 & 0 & 0 & \dots & p'(x) \end{vmatrix}.$$

Thus, $D_{m,1}(x) = p'(x)^m$ and $B_{m,1}(x) = x - p(x)/p'(x) = B_2(x)$, that is, all members of the Truncated Basic Family of order 1 are identical to Newton's method.

For $t = 2$,

$$D_{m,2}(x) = \begin{vmatrix} p'(x) & \frac{p''(x)}{2!} & 0 & \dots & 0 \\ p(x) & p'(x) & \frac{p''(x)}{2!} & \ddots & 0 \\ 0 & p(x) & p'(x) & \ddots & \vdots \\ \vdots & \vdots & \ddots & \ddots & \frac{p''(x)}{2!} \\ 0 & 0 & 0 & \dots & p'(x) \end{vmatrix},$$

which coincides with determinant of A_m (see (8.1)). Moreover, $D_{m,2}(x)$ satisfies the following recurrence:

$$D_{0,2}(x) = 1, \quad D_{1,2}(x) = p'(x),$$

$$D_{m,2}(x) = p'(x)D_{m-1,2}(x) - \frac{1}{2}p(x)p''(x)D_{m-2,2}(x), \quad \forall m \geq 2,$$

which is equivalent to (8.2).

Using the theory of symmetric functions the following theorem can be proved. We state this without proof and only give appropriate reference:

Theorem 8.2 (Jin and Kalantari (2005b)). *Let $t \geq 2, t \in \mathbb{Z}$, then for each integer $s \geq t+2$, $B_{s,t}(x)$ has an order of convergence $t+1$ for a simple root of p, say θ_1, and*

$$\lim_{x \to \infty} \frac{B_{s,t}(x) - \theta_1}{(x - \theta_1)^{t+1}} = \frac{(-1)^{t+1}}{(t+1)!} \frac{p^{(t+1)}(\theta_1)}{p'(\theta_1)}.$$ □

8.4 Applications

In Chapter 9 we will formally prove:

Theorem 8.3. *Given a polynomial $p(x)$ and a point x_0 in the Voronoi region of a root θ, the Basic Sequence $\{B_m(x_0)\}_{m=2}^{\infty}$ will converge to θ.* □

Theorem 8.3 offers a new method for polynomial root-finding. Given a good approximation x_0 to a root of the given polynomial, we make use of the sequence $\{B_m(x_0)\}_{m=2}^{\infty}$ to produce better and better approximations. Once the *normalized derivatives*, $p^{(j)}(x_0)/j!$, $j = 0, \dots, n$, are known there is no need to do additional function/derivative evaluations. Furthermore, once $B_m(x_0)$ is evaluated, the evaluation of $B_{m+1}(x_0)$ can be achieved in $O(mn)$ arithmetic operations (and possibly even faster). This follows from the recursive formula for D_m. In particular, for fixed polynomial degree, the evaluation of $B_{m+1}(x_0)$ can be achieved in $O(m)$ time. The entries of $B_m(x_0)$ can be evaluated efficiently since the evaluation of $p^{(j)}(x_0)/j!$, $j = 0, \ldots, n$ can be established in $O(n \log^2 n)$ arithmetic operations, see Kung (1974). The recursive formula of D_m is also attractive from the point of view of parallel evaluation of successive elements of the sequence $\{B_m(x_0)\}_{m=2}^{\infty}$.

Aside from polynomial root-finding, the convergence property in Theorem 8.3 together with the Truncated Basic Family motivates a new strategy for the approximation of roots of more general functions. The strategy is as follows: given the current input x_0, one approximates a root of the corresponding n-th degree Taylor polynomial, for an appropriately selected degree n, and via the above mentioned sequence. The new approximation replaces the old one, and the process is repeated.

We will now formalize an algorithm. Consider $f(x)$, say a real or (complex) valued function having n continuously differentiable derivatives in a neighborhood of a root. Now consider the following algorithm for approximation of a root $f(x)$:

Root-finding(n):

Step 1. Given x_0, compute $P_n(x)$, the n-th degree Taylor Polynomial of f at x_0.

Step 2. Approximate a root of $P_n(x)$ using the sequence $\{B_{m,n}(x_0)\}_{m=n+1}^{\infty}$:

Strategy I. Let $x_1 = B_{n+1,n}(x_0)$.

Strategy II. Fix $\epsilon > 0$. Let $x_1 = B_{m,n}(x_0)$, where m is the least $m \geq n+2$ such that

$$|B_{m,n}(x_0) - B_{m-1,n}(x_0)| \leq \epsilon.$$

Step 3. Replace x_0 with x_1. Go to Step 1.

When $n = 1$, the above algorithm is simply Newton's method. For $n \geq 2$, the approximation of the zero of the Taylor polynomial in Step 2 is via a single approximation $B_{n+1,n}(x_0)$, according to Strategy I, or the sequence of approximations $\{B_{m,n}(x_0)\}_{m=n+1}^{\infty}$, according to Strategy II. The justification in utilizing the sequence $\{B_{m,n}(x_0)\}_{m=n+1}^{\infty}$ lies in the fact that when x_0 is a reasonably good approximation to a root of $P_n(x)$, then the sequence converges to that root. Note that the sequence $\{B_{m,n}(x_0)\}_{m=n+1}^{\infty}$ corresponds to the values of the Basic Family as written with respect to the Taylor polynomial $P_n(x)$.

Strategy II of Step 2 allows the possibility of refining the initial approximation, $B_{n+1,n}(x_0)$, by performing only elementary operations on x_0, and the computed normalized derivatives, $P_n^{(j)}(x_0)/j!$, $j = 0, \ldots, n$.

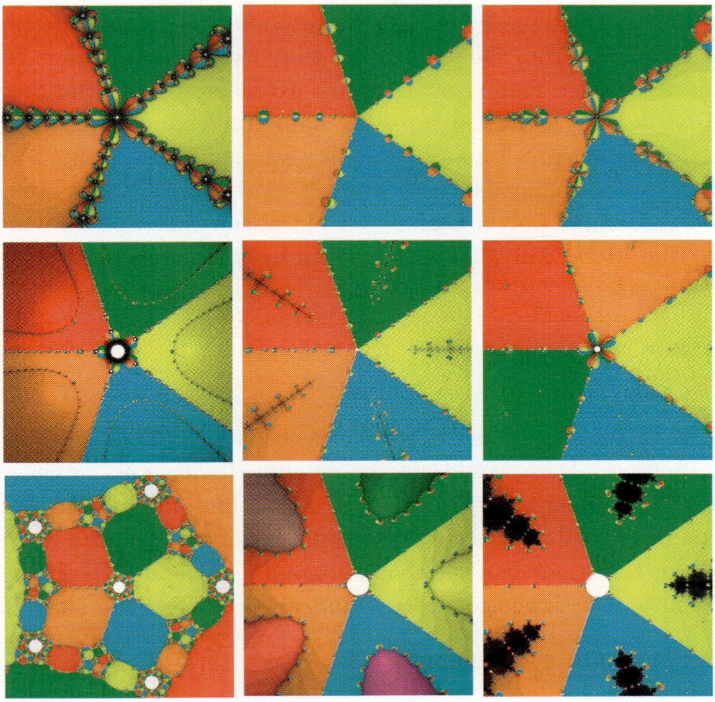

Fig. 8.1 Polynomiographs of $z^5 - 1$ with Halley Family $m = 2, 3, 4$ first row; $m = 6, 7, 8$ second row; and $m = 14, 35, 45$ third row.

Fig. 8.2 Polynomiographs of $z^9 - 1$ for $m = 2, 4, 40$.

8.5 Polynomiography with the Truncated Basic Family

In this section we offer some polynomiography with the truncated Basic Family and contrast these with the corresponding polynomiography with the Basic Family. From the point of view of visualization and art certainly the truncated Basic Family offers a rich extension. From the theoretical point of view itself it does give rise to many interesting problems. In fact the images in Figures 8.1 and 8.2 are all restricted to the Halley Family as applied to two polynomials $z^5 - 1$ and $z^9 - 1$. It appears that even in this case and the simple polynomial $z^5 - 1$ extraneous fixed points may arise. We note that Mandelbrot-like sets do seem to appear, see right-most image in the third row.

8.6 Conclusions

Clearly, on the one hand the Truncated Basic Family offers many algorithms for root-finding. These could also be combined with some other existing methods to give rise to novel practical algorithms for root-finding. The actual implementation of these is certainly a worthy task to pursue in the future. On the other hand, polynomiography with the Truncated Basic offers a whole new dimension into visualization of a polynomial equation.

Chapter 9

Characterizations of Solutions of Homogeneous Linear Recurrence Relations via the Basic Family and Polyhedral Representations *

There is a vast literature on homogeneous linear recurrence relations with constant coefficients. However, their connections to polynomial root-finding do not seem to have been fully investigated or discovered. In this chapter we develop a strong relationship between a homogeneous linear recurrence relation and the Basic Family of iteration functions, in particular the Basic Sequence. Some of these results were initially shown in Kalantari (2004b). We prove many properties of the Basic Sequence, including its convergence properties and connections with Voronoi regions of polynomial roots. In doing so we gain much insights into the solutions to homogeneous linear recurrence relations such as determinantal representations, polyhedral representation, bounds and more. Conversely, we gain valuable insights on the Basic Family itself. These connections not only offer new interpretations, but give rise to algorithms for root-finding as well as visualizations associated with a homogeneous linear recurrence relation through polynomiography. These connections even suggest new insights on such well-known recurrences as the Fibonacci and Lucas numbers and their generalized versions. We offer polyhedral representation of solutions of a homogeneous linear recurrence relation. These call for novel applications of linear programming or integer programming techniques, over real or complex numbers. In particular, the polyhedral approach gives rise to the definition of a zero-one Fibonacci polytope from which many identities on Fibonacci and Lucas numbers can be derived mechanically. In a subsequent chapter we

*Part of this chapter has been reprinted from On Homogeneous Linear Recurrence Relations and Approximation of Zeros of Complex Polynomials, in M. B. Nathanson (ed.), *Unusual Applications in Number Theory, DIMACS Series in Discrete Mathematics and Theoretical Computer Science*, Vol. 64, pp. 125–143 (2004); with permission from the American Mathematical Society.

will give polynomiography of several special recurrence relations, including the generalized Fibonacci sequences.

9.1 Introduction

There is a close relationship between polynomial root-finding through the Basic Family and a homogeneous recurrence relation. This relationship offers mutual theoretical and practical insights on the two problems which we will exhibit here in this chapter. A summary of these connections is as follows.

On the one hand given a polynomial equation $p(z) = 0$ of degree n, and the corresponding Basic Family $\{B_m(z) = z - p(z)D_{m-2}/D_{m-1}(z)\}_{m=2}^{\infty}$, for each complex number w we may associate a linear homogeneous recurrence relation of degree n, that generates $\{D_m(w)\}_{m=2}^{\infty}$, satisfying the initial consitions $D_0(w) = 1$, $D_j(w) = 0$ for $j = -n+1, \ldots, -1$. This gives rise to the *Basic Sequence*,

$$\left\{B_m(w) \equiv w - p(w)\frac{D_{m-2}(w)}{D_{m-1}(w)}\right\}_{m=2}^{\infty}.$$

In this chapter we formally prove the property that for each complex number w in the Voronoi region of a root θ of $p(z)$ (i.e. the set of points that are closer to θ than any other root), the corresponding Basic Sequence converges to θ. We thus call homogeneous linear recurrence relations corresponding to points within the same Voronoi region as *conjugate linear homogeneous relations.*

The Basic Sequence and its convergence properties give rise to a technique within polynomiography for visualization of polynomial equations. The justification in viewing this visualization as polynomiography lies in the fact that the Basic Sequence corresponds to the pointwise evaluation of the Basic Family $\{B_m(z)\}_{m=2}^{\infty}$, with respect to the underlying polynomial.

On the other hand, to every homogeneous linear recurrence relation of degree n, $\{a_m\}$, with the particular initial conditions $a_0 = 1$, $a_j = 0$ for $j = -n+1, \ldots, -1$, we may associate a polynomial equation $p(z) = 0$ corresponding to the evaluation of reciprocal of its characteristic equation. We may thus interpret a_m as that of evaluation of $D_m(a)$ with respect to $p(z)$ at $z = 0$. The corresponding Basic Sequence is the ratio $-p(0)a_{m-2}/a_{m-1}$ and it converges to the least modulus root of $p(z)$, if it happens to be uniquely defined.

9.2 Homogeneous Linear Recurrence Relations

Consider a general homogeneous linear recurrence relation (HLRR) of degree n with constant coefficients:

$$a_m = c_1 a_{m-1} + c_2 a_{m-2} + \cdots + c_n a_{m-n}, \tag{9.1}$$

where $c_1, \ldots, c_n$ are given numbers in $\mathbb{C}$, and $c_n \neq 0$.

We shall refer to the following set of initial conditions as the *Basic Initial Conditions*:

$$a_0 = 1, \qquad a_{-1} = a_{-2} = \cdots = a_{-n+1} = 0. \tag{9.2}$$

We shall refer to the solution $\{a_m\}_{m=1}^{\infty}$ of (9.1) and (9.2) as the *Fundamental Solution* of HLRR. The significance of this particular set of initial conditions and the Fundamental Solution will become evident as we unveil connections to polynomial root-finding and also to linear programming.

Example 9.1. Consider the HLRR $a_m = a_{m-1} + a_{m-2}$. The Fundamental Solution of this HLRR is the sequence of Fibonacci numbers

$$F_m = F_{m-1} + F_{m-2}, \quad F_0 = 1, F{-1} = 0.$$

However, the Lucas numbers are not Fundamental Solutions:

$$L_m = L_{m-1} + L_{m-2}, \quad L_0 = 1, L{-1} = 2,$$

and that is really their main difference with Fibonacci sequence.

The *characteristic polynomial* of the sequence $\{a_m\}$ is the polynomial

$$q(z) = z^n - (c_1 z^{n-1} + c_2 z^{n-2} + \cdots + c_{n-1} z + c_n). \tag{9.3}$$

The *characteristic equation* refers to the equation

$$q(z) = 0. \tag{9.4}$$

It is well known that a_m can be represented in terms of the roots of $q(z)$. More precisely, let $\eta_1, \ldots, \eta_t$ be the set of distinct roots of $q(z)$ having multiplicities $n_1, \ldots, n_t$. Then for each $i = 1, \ldots, t$ there exists a polynomial $\alpha_i(z)$, such that either $\alpha_i(z)$ is identically zero, or it is a polynomial of degree $\leq n_i - 1$, such that

$$a_m = \alpha_1(m)\eta_1^m + \cdots + \alpha_t(m)\eta_t^m, \quad \forall\, m \geq -n+1. \tag{9.5}$$

Remark 9.1. In fact the above representation could be stated for all $m \in \mathbb{Z}$, thus extending to negative infinity as well.

We now prove some properties of the HLRR with the Basic Initial Conditions, (9.2). These initial conditions actually arise in the context of Bernoulli method for computing the dominant root of a polynomial, when the dominant root is uniquely defined, but possibly with multiplicity higher than one. In particular, the following is well-known in the literature of homogeneous recurrence relations (e.g. Hildebrand (1974), Householder (1970)):

Proposition 9.1. *Consider the homogeneous linear recurrence relation (9.1) with the Basic Initial Conditions (9.2). If η_r is the unique dominant root of $q(z)$, then $\alpha_r(z)$ is not identically zero. Moreover,*

$$\lim_{m\to\infty} \frac{a_m}{a_{m-1}} = \eta_r.$$

□

Remark 9.2. In fact it can be shown that given the initial condition (9.2), $\alpha_i(z)$ is not identically zero for all $i = 1, \ldots, t$, regardless of the uniqueness of the dominant root of $q(z)$. Thus the convergence of the ratio a_m/a_{m-1} to the dominant root is an unrelated matter altogether. This fact appears to have been overlooked in the vast literature on homogeneous linear recurrence relations, including even recent ones. In what follows we will first prove a weaker version of the above proposition. Next using its proof we will derive an explicit formula for the coefficients $\alpha_i(z)$ when the roots are all simple. Later in the chapter, (9.4), we will argue why $\alpha_i(z)$ in representing the terms of an HLRR, see (9.5), will be nonzero for all i, regardless of the dominance of the modulus.

Proposition 9.2. *Consider the homogeneous linear recurrence relation (9.1) with the Basic Initial Conditions (9.2). Assume that $|\eta_i| \neq |\eta_j|$, if $i \neq j$. Let r be the index satisfying*

$$|\eta_r| = \max\{|\eta_i| : \alpha_i(z) \not\equiv 0, \quad i = 1, \ldots, t\},$$

i.e. the index of the root with largest modulus whose coefficient polynomial is not identically zero. Then r is well-defined and

$$\lim_{m\to\infty} \frac{a_m}{a_{m-1}} = \eta_r.$$

Moreover, if all the roots of $q(z)$ are simple, then $\alpha_i(z)$ is a nonzero constant for all $i = 1, \ldots, n$. In particular, η_r is the root of $q(z)$ having the largest modulus.

Proof. Since $a_0 = 1$ we conclude that not all $q_i(z)$'s are zero. Hence r is well-defined. From the assumption that all distinct roots of $Q(z)$ have different modulus and the definition of η_r, for any $i \neq r$ such that $\alpha_i(z) \not\equiv 0$, we have $|\eta_i/\eta_r| < 1$. Since for any such index $\alpha_i(z)$ is a polynomial, it follows that

$$\lim_{m\to\infty} \alpha_i(m)\left(\frac{\eta_i}{\eta_r}\right)^m = 0. \tag{9.6}$$

First we claim that the sequence $\{a_m\}$ cannot have a subsequence $\{a_{m_k}\}$ where $a_{m_k} = 0$ for all k. Suppose such a subsequence exists. If all but one of the $\alpha_i(z)$s are identically zero, then from (9.1) we have

$$a_m = \alpha_l(m)\eta_l^m,$$

for some l and $\alpha_l(m) \not\equiv 0$. Thus, $a_m \neq 0$ for all $m \geq 0$. Hence under the assumption that $\{a_{m_k} = 0\}$ there must exist at least two indices for which $\alpha_i(z)$ is not identically zero.

Since any $a, b \in \mathbb{C}$, $|a - b| \geq |a| - |b|$, we have

$$\left|\frac{a_{m_k}}{\eta_r^{m_k}}\right| = \left|\sum_{i=1}^{t} \alpha_i(m_k)\left(\frac{\eta_i}{\eta_r}\right)^{m_k}\right|$$

$$\geq \left|\alpha_r(m_k)\right| - \sum_{i=1,\ i\neq r,\ \alpha_i(z)\not\equiv 0}^{t} \left|\alpha_i(m_k)\left(\frac{\eta_i}{\eta_r}\right)^{m_k}\right|. \tag{9.7}$$

Taking the limit in (9.7) and using (9.6), we get

$$0 \geq \lim_{k\to\infty} |\alpha_r(m_k)|. \tag{9.8}$$

But the limit in (9.8) is either a nonzero number, or infinity. This is a contradiction. Thus a_m can be zero at most for a finite set of indices m.

Since $\alpha_i(z)$ is a polynomial, if it is not identically zero then we have

$$\lim_{m\to\infty} \frac{\alpha_i(m)}{\alpha_i(m-1)} = 1. \tag{9.9}$$

Thus for m large enough the ratio a_m/a_{m-1} is well-defined and is given by:

$$\frac{a_m}{a_{m-1}} = \frac{\eta_r^m \sum_{i=1}^t \alpha_i(m)\left(\frac{\eta_i}{\eta_r}\right)^m}{\eta_r^{m-1} \sum_{i=1}^t \alpha_i(m-1)\left(\frac{\eta_i}{\eta_r}\right)^{m-1}}. \tag{9.10}$$

Now using (9.6) and (9.9) it follows that the ratio in (9.10) converges to η_r.

Next we assume that all the roots of $q(z)$ are simple. In this case $t = n$, and $\alpha_i(z)$ is a constant, say $\alpha_i \in \mathbb{C}$, for all $i = 1, \ldots, n$. We wish to show that $\alpha_i \neq 0$ for all $i = 1, \ldots, n$.

From the initial condition (9.2), while substituted into (9.5), we get the following system of linear equations:

$$\begin{pmatrix} \eta_1^{-n+1} & \eta_2^{-n+1} & \cdots & \eta_n^{-n+1} \\ \eta_1^{-n+2} & \eta_2^{-n+2} & \cdots & \eta_n^{-n+2} \\ \vdots & \vdots & \cdots & \vdots \\ \eta_1^{-1} & \eta_2^{-1} & \cdots & \eta_n^{-1} \\ 1 & 1 & \cdots & 1 \end{pmatrix} \begin{pmatrix} \alpha_1 \\ \alpha_2 \\ \vdots \\ \alpha_{n-1} \\ \alpha_n \end{pmatrix} = \begin{pmatrix} 0 \\ 0 \\ \vdots \\ 0 \\ 1 \end{pmatrix}. \tag{9.11}$$

Suppose that some α_i is zero. Without loss of generality we may assume $\alpha_1 = 0$. Substituting $\alpha_1 = 0$ in the above system of linear equations gives:

$$\begin{pmatrix} \eta_2^{-n+1} & \eta_3^{-n+1} & \cdots & \eta_n^{-n+1} \\ \eta_2^{-n+2} & \eta_3^{-n+2} & \cdots & \eta_n^{-n+2} \\ \vdots & \vdots & \cdots & \vdots \\ \eta_2^{-1} & \eta_3^{-1} & \cdots & \eta_n^{-1} \end{pmatrix} \begin{pmatrix} \alpha_2 \\ \alpha_3 \\ \vdots \\ \alpha_n \end{pmatrix} = \begin{pmatrix} 0 \\ 0 \\ \vdots \\ 0 \end{pmatrix}.$$

But the coefficient matrix is a product of a Vandermonde matrix in $\eta_2, \ldots, \eta_n$ and the diagonal matrix $\text{diag}(\eta_1^{-2}, \cdots, \eta_n^{n-1})$. Since the η_i's are distinct, the Vandermonde matrix is invertible. Hence if $\alpha_1 = 0$, then $\alpha_2 = \cdots = \alpha_n = 0$. But this contradicts that $a_0 \neq 0$. □

9.3 Explicit Representation of the Fundamental Solution

In this section we develop an explicit representation of the Fundamental Solution of an HLRR when the roots of the characteristic polynomial are simple. We will then use this in the next section to give a further refinement.

Theorem 9.1. *Consider HLRR (9.1) with the Basic Initial Conditions (9.2) and assume the characteristic roots are simple. We have*

$$\alpha_i = \frac{\eta_i}{\prod_{j=1}^{n} \eta_j} \; \frac{1}{\prod_{j=1, j\neq i}^{n} (\eta_j^{-1} - \eta_i^{-1})}. \tag{9.12}$$

Proof. Substituting the initial conditions (9.2) into (9.11) and also letting

$$\theta_i = \frac{1}{\eta_i}, \quad i = 1, \ldots, n,$$

we get the following system of linear equations:

$$\begin{pmatrix} 1 & 1 & \cdots & 1 \\ \theta_1 & \theta_2 & \cdots & \theta_n \\ \vdots & \vdots & \ddots & \vdots \\ \theta_1^{n-1} & \theta_2^{n-1} & \cdots & \theta_n^{n-1} \end{pmatrix} \begin{pmatrix} \alpha_1 \\ \alpha_2 \\ \vdots \\ \alpha_n \end{pmatrix} \begin{pmatrix} 1 \\ 0 \\ \vdots \\ 0 \end{pmatrix}$$

From Cramer's rule each α_i is the quotient of two determinants, Δ_1/Δ_2, where Δ_2 is the determinant of an $n \times n$ Vandermonde coefficient matrix, we denote by $\Delta(\theta_1, \ldots, \theta_n)$. Δ_1 is the determinant of the coefficient matrix whose i-th column is replaced with the vector $(1, 0, \cdots, 0)^T$.

We shall compute the formula for α_1 and by symmetry this can be modified to give the formula for all α_i. It is well-known that

$$\Delta(\theta_1, \cdots, \theta_n) = \prod_{1 \le i < j \le n} (\theta_j - \theta_i).$$

Expanding Δ_1 about its first column, the vector $(1, 0, \ldots, 0)^T$, we get

$$\Delta_1 = \begin{pmatrix} \theta_2 & \theta_3 & \cdots & \theta_n \\ \theta_2^2 & \theta_3^2 & \cdots & \theta_n^2 \\ \vdots & \vdots & \ddots & \vdots \\ \theta_2^{n-1} & \theta_3^{n-1} & \cdots & \theta_n^{n-1} \end{pmatrix} = \begin{pmatrix} 1 & 1 & \cdots & 1 \\ \theta_2 & \theta_3 & \cdots & \theta_n \\ \vdots & \vdots & \ddots & \vdots \\ \theta_2^{n-2} & \theta_3^{n-2} & \cdots & \theta_n^{n-2} \end{pmatrix} \begin{pmatrix} \theta_2 & 0 & \cdots & 0 \\ 0 & \theta_3 & \cdots & 0 \\ \vdots & \vdots & \ddots & \vdots \\ 0 & 0 & \cdots & \theta_n \end{pmatrix}$$

$$= \Delta(\theta_2, \cdots, \theta_n) \prod_{i=2}^{n} \theta_i.$$

Thus,

$$\alpha_1 = \frac{\Delta(\theta_2, \cdots, \theta_n) \prod_{i=2}^n \theta_i}{\Delta(\theta_1, \cdots, \theta_n)} = \frac{\Pi_{j=2}^n \theta_i}{\Pi_{j=2}^n (\theta_j - \theta_1)} = \frac{\Pi_{j=1}^n \theta_i}{\theta_1} \frac{1}{\Pi_{j=2}^n (\theta_j - \theta_1)}.$$

But this coincides with (9.12). Hence the proof of the theorem. □

9.4 Explicit Representation Via Characteristic Polynomial

In this section we shall gain more insight by connecting HLRR to polynomial root-finding.

Definition 9.1. Define the negative reciprocal of the characteristic polynomial $q(z)$ to be

$$p(z) = -z^n q(\frac{1}{z}) = c_n z^n + c_{n-1} z^{n-1} + \cdots + c_1 z - 1. \tag{9.13}$$

We will now consider homogeneous linear recurrence relation defined in more generality. Given a complex number w, we define a homogeneous linear recurrence relation by HLRR(w), where for each $m \geq 1$ we set

$$a_m(w) = \sum_{i=1}^{n} c_i(w) a_{m-i}(w), \tag{9.14}$$

with

$$c_i(w) = (-1)^{i-1} p^{i-1}(w) \frac{p^{(i)}(w)}{i!}.$$

We consider HLRR(w) with its Basic Initial conditions

$$a_0(w) = 1, \quad a_j(w) = 0, \quad \forall\ j = -1, \ldots, -n+1. \tag{9.15}$$

The characteristic polynomial associated with the sequence $\{a_m(w)\}$ is the polynomial

$$q_w(z) = z^n - (c_1(w) z^{n-1} + c_2(w) z^{n-2} + \cdots + c_{n-1}(w) z + c_n(w)). \tag{9.16}$$

The corresponding negative reciprocal is

$$p_w(z) = z^n q_w(\frac{1}{z}) = c_n(w) z^n + \cdots + c_1(w) z - 1.$$

At $w = 0$, $p(0) = 1$ and $p^{(i)}(0)/i! = c_i$. Thus we get

$$c_i(0) = (-1)^{i-1}(-1)^{i-1} c_i = c_i.$$

Hence

$$a_m(0) = a_m, \quad q_0(z) = q(z), \quad p_0(z) = p(z),$$

so that HLRR(0) coincides with HLRR.

Definition 9.2. We define the Universal Class of HLRR associated with the reciprocal polynomial $p(z) = c_n z^n + \cdots + c_1 z - 1$ to be

$$U(p(z)) = \{HLRR(w) : w \in \mathbb{C}\}.$$

The significance of $U(p(z))$ lies in the following connection between HLRR(w) and HLRR(0):

Theorem 9.2. *If θ is a root of the negative reciprocal polynomial, $p(z)$, then $\eta(w) = p(w)/(w - \theta)$ is a root of $q_w(z) = z^n - \sum_{i=1}^{n-1} c_i(w) z^{n-i}$. Conversely, if $\eta(w)$ is a root of $q_w(z)$, then $\theta = w - p(w)/\eta(w)$ is a root of $p(z)$.*

Equivalently, θ is a root of $p(z)$ if and only if $\theta(w) = (w - \theta)/p(w)$ is a root of $p_w(z) = -z^n q_w(1/z)$.

Proof. Suppose that $\eta(w)$ is a root of $q_w(z)$, i.e.

$$\eta(w)^n - \sum_{i=1}^{n}(-1)^{i-1}p^{i-1}(w)\frac{p^{(i)}(w)}{i!}\eta^{n-i} = 0. \tag{9.17}$$

Multiplying both sides by $p(w)/\eta(w)^n$ we get

$$p(w) - \sum_{i=1}^{n}(-1)^{i-1}p^{i}(w)\frac{p^{(i)}(w)}{i!}\eta(w)^{-i} = 0. \tag{9.18}$$

From the definition of θ we get

$$\eta(w) = \frac{p(w)}{(w-\theta)} = -\frac{p(w)}{(\theta - w)}. \tag{9.19}$$

Substituting (9.19) into (9.18) gives

$$\sum_{i=0}^{n}\frac{p^{(i)}(w)}{i!}(\theta - w)^i = 0. \tag{9.20}$$

But from Taylor's theorem the left-hand side of the above coincides with $p(\theta)$. Hence θ is a root of $p(z)$. The converse follows simply by reversing the steps. □

Corollary 9.1. *Let the distinct roots of $q_w(z)$ be $\eta_i(w)$, $i = 1, \ldots, t$ with multiplicities n_i. Then for all $m \in \mathbb{Z}$ we have:*

$$a_m(w) = \sum_{i=1}^{t}\alpha_{i,w}(m)\eta_i^m(w) = \sum_{i=1}^{t}\alpha_{i,w}(m)(\frac{p(w)}{(w-\theta_i)})^m,$$

where θ_i, $i = 1, \ldots, t$ is the set of distinct zeros of the negative reciprocal reciprocal $p(z)$, and $\alpha_{i,w}(z)$ is either identically zero or a polynomial of degree at most $n_i - 1$. □

Theorem 9.3. *If the roots of the negative reciprocal polynomial $p(z)$ are simple, then $\alpha_{i,w}(z)$ is independent of z:*

$$\alpha_{i,w}(z) = \frac{p(w)}{(w-\theta_i)}\frac{1}{p'(\theta_i)} \equiv \alpha_i(w).$$

Thus, for all $m \in \mathbb{Z}$ we have

$$a_m(w) = \sum_{i=1}^{n}(\frac{p(w)}{(w-\theta_i)})^{m+1}\frac{1}{p'(\theta_i)}.$$

Proof. From Theorem 9.1 we have

$$\alpha_i(w) = \frac{\eta_i(w)}{\prod_{j=1}^n \eta_j(w)} \frac{1}{\prod_{j=1,j\neq i}^n (\eta_j^{-1}(w) - \eta_i^{-1}(w))}. \tag{9.21}$$

We have

$$p(z) = c_n z^n + \cdots + c_1 z - 1 = (z - \theta_1) \cdots (z - \theta_n).$$

Thus

$$\prod_{j=1}^n \eta_j(w) = \prod_{j=1}^n \frac{p(w)}{(w - \theta_i)} = p(w)^{n-1}.$$

Also,

$$\prod_{j=1,j\neq i}^n (\eta_j^{-1}(w) - \eta_i^{-1}(w)) = \frac{\prod_{j=1,j\neq i}^n (\theta_i - \theta_i)}{p(w)^{n-1}} = \frac{p'(\theta)}{p(w)^{n-1}}.$$

Substituting these into (9.21) proves the formula for $\alpha_i(w)$, and hence for $a_m(w)$. □

In particular, when $w = 0$ we get the following statement on the original HLRR considered in the chapter.

Corollary 9.2. *Consider HLRR (9.1) with Basic Initial Conditions (9.2). If the characteristic polynomial has simple roots* $\eta_1, \ldots, \eta_n$, *then*

$$a_m = \sum_{i=1}^n \alpha_i \eta_i^m = \sum_{i=1}^n \frac{1}{\theta_i^{m+1}} \frac{1}{p'(\theta_i)},$$

i.e.,

$$\alpha_i = \frac{1}{\theta_i} \frac{1}{p'(\theta_i)}.$$

□

In particular, when $w = 0$

$$\alpha_i(0) = \alpha_i = \frac{1}{\theta_i} \frac{1}{p'(\theta_i)}$$

and we have

$$a_m(0) = a_m = \sum_{i=1}^n \alpha_i \eta_i^m = \sum_{i=1}^n \frac{1}{\theta_i^{m+1}} \frac{1}{p'(\theta_i)}.$$

We make an observation based on Theorem 9.3. First a definition.

Definition 9.3. For a given complex number w we call the *shifted Fundamental Solution* to the sequence

$$\{\widehat{a}_m(w) = a_{m-1}(w)\},$$

where $\{a_m(w)\}$ is the Fundamental solution of the corresponding HLRR.

Corollary 9.3. *If the roots of $p(z)$ are simple, then we have*

$$\widehat{a}_m(w) = \sum_{i=1}^{n} (\frac{p(w)}{(w-\theta_i)})^m \frac{1}{p'(\theta_i)}.$$

Thus the coefficient polynomials in representing the shifted Fundamental Solution are $1/p'(\theta_i)$, $i = 1, \ldots, n$, independent of w. □

Example 9.2. Consider the Fibonacci sequence $F_m = F_{m-1} + F_{m-2}$, $F_0 = 1$, $F_{-1} = 0$. Then the characteristic polynomial and its negative reciprocal are

$$q(z) = z^2 - z - 1, \quad p(z) = z^2 q(1/z) = z^2 + z - 1.$$

$p'(z) = 2z + 1$, and the roots are $\theta_i = \frac{1}{2}(-1 \pm \sqrt{5})$, $i = 1, 2$. $p'(\theta_i) = \pm\sqrt{5}$. If $\phi = \frac{1}{2}(1 + \sqrt{5})$, the golden ration, then $1/\theta_1 = \phi$, and $1/\theta_2 = 1 - \phi$. Thus

$$F_m = \frac{\phi^{m+1} - (1-\phi)^{m+1}}{\sqrt{5}}, \quad m \in \mathbb{Z}.$$

The shifted Fibonacci sequence, $\{\widehat{F}_m\}$, satisfies $\widehat{F}_1 = 1$, $\widehat{F}_2 = 1$ which coincides with the usual definition of Fibonacci sequence.

9.5 Approximation of Polynomial Roots Using HLRR

Given an HLRR and its Basic Initial Conditions, we have seen that given any complex number w, we can associate a homogeneous linear recurrence relation defined with respect to w and the roots of its negative reciprocal polynomial $p(z) = c_n z^n + \cdots + c_1 z - 1$.

In this section we reverse the relationship and associate homogeneous linear recurrence relations to an arbitrary polynomial with the aim of relating it to the approximation of polynomial roots.

Let

$$p(z) = p_n z^n + \cdots + p_1 z + p_0 \tag{9.22}$$

be a polynomial with coefficients $p_i \in \mathbb{C}$, $i = 1, \ldots, n$, and $p_0 p_n \neq 0$.

For each $w \in \mathbb{C}$ we associate a sequence $\{d_m(w)\}$ defined via the following homogeneous linear recurrence relation of degree n:

$$d_m(w) = \sum_{i=1}^{n} (-1)^{i-1} p^{i-1}(w) \frac{p^{(i)}(w)}{i!} d_{m-i}(w), \tag{9.23}$$

where

$$d_0(w) = 1, \quad d_j(w) = 0, \quad \forall \ j = -1, \ldots, j = -n + 1. \tag{9.24}$$

The corresponding characteristic polynomial, denoted by $q_w(z)$ is

$$q_w(z) = z^n - \sum_{i=1}^{n} (-1)^{i-1} p^{i-1}(w) \frac{p^{(i)}(w)}{i!} z^{n-i}.$$

As before if $\eta_1, \ldots, \eta_t$ is the set of distinct roots of $q_w(z)$ having multiplicities $n_1, \ldots, n_t$, then

$$d_m(w) = \alpha_1(m)\eta_1^m + \cdots + \alpha_t(m)\eta_t^m, \quad \forall \;\; m \geq -n+1. \tag{9.25}$$

Since the roots of $p(z)$ and $-p(z)/p_0$ are identical, Theorem 9.2 applies to arbitrary $p(z)$, (9.22) so that we may conclude: If $\eta(w)$ is a root of $q_w(z)$, then $\theta = w - p(w)/\eta(w)$ is a root of $p(z)$. Conversely, if θ is a root of $p(z)$, then $\eta(w) = -p(w)/(\theta - w)$ is a root of $q_w(z)$.

Let

$$R_p = \{\theta_1, \ldots, \theta_t\} \tag{9.26}$$

be the set of distinct roots of $p(z)$. The elements of R_p partition the Euclidean plane into Voronoi regions and their boundaries. The Voronoi region of a root θ is a convex polygon defined by the locus of points which are closer to this root than to any other root. More precisely, the Voronoi region of a root θ is

$$V(\theta) = \{z \in \mathbb{C} : |z - \theta| < |z - \theta'|, \;\; \theta' \in R_p, \;\; \theta' \neq \theta\}. \tag{9.27}$$

Definition 9.4. Given $w \in \mathbb{C}$ the *Basic Sequence* at w is defined as

$$b_m(w) = w - p(w)\frac{d_{m-2}(w)}{d_{m-1}(w)}, \qquad m = 2, 3, \ldots. \tag{9.28}$$

Theorem 9.4. *Given $w \in V(\theta)$ for some root θ of $p(z)$, the corresponding Basic Sequence is well-defined satisfying*

$$\lim_{m \to \infty} b_m(w) = \theta.$$

Proof. From Theorem 9.2 for any $w \in \mathbb{C}$, different from a root of $p(z)$, the set of distinct roots of $q_w(z)$ is

$$\{\eta_1(w) = -\frac{p(w)}{\theta_1 - w}, \ldots, \eta_t = -\frac{p(w)}{\theta_t - w}\}. \tag{9.29}$$

Since $w \in V(\theta)$, then $\eta(w) = -p(w)/(\theta - w)$ is the dominant root of $q_w(z)$. Then from Proposition 9.1 it follows that

$$\lim_{m \to \infty} \frac{d_m(w)}{d_{m-1}(w)} = \eta(w).$$

□

Definition 9.5 (Conjugate HLRRs). *Given two complex numbers w_1 and w_2 we shall say their corresponding homogeneous linear recurrence relations, $\{d_m(w_1) : m \geq 2\}$ and $\{d_m(w_2) : m \geq 2\}$ are conjugate if w_1 and w_2 belong to the same Voronoi region.*

Remark 9.3. The notion of conjugacy says that when considering a homogeneous linear recurrence relation as a tool for approximating the roots of a polynomial, two conjugate homogeneous linear recurrences are really attempting to accomplish the same task. For instance if we consider the Fibonacci sequence as a means for approximating the golden ratio, then there are infinitely many conjugate recurrences which do just as well as the Fibonacci sequence.

Remark 9.4. For a point w that belongs to the boundary of the Voronoi regions, thus equidistant to two distinct roots of $p(z)$ the convergence of $b_m(w)$ to a root may not occur. For instance, for $p(z) = z^2 - 2$ the origin is a point on the boundary of Voronoi regions since it is equidistant to $\sqrt{2}$ and $-\sqrt{2}$. It is easy to see that for $a = 0$ the corresponding sequence, $\{d_m(0)\}$, has a subsequence that consists of zeros. Thus, the Basic Sequence is undefined. In the next example we consider a case where the Basic Sequence at a point on the Voronoi boundary could converge to a root.

Example 9.3. Consider the case of $p(z) = (z+1)(z^2-1)$, $a = 0$. In this case we show that the Basic Sequence $B_m(0)$ converges to the root $\theta = -1$. An intuitive explanation of this behavior could be due to the fact that one can attribute more *weight* to the root $\theta = -1$ in attracting the origin by virtue of its multiplicity.

Since $p'(z) = 3z^2 + 2z - 1$, $p''(z) = 6z + 2$, $p'''(z) = 6$. Thus

$$B_m(0) = \frac{D_{m-2}(0)}{D_{m-1}(0)},$$

where

$$D_m(0) = p'(0)D_{m-1}(0) - p(0)\frac{p''(0)}{2!}D_{m-2}(0) + p^2(0)\frac{p'''(0)}{3!}D_{m-3}(0).$$

Substituting,

$$D_m(0) = -D_{m-1}(0) + D_{m-2}(0) + D_{m-3}(0),$$

where $D_0(0) = 1$, $D_{-1}(0) = D_{-2}(0) = 0$. Computing the sequence $\{D_m(0)\}$ we get

$$\{B_m(0)\}_{m=2}^{\infty} = \{-1, -1/2, -1, -2/3, -1, -3/4, \dots\}.$$

More specifically,

$$B_m(0) = \begin{cases} -1, & \text{if } m \text{ is even;} \\ -\frac{m-1}{m+1}, & \text{if } m \text{ is odd.} \end{cases}$$

9.6 Basic Sequence and Connection to the Basic Family

As before let

$$p(z) = p_n z^n + \cdots + p_1 z_1 + p_0,$$

with $n \geq 2$ with complex coefficients, and $p_0 p_n \neq 0$.

We recall the definition and some properties of $B_m(z)$. For each $m \geq 2$,

$$B_m(z) \equiv z - p(z)\ \frac{D_{m-2}(z)}{D_{m-1}(z)},$$

where $D_0(z) \equiv 1$, and for each natural number $m \geq 1$,

$$D_m(z) = \det \begin{pmatrix} p'(z) & \frac{p''(z)}{2!} & \dots & \frac{p(m-1)(z)}{(m-1)!} & \frac{p^{(m)}(z)}{(m)!} \\ p(z) & p'(z) & \ddots & \ddots & \frac{p(m-1)(z)}{(m-1)!} \\ 0 & p(z) & \ddots & \ddots & \vdots \\ \vdots & \vdots & \ddots & \ddots & \frac{p''(z)}{2!} \\ 0 & 0 & \dots & p(z) & p'(z) \end{pmatrix}.$$

Then for each $i = m+1, \dots, n+m-1$

$$\widehat{D}_{m,i}(z) = \det \begin{pmatrix} \frac{p''(z)}{2!} & \frac{p'''(z)}{3!} & \dots & \frac{p^{(m)}(z)}{(m)!} & \frac{p^{(i)}(z)}{i!} \\ p'(z) & \frac{p''(z)}{2!} & \ddots & \frac{p^{(m-1)}(z)}{(m-1)!} & \frac{p^{(i-1)}(z)}{(i-1)!} \\ p(z) & p'(z) & \ddots & \vdots & \vdots \\ \vdots & \vdots & \ddots & \frac{p''(z)}{2!} & \frac{p^{(i-m+2)}(z)}{(i-m+2)!} \\ 0 & 0 & \dots & p'(z) & \frac{p^{(i-m+1)}(z)}{(i-m+1)!} \end{pmatrix}.$$

Some basic properties of the Basic Family are

(1) For all $m \geq 1$ we have,

$$D_m(z) = \sum_{i=1}^{n} (-1)^{i-1} \frac{p^{i-1}(z) p^{(i)}(z)}{i!} D_{m-i}(z),$$

where $D_0(z) = 1, \quad D_{-1}(z) = D_{-2}(z) = \cdots = D_{-n+1}(z) = 0.$

(2) If θ is a simple root of $p(z)$, then

$$B_m(z) = \theta + \sum_{i=m}^{m+n-2} (-1)^m \frac{\widehat{D}_{m-1,i}(z)}{D_{m-1}(z)} (z-\theta)^i.$$

(3) There exists $r > 0$ such that given any $a_0 \in N_r(\theta) = \{z : |z-\theta| \leq r\}$, the fixed-point iteration $a_{k+1} = B_m(a_k)$ is well-defined, it converges to θ having order m. Specifically,

$$\lim_{k\to\infty} \frac{(\theta - a_{k+1})}{(\theta - a_k)^m} = (-1)^{m-1} \frac{\widehat{D}_{m-1,m}(\theta)}{D_{m-1}(\theta)} = (-1)^{m-1} \frac{\widehat{D}_{m-1,m}(\theta)}{p'(\theta)^{m-1}}.$$

We now prove:

Theorem 9.5. *Given $p(z)$, for any input $w \in \mathbb{C}$ the associated sequences $\{d_m(w)\}$ (9.24) and the corresponding Basic Sequence $\{b_m(w)\}$ (9.28) satisfy:*

$$d_m(w) = D_m(w), \qquad b_m(w) = B_m(w), \quad \forall \ m \geq 0.$$

In particular, the following hold:

(1) $D_m(w) = d_m(w)$. Moreover when the roots of $p(z)$ are simple we have

$$D_m(w) = \sum_{i=1}^{n} (\frac{p(w)}{w-\theta_i})^{m+1} \frac{1}{p'(\theta_i)}.$$

(2) $d_m(w)$ can be represented as matrix $m \times m$ matrix corresponding to $D_m(w)$.

(3) For each simple root θ of $p(z)$ there exists an open neighborhood such that given any fixed $m \geq 2$, and any w_0 within this neighborhood, the sequence $\{w_k\}$ defined by

$$w_{k+1} = b_m(w_k),$$

converges to θ having order of convergence equal to m.

(4) $D_m(w)$ and hence $B_m(w)$ can be computed in $O(n \log n \log m)$ operations.

Proof. The equality $d_m(w) = D_m(w)$ is immediate since they correspond to the same homogeneous recurrence relations satisfying the same initial conditions. Except for the last part of the proof the remaining parts follow from earlier results in this section and the connections to the Basic Family.

To prove the last part of the theorem. We review some computational properties of the terms a_m of a general HLRR. From (9.1) and the identity equations

$$a_i = a_i, \quad i = 1, \ldots, m,$$

we get

$$\begin{pmatrix} a_m \\ a_{m-1} \\ \vdots \\ a_{m-n} \end{pmatrix} = \begin{pmatrix} c_1 & c_2 & \cdots & c_{n-1} & c_n \\ 1 & 0 & \cdots & 0 & 0 \\ \vdots & \vdots & \cdots & \vdots & \vdots \\ 0 & 0 & \cdots & 1 & 0 \end{pmatrix} \begin{pmatrix} a_{m-1} \\ a_{m-2} \\ \vdots \\ a_{m-n-1} \end{pmatrix}.$$

Thus, if for each $k \geq 0$ we let $v_k = (a_k, \cdots, a_{k-n})^T$, and let A denote the coefficient matrix above, then

$$v_m = Av_{m-1} = \cdots = A^m v_0, \qquad v_0 = (1, 0, \ldots, 0)^T. \tag{9.30}$$

Using (9.30) it can be shown that v_m and hence a_m can be computed in $O(n^2 \log m)$ arithmetic operations, see Gries and Levin (1980). It is also possible to compute a_m in $O(n \log n \log m)$ arithmetic operations, see Fiduccia (1985).

The complexity of computing $D_m(w)$ and hence $B_m(w)$ follows from two results. Firstly, given any w the evaluation of $p^{(j)}(w)/j!$, $j = 0, \ldots, n$, called the *normalized derivatives*, can be established in $O(n \log^2 n)$ arithmetic operations, see Kung (1974). Secondly, that the m-th term of a sequence defined via a homogeneous linear recurrence relation can be computed in $O(n \log n \log m)$ arithmetic operations. Combining these two results the claimed complexity follows. □

We know that the basic sequence for a point in the Voronoi region of a root converges to that root. The next result gives an explicit representation of the error for the special case when all the roots are simple and arbitrary point w.

Proposition 9.3. *Assume that $p(z)$ has simple roots $\theta_i, i = 1, \ldots, n$. Let $w \in \mathbb{C}$ be an arbitrary point different from the zeros of $p(z)$. For $i = 1, \ldots, n$ let $r_i(w) = (w - \theta_1)/(w - \theta_i)$, and $\rho_i = p'(\theta_1)/p'(\theta_i)$. Then*

$$B_m(w) - \theta_1 = \frac{\sum_{i=2}^n r_i^m(w)(\theta_i - \theta_1)\rho_i}{1 + \sum_{i=2}^n r_i^m(w)\rho_i}.$$

In particular, if w lies in the Voronoi region of θ_1,

$$\lim_{m \to \infty} B_m(w) = \theta_1.$$

More specifically, suppose $l \leq |\theta_i - \theta_j| \leq u$ for all $1 \leq i \neq j \leq n$. Suppose that $|r_i(w)| \leq (1 - \epsilon)$ for all $i = 1, \ldots, n$, $\epsilon \in [0, 1]$. Let m_0 be the smallest m such that $(1 - \epsilon)^m (n - 1)(u/l)^{n-2} < 1$. Then for all $m \geq m_0$ we have

$$|B_m(w) - \theta_1| \leq \frac{(1 - \epsilon)^m (n - 1)(\frac{u}{l})^{n-2} u}{1 - (1 - \epsilon)^m (n - 1)(\frac{u}{l})^{n-2}}.$$

Proof. We have

$$D_k(w) = p(w)^{k+1} \sum_{i=1}^{n} \frac{1}{(w-\theta_i)^{k+1}} \frac{1}{p'(\theta_i)}.$$

Substituting into the error and simplifying we get

$$B_m(w) - \theta_1 = (w-\theta_1) - p(w)\frac{D_{m-2}(w)}{D_{m-1}(w)}.$$

Multiplying and dividing $p(w)D_{m-2}(w)/D_{m-1}(w)$ by $(w-\theta_1)^m p'(\theta_1)$, substituting for $D_k(w)$, and simplifying we get

$$p(w)\frac{D_{m-2}(w)}{D_{m-1}(w)} = (w-\theta_1)\frac{1+\sum_{i=2}^{n} r_i^{m-1}(w)\rho_i}{1+\sum_{i=2}^{n} r_i^{m}(w)\rho_i}.$$

Thus

$$B_m(w) - \theta_1 = (w-\theta_1)\frac{\sum_{i=2}^{n} r_i^{m-1}(w)(r_i(w)) - 1)\rho_i}{1+\sum_{i=2}^{n} r_i^{m}(w)\rho_i}.$$

But

$$(w-\theta_1)(r_i(w)-1) = (w-\theta_1)\frac{(w-\theta_1)-(w-\theta_i)}{(w-\theta_i)} = r_i(w)(\theta_i-\theta_1).$$

Hence we have the claimed formula for $B_m(w) - \theta_1$. Since $|r_i| < 1$ and $|\rho_i|$ is bounded for all i, the convergence to θ_1 follows trivially.

To get the upper bound on the gap we simply need to observe that

$$\rho_i = \frac{p'(\theta_1)}{p'(\theta_i)} = \frac{\prod_{j=2}^{n}(\theta_1-\theta_j)}{\prod_{j=1, j\neq i}^{n}(\theta_i-\theta_j)}.$$

Hence

$$(\frac{l}{u})^{n-2} \leq |\rho_i| \leq (\frac{u}{l})^{n-2}.$$

Thus the numerator in $|B_m(w) - \theta_1|$ is bounded above by $(1-\epsilon)^m(n-1)u^{n-1}/l^{n-2}$. On the other hand

$$|1+\sum_{i=2}^{n} r_i(w)^m\rho_i| \geq 1 - |\sum_{i=2}^{n} |r_i(w)|^m|\rho_i|| \geq 1-(1-\epsilon)^m(n-1)(\frac{u}{l})^{n-2}.$$

□

Finally, here we will attempt to justify why the coefficient functions $\alpha_i(z)$ in representing the terms of an HLRR with its Basic Initial Conditions with respect to distinct roots of its characteristic polynomial, $\eta_1, \ldots, \eta_t$, (see (9.5)) cannot be identically be zero. Equivalently this can be stated in terms of $\{D_m(w)\}$ as follows:

Proposition 9.4. *Given a polynomial $p(z)$ of degree $n \geq 2$ with distinct roots $\theta_1, \ldots, \theta_t$ of multiplicities $n_1, \ldots, n_t$. Let w be a given complex number different from the roots of $p(z)$ and consider $\{D_m(w)\}$. Let $\eta_i = p(w)/(w - \theta_i)$, $i = 1, \ldots, t$. For $i = 1, \ldots, t$ there exists a polynomial $\alpha_i(z)$, not identically zero, of degree $\leq n_i - 1$, such that*

$$D_m(w) = \alpha_1(m)\eta_1^m + \cdots + \alpha_t(m)\eta_t^m, \quad \forall\ m \geq -n + 1. \tag{9.31}$$

If $t = n$ then $\alpha_i(z) = \alpha_i = \eta_i / p'(\theta)$ and

$$D_m(w) = p(w)^{m+1} \sum_{i=1}^{n} \frac{1}{(w - \theta_i)^{m+1}} \frac{1}{p'(\theta_i)}.$$

We will only prove here a special case that gives the flavor for proving the general case.

We give polyhedral representation of solutions of a homogeneous linear recurrence relation. These findings, inspired by connections to polynomial root-finding, call for novel applications of linear or integer programming techniques, over real or complex numbers. These also allow computing bounds on solutions. In particular, the polyhedral approach gives rise to the definition of a zero-one Fibonacci polytope from which many identities on Fibonacci and Lucas numbers can be derived mechanically.

Proof. [Jin (2005b), a special case] We will prove this for the special case when one of the roots has multiplicity 2 and the other roots are simple. If all the roots of $p(z)$ are simple we have seen that

$$D_m(w) = \sum_{i=1}^{n} \alpha_i \eta_i^m = p(w)^{m+1} \sum_{i=1}^{n} \frac{1}{(w - \theta_i)^{m+1}} \frac{1}{p'(\theta_i)}.$$

In the general case

$$D_m(w) = \sum_{i=1}^{t} \alpha_i(m) \eta_i^m.$$

We would like to consider the case when $t = n - 1$ and argue that $\alpha_i(m)$ is not identically zero.

Assume that θ_i and θ_j are distinct simple roots. We will analyze the coefficients when θ_j approaches θ_i. From the assumption of simplicity of α_i and α_j we have

$$p(z) = (z - \theta_i)(z - \theta_j)g(z),$$

where

$$g(\theta_i) \neq 0,\ g(\theta_j) \neq 0.$$

Then

$$p'(z) = (z - \theta_j)g(z) + (z - \theta_i)g(z) + (z - \theta_i)(z - \theta_j)g'(z).$$

Hence,

$$p'(\theta_i) = (\theta_i - \theta_j)g(\theta_i),$$
$$p'(\theta_j) = (\theta_j - \theta_i)g(\theta_j).$$

We have

$$\begin{aligned}
&\beta_i\eta_i^m + \beta_j\eta_j^m \\
&= p(w)^{m+1}\left[\frac{1}{(w-\theta_i)^{m+1}(\theta_i-\theta_j)g(\theta_i)} + \frac{1}{(w-\theta_j)^{m+1}(\theta_j-\theta_i)g(\theta_j)}\right] \\
&= \frac{p(w)^{m+1}}{(w-\theta_i)^{m+1}}\left[\frac{(w-\theta_j)^{m+1}g(\theta_j) - (w-\theta_i)^{m+1}g(\theta_i)}{(\theta_i-\theta_j)g(\theta_i)g(\theta_j)(w-\theta_j)^{m+1}}\right].
\end{aligned}$$

Defining

$$G(z) = (w - z)^{m+1}g(z)$$

we have

$$\begin{aligned}
&\lim_{\theta_j\to\theta_i} \frac{(w-\theta_j)^{m+1}g(\theta_j) - (w-\theta_i)^{m+1}g(\theta_i)}{(\theta_i-\theta_j)} \\
&= \lim_{\theta_j\to\theta_i} \frac{G(\theta_j) - G(\theta_i)}{(\theta_i-\theta_j)} \\
&= -G'(\theta_i) \\
&= (m+1)(w-\theta_i)^m g(\theta_i) - (w-\theta_i)^{m+1}g'(\theta_i).
\end{aligned}$$

Thus,

$$\begin{aligned}
&\lim_{\theta_j\to\theta_i} \beta_i\eta_i^m + \beta_j\eta_j^m \\
&= \frac{p(w)^{m+1}}{(w-\theta_i)^{m+1}}\left[\frac{(m+1)(w-\theta_i)^m g(\theta_i) - (w-\theta_i)^{m+1}g'(\theta_i)}{g(\theta_i)^2(w-\theta_i)^{m+1}}\right] \\
&= \eta_i^{m+1}\left[\frac{m}{g(\theta_i)(w-\theta_i)} + \frac{1}{g(\theta_i)(a-\theta_i)} - \frac{g'(\theta_i)}{g(\theta_i)^2}\right] \\
&= \eta_i^m\left[\frac{\eta_i}{g(\theta_i)(w-\theta_i)}m + \frac{\eta_i}{g(\theta_i)(w-\theta_i)} - \frac{\eta_i g'(\theta_i)}{g(\theta_i)^2}\right].
\end{aligned}$$

Note that

$$g(\theta_i) = \frac{p''(\theta_i)}{2} \neq 0.$$

Hence we conclude that the coefficient polynomial $\alpha_i(z)$ is not zero. More specifically, $\alpha_i(z)$ is a linear polynomial in z. Thus $\alpha_i(m)$ is equal to zero for at most one value of m. □

The proof of the general case, as can be imagined from this proof, is more complicated. Moreover, the closed form of the actual coefficients, $\alpha_i(z)$, would be more complex.

9.7 The Basic Sequence and the Bernoulli Method

Here we make a connection between the Basic Sequence and a classic method known as the *Bernoulli iteration* or the Bernoulli method due to Daniel Bernoulli, see Hildebrand (1974), Householder (1970).

Consider again a complex polynomial $p(z) = p_n z^n + \cdots + p_1 z + p_0$ of degree n, with $p_0 \neq 0$, having t distinct roots $R_p = \{\theta_1, \ldots, \theta_t\}$, of multiplicity $n_1, \ldots, n_t$, respectively. We can consider $p(z)$ as the characteristic polynomial of the recurrence relation

$$p_n u_m + p_{n-1} u_{m-1} + \cdots + p_0 u_{m-n} = 0. \tag{9.32}$$

For any set of initial conditions we get

$$u_m = \beta_1(m)\theta_1 + \cdots + \beta_t(m)\theta_t, \tag{9.33}$$

where for each $i = 1, \ldots, t$, $\beta_i(z)$ is either identically zero or a polynomial of degree $\leq n_i - 1$. It thus follows, as in the proof of Proposition 9.2, that if the coefficient polynomial corresponding to the dominant root is nonzero, then the ratio u_m/u_{m-1} converges to the dominant root of $p(z)$. Different schemes for selecting the initial conditions so that under the assumption of simplicity of the roots convergence of ratio u_m/u_{m-1} to the dominant root is guaranteed, are discussed in several numerical analysis books, see e.g. Hildebrand (1974).

In order to construct sequences convergent to the least modulus root, one can consider the polynomial

$$g(y) = p_0 y^n + p_1 y^{n-1} + \cdots + p_n. \tag{9.34}$$

Clearly, θ is a root of $p(z)$ if and only if $1/\theta$ is a root of $g(y)$. Thus if the sequence $\{v_m\}$ is defined via the homogeneous recurrence relation

$$p_0 v_m + p_1 v_{m-1} + \cdots + p_n v_{m-n} = 0, \tag{9.35}$$

then, under some regularity conditions, the ratio v_{m-1}/v_m converges to the least root of $p(z)$. The approximation of the extreme roots via the ratios u_m/u_{m-1} and v_{m-1}/v_m is what is known as the Bernoulli method.

In what follows we will show that through the use of Taylor's theorem the notion of least modulus and largest modulus is only relative to the input with respect to which the polynomial $p(z)$ is being represented. Thus we may define the ratios u_m/u_{m-1} and v_{m-1}/v_m in more generality and produce convergent sequences to any root. Finally, we establish a relationship between the latter generalization and the Basic Sequence. This results in a deeper understanding of the Bernoulli method as well as the existence of a precise error formula for its iterates.

Let $a \in \mathbb{C}$ be any complex number that is not a root of $p(z)$, and θ a root of $p(z)$. From Taylor's theorem we have

$$\sum_{i=0}^{n} \frac{p^{(n-i)}(a)}{(n-i)!}(\theta - a)^{n-i} = 0 = \sum_{i=0}^{n} \frac{p^{(i)}(a)}{i!}(\theta - a)^{i}. \tag{9.36}$$

Letting $z = \theta - a$, from the left-hand side of (9.36) we get

$$\sum_{i=0}^{n} \frac{p^{(n-i)}(a)}{(n-i)!} z^{n-i} = 0, \tag{9.37}$$

which can be viewed as the characteristic equation of the homogeneous recurrence relation

$$\sum_{i=0}^{n} \frac{p^{(n-i)}(a)}{(n-i)!} u_{m-i}(a) = 0. \tag{9.38}$$

Also by dividing the right-hand side of (9.36) by $(\theta - a)^n$ and letting $w = 1/(\theta - a)$, we get

$$\sum_{i=0}^{n} \frac{p^{(i)}(a)}{i!} w^{n-i} = 0. \tag{9.39}$$

The above can be viewed as the characteristic equation of the homogeneous recurrence relation

$$\sum_{i=0}^{n} \frac{p^{(i)}(a)}{i!} v_{m-i}(a) = 0. \tag{9.40}$$

Definition 9.6. Let the Bernoulli Sequence associated with the farthest root from a be defined as

$$\alpha_m(a) = a + \frac{u_{m-1}(a)}{u_{m-2}(a)},$$

and the Bernoulli Sequence associated with the closet root to a be defined as

$$\beta_m(a) = a + p(a)\frac{v_{m-2}(a)}{v_{m-1}(a)}.$$

In particular, if $a = 0$ the corresponding Bernoulli Sequence coincides with the usual iterates of the Bernoulli method.

Theorem 9.6. *Given $p(z)$, let a be any complex number not equidistant to two roots of $p(z)$. Consider the Bernoulli Sequences with the initial conditions $u_1(a) = 1$, $u_j(a) = 0$, for all $j < 0$; and $v_1(a) = 1$, $v_j(a) = 0$, for all $j < 0$. We have*

(1)

$$\lim_{m\to\infty} \alpha_m(a) = \theta, \qquad \lim_{m\to\infty} \beta_m(a) = \theta',$$

where θ and θ' are roots of $p(z)$.

(2) If all the roots of $p(z)$ are distinct, then θ is the farthest root from a, and θ' the nearest root to a.

(3)

$$v_m(a) = (-1)^m \frac{d_m(a)}{p(a)^m}.$$

(4)

$$\beta_m(a) = b_m(a) = B_m(a),$$

(i.e. the Bernoulli Sequence associated with the nearest root to a coincides with the Basic Sequence associated with a, and hence the pointwise evaluation of the Basic Family at a).

Proof. Define

$$\overline{Q}_a(z) = \sum_{i=0}^{n} \frac{p^{(i)}(a)}{i!} z^{n-i}, \qquad \widehat{Q}_a(z) = \sum_{i=0}^{n} \frac{p^{(n-i)}(a)}{(n-i)!} z^{n-i}. \tag{9.41}$$

The set of distinct roots of $\overline{Q}_a(z)$ and $\widehat{Q}_a(z)$ are, respectively,

$$\overline{R} = \{\overline{\eta}_1 = \theta_1 - a, \ldots, \overline{\eta}_t = \theta_t - a\}, \quad \widehat{R} = \{\widehat{\eta}_1 = \frac{1}{\theta_1 - a}, \ldots, \widehat{\eta}_t = \frac{1}{\theta_t - a}\}. \tag{9.42}$$

If a is not equidistant two roots of $p(z)$, $|\overline{\eta}_i| \neq |\overline{\eta}_j|$, for $i \neq j$. From Proposition 9.2 it follows that $\alpha_m(a)$ converges to a root of $p(z)$. Similarly, it follows that $\beta_m(a)$ converges to a root of $p(z)$. From Proposition 9.2 it also follows that if all the roots are simple, then $\alpha_m(a)$ converges to the farthest root from a, and $\beta_m(a)$ converges to the nearest root to a.

Next we establish the equivalence of $\beta_m(a)$, and $b_m(a)$. First we prove by induction that

$$v_m(a) = (-1)^m \frac{d_m(a)}{p(a)^m}.$$

The above is true for $m = 0$. Assume that it is true for all $k = 0, \ldots, m-1$. Substituting $v_k(a) = (-1)^k d_k(a)/p(a)^k$ for $k = 0, \ldots, m-1$ in (9.40), we get

$$p(a)v_m(a) = -\sum_{i=1}^{n} \frac{p^{(i)}(a)}{i!} (-1)^{m-i} \frac{d_{m-i}(a)}{p(a)^{m-i}}. \tag{9.43}$$

Multiplying the above by $p(a)^{m-1}$ and since $(-1)^{-i+1} = (-1)^{i-1}$ we get

$$p(a)^m v_m(a) = (-1)^m \sum_{i=1}^{n} (-1)^{i-1} \frac{p^{(i)}(a)}{i!} d_{m-i}(a) p(a)^{i-1}. \tag{9.44}$$

But from (9.23) the right-hand side of the above coincides with $(-1)^m d_m(a)$. From this it easily follows that $\beta_m(a) = b_m(a)$. The equivalence of $\beta_m(a)$ and $B_m(a)$ was already established in Theorem 9.5. □

9.8 Determinantal Representation of Fundamental Solution

In this section we consider the general homogeneous linear recurrence relation with its Basic Initial conditions, i.e.,

$$a_m = c_1 a_{m-1} + c_2 a_{m-2} + \cdots + c_n a_{m-n}. \tag{9.45}$$

Theorem 9.7. *Let $\{a_m\}$ be the Fundamental Solution to HLRR. Then for each $m \geq 1$ we have*

$$a_m = \det \begin{pmatrix} c_1 & c_2 & \dots & c_{m-1} & c_m \\ -1 & c_1 & \ddots & \ddots & c_{m-1} \\ 0 & -1 & \ddots & \ddots & \vdots \\ \vdots & \vdots & \ddots & \ddots & c_2 \\ 0 & 0 & \dots & -1 & c_1 \end{pmatrix}$$

$$= \det \begin{pmatrix} c_1 & -c_2 & \dots & (-1)^{m-2} c_{m-1} & (-1)^{m-1} c_m \\ 1 & c_1 & \ddots & \ddots & (-1)^{m-2} c_{m-1} \\ 0 & 1 & \ddots & \ddots & \vdots \\ \vdots & \vdots & \ddots & \ddots & -c_2 \\ 0 & 0 & \dots & 1 & c_1 \end{pmatrix}$$

where

$$c_i = 0, \quad \forall \quad i > n.$$

Proof. Consider

$$p(z) = c_n z^n + \cdots + c_1 z - 1,$$

the negative reciprocal corresponding to HLRR. We have

$$c_i = (-1)^{i-1} p^{i-1}(0) \frac{p^{(i)}(0)}{i!} = (-1)^{i-1} (-1)^{i-1} \frac{p^{(i)}(0)}{i!}, \qquad i = 1, \dots, n. \tag{9.46}$$

From this and Theorem 9.5 it follows that a_m coincides with the quantity

$$D_m(0) = \sum_{i=1}^{n} (-1)^{i-1} p^{i-1}(z) \frac{p^{(i)}(z)}{i!}$$

corresponding to this polynomial. Hence the first claimed determinantal representation of a_m.

Next we can take

$$p(z) = p_n z^n + \cdots + p_1 z + 1,$$

where

$$p_i = (-1)^{i-1} c_i, \quad i = 1, \ldots, n.$$

Then

$$c_i = (-1)^{i-1} p^{i-1}(0) \frac{p^{(i)}(0)}{i!} = (-1)^{i-1} \frac{p^{(i)}(0)}{i!}.$$

Again from this, and Theorem 9.5 and the determinantal representation of a_m the second claimed determinantal representation follows. □

Theorem 9.8. *For each $m \geq 1$, we have*

$$|a_m| \leq \left(1 + \sum_{i=1}^{n} |c_i|\right)^{m/2}.$$

Proof. From Hadamard inequality (see e.g. Marcus and Minc (1964)), given an $m \times m$ matrix $A = (a_{ij})$, we have

$$\det(A) \leq \left(\prod_{i=1}^{m} \sum_{i=1}^{m} |a_{ij}|^2\right)^{1/2}.$$

But this implies the claimed bound. □

9.9 Application to Fibonacci Sequence and Generalizations

Let us consider again the Fibonacci sequence $F_m = F_{m-1} + F_{m-2}$, $F_0 = 1$, $F_{-1} = 0$. The characteristic polynomial and its negative reciprocal are

$$q(z) = z^2 - z - 1 = p(z) = z^2 + z - 1.$$

The roots are $\frac{1}{2}(-1 \pm \sqrt{5})$. Thus the closer root to the origin is $\theta = (\sqrt{5} - 1)/2$.

From the expansion formula for $B_m(z)$ as applied to $p(z) = z^2 - z - 1$ at $z = 0$ we get:

$$B_m(0) - \theta = \frac{D_{m-2}}{D_{m-1}} = (-1^m)\frac{\widehat{D}_{m-1,m}}{D_{m-1}}\theta^m.$$

Here

$$D_m = \det\begin{pmatrix} 1 & 1 & \dots & 0 & 0 \\ -1 & 1 & \dots & 0 & 0 \\ 0 & -1 & \dots & \vdots & \vdots \\ \vdots & \vdots & \ddots & 1 & 1 \\ 0 & 0 & \dots & -1 & 1 \end{pmatrix}, \quad \widehat{D}_{m,m-1}(z) = \det\begin{pmatrix} 1 & 0 & \dots & 0 & 0 \\ 1 & 1 & \dots & 0 & 0 \\ -1 & 1 & \ddots & \vdots & \vdots \\ \vdots & \vdots & \ddots & 1 & 0 \\ 0 & 0 & \dots & 1 & 1 \end{pmatrix} = 1.$$

Thus we get

Theorem 9.9.

$$\frac{F_{m-2}}{F_{m-1}} - \theta = (-1)^m \frac{1}{F_{m-1}}\theta^m. \tag{9.47}$$

□

We state the following corollary which bounds the convergence of the ratio to θ:

Corollary 9.4.

$$|\frac{F_{m-2}}{F_{m-1}} - \theta| \leq \theta^{2m-2}.$$

Proof. From an easy induction (see e.g. Knuth (1973), p. 84) we have the lower bound

$$F_m \geq 1/|\phi^{m-1}|,$$

where $\phi = 1/\theta = (\sqrt{5}+1)/2$ is the Golden ratio. Substituting into the formula in Theorem 9.9, we get the desired bound. □

Theorem 9.9 also proves the oscillating behavior of convergence of the ratio of Fibonacci numbers F_m/F_{m-1} to the golden ratio. Replacing m with $m+1$, Theorem 9.9 gives:

$$\frac{F_{m-1}}{F_m} - \theta = (-1)^{m+1}\frac{1}{F_m}\theta^{m+1}. \tag{9.48}$$

Subtracting (9.48) from (9.47) and since

$$F_m = \frac{(\phi^{m+1} - \hat{\phi}^{m+1})}{\sqrt{5}}, \tag{9.49}$$

where $\hat{\phi} = (-\sqrt{5}+1)/2 = -\theta$, we get

$$\frac{F_{m-2}}{F_{m-1}} - \frac{F_{m-1}}{F_m} = (-1)^m \theta^m \Big(\frac{1}{F_{m-1}} + \frac{1}{F_m}\theta\Big) = \frac{(-1)^m}{F_{m-1}F_m} \frac{1}{\phi^{m+1}} \big(\phi F_m + F_{m-1}\big).$$

It is easy to show that

$$\frac{1}{\phi^{m+1}} \big(\phi F_m + F_{m-1}\big) = 1.$$

Substituting this in the previous equation implies

$$F_m F_{m-2} - F_{m-1}^2 = (-1)^m \tag{9.50}$$

which is a well-known identity going back to J.D. Cassini, see [Knuth (1973)] and the book of Koshy (2001).

Consider the generalized Fibonacci numbers of order n (see Miles (1960))

$$F_m^{(n)} = F_{m-1}^{(n)} + F_{m-2}^{(n)} + \cdots + F_{m-n}^{(n)},$$

with their Basic Initial Conditions

$$F_0^{(n)} = 1, \quad F_j^{(n)} = 0, \quad j = -1, \ldots, -n+1.$$

Applying Theorem 9.8 to the determinantal formula for $F_m^{(n)}$ gives:

Corollary 9.5.

$$F_m^{(n)} \leq (n+1)^{m/2}.$$

□

When $m = 2$, a well-known bound on the ordinary Fibonacci numbers is:

$$F_m \leq [(1+\sqrt{5})/2]^m \approx 1.618^m.$$

Our bound gives:

$$F_m \leq (\sqrt{3})^m \approx 1.732^m$$

which is comparable.

The determinantal representation of the ordinary Fibonacci numbers (i.e. $n = 2$) appears as an exercise in Knuth (1973) (see p. 84). However, we are not aware of any other determinantal representation of the generalized Fibonacci numbers.

9.10 Experimental Results Via Polynomiography

In this section we present some preliminary experimentation with polynomiography of which depicts the convergence behavior of the Basic Family $B_m(z)$, as applied to a given polynomial, for a fixed m. In subsequent chapters we will exhibit many more polynomiographies with respect to the convergence behavior of the Basic Sequence $\{B_m(a)\}_{m=2}^{\infty}$, hence visualizations associated with a homogeneous recurrence relations.

Consider a polynomial $p(z)$ and a fixed natural number $m \geq 2$. As already thoroughly explained in earlier chapters, the basins of attraction of a root of $p(z)$ with respect to the iteration function $B_m(z)$ are regions in the complex plane such that given an initial point within them the corresponding sequence $a_{k+1} = B_m(a_k)$ will converge to that root. The fractal nature and the images of the basins of attractions of Newton's method are now quite familiar for some special polynomials such as $p(z) = z^n - 1$. However, this is not the case as far as the behavior of other members of the Basic Family is concerned.

In Figure 9.10, first row, we consider a polynomial with a random set of roots, depicted as dots. The figures show the evolution of the basins of attraction of the roots to the Voronoi regions as m takes the values $2, 4, 10$, and 50. In the remaining rows the polynomiographs demonstrate the growth of the basin of attractions for the polynomials $p(z) = z^3 - 1$, $P(z) = z^4 - 1$, and $p(z) = z^9 - 1$; corresponding to different values of m. Note that in all the images as m increases the regions with chaotic behavior dramatically decrease to merely the boundaries of the Voronoi regions. Note also that, as predicted by the theory, for each root the basins of attractions converge to the Voronoi region of that root. The roots of $p(z) = z^n - 1$ are the roots of unity and hence the Voronoi regions are completely symmetric. In all the figures in the case of $m = 2$, i.e. Newton's method, the basins of attractions are chaotic. However, these regions rapidly improve by increasing m.

9.11 A Representation Theorems for Arbitrary Solutions

Here we use the Fundamental Solution of HLRR and also linear programming to represent solutions of HLRR with arbitrary set of initial conditions. We prove

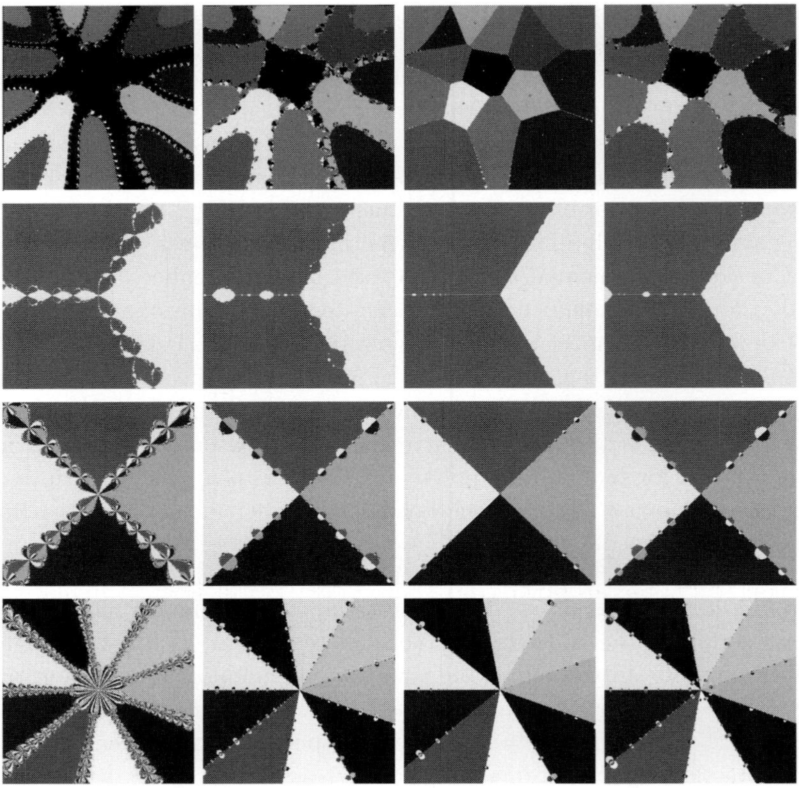

Fig. 9.1 Evolution of basins of attraction to Voronoi regions via polynomiographs of $B_m(z)$: random points, $m = 2, 4, 10, 50$ (first row); corresponding to $p(z) = z^3 - 1$, $m = 2, 3, 10, 50$ (second row); corresponding to $p(z) = z^4 - 1$, $m = 2, 3, 4, 50$ (third row); $p(z) = z^9 - 1$, $m = 2, 3, 10, 50$.

Theorem 9.10 (Representation Theorem for Solution HLRR). *Let $\{b_m\}$ be a sequence satisfying the recurrence relation (9.1) but with an arbitrary given set of initial conditions $b_0, b_1, \ldots, b_{n-1}$. Then for all $m \geq 0$*

$$b_m = \alpha_{n-1}a_m + \alpha_{n-2}a_{m-1} + \cdots + \alpha_0 a_{m-n+1}, \tag{9.51}$$

where α_i satisfies the matrix equation $A\alpha = b$, where

$$A = \begin{pmatrix} a_0 & 0 & \ldots & 0 \\ a_1 & a_0 & \ldots & 0 \\ \vdots & \vdots & \ddots & \vdots \\ a_{n-1} & a_{n-2} & \ldots & a_0 \end{pmatrix}, \quad \alpha = \begin{pmatrix} \alpha_{n-1} \\ \alpha_{n-2} \\ \vdots \\ \alpha_0 \end{pmatrix}, \quad b = \begin{pmatrix} b_0 \\ b_1 \\ \vdots \\ b_{n-1} \end{pmatrix}.$$

Conversely, if $b_m = \alpha_{n-1}a_m + \alpha_{n-2}a_{m-1} + \cdots + \alpha_0 a_{m-n+1}$ *for a given set of* α_i*, then* b_m *satisfies the recurrence relation (9.1).*

Proof. First we note that since $a_0 \neq 0$, A is invertible and hence the matrix equation has a solution in α. Thus (9.51) is true for $m = 0, \ldots, n-1$. To prove that it is true for $m = n$, let $c = (c_n, c_{n-1}, \cdots, c_1)^T$. Then we have:

$$c^T b = \sum_{i=1}^{n} c_{n-i+1} b_{i-1} = b_n, \quad c^T A = (a_n, a_{n-1}, \cdots, a_0)^T.$$

Thus,

$$b_n = c^T b = c^T A\alpha = \alpha_{n-1}a_n + \alpha_{n-2}1a_{n-1} + \cdots + \alpha_0 a_1.$$

Hence (9.51) is true for $m = n$. Using similar approach and induction proves (9.51) is true for general m. The proof of the converse is trivial since once the vector α is fixed, b is then defined as $A\alpha$. □

According to the representation theorem, any solution of HLRR with an arbitrary set of initial conditions can be written as the linear combination of the fundamental solution where the coefficients can be explicitly computed. In particular if we are given the representations of a_m in terms of the characteristic roots we can also compute the representation of b_m in terms of these roots.

Example 9.4. As an example of applications of the representation theorem consider Lucas numbers, $L_{m+2} = L_{m+1} + L_m$, with $L_0 = 2$, $L_1 = 1$. It is easy to show that $L_m = 2F_m - F_{m-1}$.

Next we prove a more general representation theorem. For a given fixed index $t \geq n$ set

$$A^t = \begin{pmatrix} a_{t-n} & \cdots & a_1 & a_0 & 0 & \ldots & 0 \\ a_{t-n+1} & \cdots & a_2 & a_1 & a_0 & \ldots & 0 \\ \vdots & \cdots & \vdots & \vdots & \vdots & \ddots & \vdots \\ a_{t-1} & \cdots & a_n & a_{n-1} & a_{n-2} & \ldots & a_0 \end{pmatrix}, \tag{9.52}$$

$$\beta^t = \begin{pmatrix} \beta_{t-1} \\ \beta_{t-2} \\ \vdots \\ \beta_0 \end{pmatrix}, \quad b^k = \begin{pmatrix} b_k \\ b_{k+1} \\ \vdots \\ b_{n+k-1} \end{pmatrix}. \tag{9.53}$$

When $t = n$ and $k = 0$ this coincides with the previous system, $A\alpha = b$. Next we would like to define a polyhedral set, over the real inputs or complex ones.

Definition 9.7. Let A be an $n \times t$ matrix with complex entries and rank n, and let $b \in \mathbb{C}^n$ be given. We define the *complex polyhedron* associated with A, b to be the set:

$$\Omega = \{\beta \in \mathbb{C}^t : A\beta = b, \beta \geq 0\},$$

where by the inequality $\beta \geq 0$ we mean that both the real and imaginary parts of each component of β is nonnegative. An invertible $n \times n$ submatrix B of A defines a *Basic Solution* determined by setting the variables not corresponding to those of B, say β_N (called non-basic variable), equal to zero and by solving for the remaining variables, say β^t_B (called basic variables), through the matrix equation $B\beta_B = b$. The corresponding solution for β is said to be a Basic Solution. If $\beta \geq 0$, we call this solution a Basic Feasible Solution.

The above definition gives generalization of the case of real polyhedrons. Now consider the complex polyhedron associated with A^t, b^k defined in (9.52), and (9.53):

$$P_{t,k} = \{\beta^t \in \mathbb{C}^t : A^t\beta^t = b^k, \quad \beta^t \geq 0\}.$$

Theorem 9.11 (General Representation Theorem for HLRR). *Let $\{b_m\}$ be a sequence which satisfies the recurrence relation (9.1) but with any arbitrary given set of initial conditions $b_0, b_1, \ldots, b_{n-1}$. Consider $P_{t,k}$ for any given fixed indices $t \geq n$ and $k \geq 0$. Then $P_{t,k}$ has at least one Basic Solution. Moreover any Basic Solution gives rise to a representation of b_m in terms of $a_0, \ldots, a_{t-1}$. More precisely, if the index set of the Basic variables is $\{t_0, t_1, \cdots, t_{n-1}\}$ with $t_{i-1} < t_i$, then for all $m \geq k$ we have*

$$b_m = \sum_{i=1}^{n} \beta_{t_{n-i}} a_{m-n-k+1+t_{n-i}}. \tag{9.54}$$

Proof. When $k = 0$, and $t = n$, then $t_{n-i} = n-i$, for all $i = 0, 1, \ldots, n-1$. This implies the claimed representation is

$$b_m = \sum_{i=1}^{n} \beta_{n-i} a_{m+1-i}. \tag{9.55}$$

But this coincides with that in previous theorem. The proof of the general case is analogous to this special case and is omitted. □

An *extreme point* (or *vertex*) w of $P_{k,t}$ is a feasible point that cannot be written as a convex combination of two other points in $P_{k,t}$. Thus if $\alpha u + (1-\alpha)v = w$ where u, y, w lie in $P_{k,t}$, $\alpha \in [0,1]$, then α is either 0 or 1.

The following theorem is well-known in the case of real polyhedrons. The proof below extends it to the case of general complex polyhedron Ω, and hence to $P_{k,t}$.

Theorem 9.12. *Consider the complex polyhedron Ω. A feasible point of Ω is an extreme point (also called vertex) if and only if it is a Basic Feasible Solution.*

Proof. Suppose that $\beta = (\beta_1, \ldots, \beta_t)$ is a BFS of Ω, and $\beta = \alpha u + (1-\alpha)v$, $\alpha \in [0,1]$, $u, v \in \Omega$. If β_i is a nonbasic component of β then $\beta_i = 0 = \alpha u_i + (1-\alpha)v_i$. But since $u_i, v_i \geq 0$, we have $u_i = v_i = 0$. If B is the $n \times n$ matrix corresponding to β, after some permutation of the columns of A we can write

$$\beta = \begin{pmatrix} \beta_B \\ \beta_N \end{pmatrix}, \quad u = \begin{pmatrix} u_B \\ u_N \end{pmatrix}, \quad v = \begin{pmatrix} v_B \\ v_N \end{pmatrix}$$

where $\beta_N = u_N = v_N = 0$. But then $\beta_B = u_B = v_B$ since these are all solutions to $Bw = b$. Hence, $\beta = u = v$.

Conversely, suppose that β is an extreme point. Consider the set of columns of A for which $p_i \neq 0$. We claim that the columns are linearly independent. This then implies that it has n or fewer number of elements. If the number is exactly n we are done since we let B be the submatrix of A consisting of these n columns. If the number is less than n, then using the fact that $rank(A) = n$, we can extend these columns to n linearly independent ones, hence obtaining an invertible submatrix B. Thus it suffices to prove the claim. If the claim is false, there exists $w \in \mathbb{C}^t$ such that $Aw = 0$, $w \neq 0$, and $w_i = 0$ whenever $\beta_i = 0$. Now consider $\beta^1 = \beta + \alpha w$ and $\beta^2 = \beta - \alpha w$, where α is a real number. Then there must exist $\alpha > 0$ such that both β^1 and β^2 are feasible points of Ω. But

$$\beta = \frac{1}{2}(\beta^1 + \beta^2),$$

a contradiction. □

In the case of real inputs the above theorem allows the use of linear programming techniques and simplex method to search for Basic solutions of a particular type. For general complex input it also suggests the use of extension of such algorithms. While there are known extensions of linear programming duality theorems over the complex numbers, e.g. Ben-Israel Ben-Israel (1969), we are not aware of the extension of linear programming algorithms, say the simplex method. The study of homogeneous linear recurrence relations suggests new usage of linear programming algorithms over the complex numbers.

9.12 Applications to Fibonacci and Lucas Numbers

Consider the polyhedron $P_{3,k}$ corresponding to the Fibonacci sequence. We call this the *Fibonacci polytope*. The corresponding system is

$$\begin{pmatrix} F_2 & F_1 & F_0 & F_{-1} \\ F_3 & F_2 & F_1 & F_0 \end{pmatrix} \begin{pmatrix} \beta_3 \\ \beta_2 \\ \beta_1 \\ \beta_0 \end{pmatrix} = \begin{pmatrix} L_k \\ L_{k+1} \end{pmatrix}.$$

For $k = 0$ we get the system of equations:

$$\begin{cases} 2\alpha_3 + \alpha_2 + \alpha_1 = 2 \\ 3\alpha_3 + 2\alpha_2 + \alpha_1 + \alpha_0 = 1 \end{cases}.$$

There are 6 possible Basic Solutions corresponding to the pairs of variables (α_i, α_j), $i < j$. Solving the corresponding systems we get:

$$\begin{array}{lll} (\alpha_0, \alpha_1) = (-1, 2), & (\alpha_0, \alpha_2) = (-3, 2), & (\alpha_0, \alpha_3) = (-2, 1), \\ (\alpha_1, \alpha_2) = (3, -1), & (\alpha_1, \alpha_3) = (4, -1), & (\alpha_2, \alpha_3) = (-4, 3). \end{array}$$

These lead to the respective representations, all valid for all $n \geq 0$:

$$\begin{array}{lll} L_n = 2F_n - F_{n-1}, & L_n = 2F_{n+1} - 3F_{n-1}, & L_n = F_{n+2} - 2F_{n-1}, \\ L_n = -F_{n+1} + 3F_n, & L_n = -F_{n+2} + 4F_n, & L_n = 3F_{n+2} - 4F_{n+1}. \end{array}$$

We note that none of the Basic Solutions is a Basic Feasible Solution of $P_{3,0}$. And this can be proved for $P_{m,0}$ for any $m \geq 1$. More specifically for any pair of indices (i, j), $i < j$, the corresponding Basic Solution when solved for α_i via Cramer's rule gives $\alpha_i = u/v$ where u and v are determinants which are of opposing signs. This suggests if we want to have positive coefficients in the representation we should use the next polytope, $P_{3,1}$. This amounts to replacing the right-hand side with the vector $(L_1, L_2)^T = (1, 3)^T$ and recomputing. This gives the following Basic Solutions:

$$\begin{array}{lll} (\alpha_0, \alpha_1) = (2, 1), & (\alpha_0, \alpha_2) = (1, 1), & (\alpha_0, \alpha_3) = (\frac{1}{2}, \frac{3}{2}), \\ (\alpha_1, \alpha_2) = (-1, 2), & (\alpha_1, \alpha_3) = (-3, 2), & (\alpha_2, \alpha_3) = (3, -1). \end{array}$$

These lead to the respective representations, all valid for $n \geq 1$:

$$L_n = F_{n-1} + 2F_{n-2}, \quad L_n = F_n + F_{n-2}, \quad L_n = \tfrac{3}{2}F_{n+1} + \tfrac{1}{2}F_{n-2},$$
$$L_n = 2F_n - F_{n-1}, \quad L_n = 2F_{n+1} - 3F_{n-1}, \quad L_n = -F_{n+1} + 3F_n.$$

Consider the polytope $P_{t,k}$ with respect to Generalized Fibonacci of order n, where $b^{(k)}$ is a nonnegative integral, or even with zero-one elements. We call this the *Generalized Fibonacci polytope*. For instance, for $n = 4$, i.e.

$$F_m^{(4)} = F_{m-1}^{(4)} + F_{m-2}^{(4)} + F_{m-3}^{(4)} + F_{m-4}^{(4)},$$

where $t = 7$, $k = 4$. Suppressing the superscript we get

$$\begin{pmatrix} F_3 & F_2 & F_1 & F_0 & 0 & 0 & 0 \\ F_4 & F_3 & F_2 & F_1 & F_0 & 0 & 0 \\ F_5 & F_4 & F_3 & F_2 & F_1 & F_0 & 0 \\ F_6 & F_5 & F_4 & F_3 & F_2 & F_1 & F_0 \end{pmatrix} \begin{pmatrix} \beta_6 \\ \beta_5 \\ \beta_4 \\ \beta_3 \\ \beta_2 \\ \beta_1 \\ \beta_0 \end{pmatrix} = \begin{pmatrix} b_k \\ b_{k+1} \\ b_{k+2} \\ b_{k+3} \end{pmatrix}$$

and where we assume $\beta_i \geq 0$. Using this polyhedral representation a number of interesting problems arise, such as: generalizations of Lucas numbers and more general identities on generalized Fibonacci and Lucas numbers; the study of the properties of this polytope; the study of its integral vertices; more specifically zero-one vertices, as well as their computations.

9.13 Concluding Remarks

In this chapter we have formally shown that given a complex polynomial $p(z)$, to each complex number w one can associate the Basic Sequence, $\{B_m(w) = w - p(w)D_{m-2}(w)/D_{m-1}(w)\}$, where $D_m(w)$, dependent only on the normalized derivatives at w, is definable via a homogeneous linear recurrence relation or a determinant. Except possibly for the boundary points of the Voronoi region of roots, for any other input w the corresponding Basic Sequence converges to the root of $p(z)$ closest to w.

The evaluation of the Basic Sequence provides a very simple scheme for the approximation of a root of $p(z)$. As an example given the complex polynomial $p(z) = 1 + p_1 z + \cdots + p_n z^n$, and $w = 0$, the corresponding Basic Sequence $\{B_m(0)\}$ in many cases provides a very simple method for the approximation of a root of $p(z)$. Even if the sequence $\{D_m(0)\}$ has an infinite subsequence of zeros so that the Basic Sequence is not well-defined for all m, the Basic Sequence could still have subsequences that converge to a root of $p(z)$. For instance, consider $p(z) = (z^2 + 2)(z - \theta)$, say for $\theta = 1$ and $\theta = 4$. For $\theta = 1$ the Basic Sequence $\{B_m(0)\}$ converges to θ even though the origin is equidistant to the complex roots. Although the root $\theta = 4$ is not the nearest root to the origin, the Basic Sequence at $w = 0$ still has subsequences convergent to θ. If we now set $p(z)$ to be $p(z) = (z^2 + 2)(z - \theta)(z - \beta)^2$, where β is a real number, again at $w = 0$ we observe the convergence of its corresponding Basic Sequence, or a subsequence of it, to a root of $p(z)$.

The Basic Sequence can be efficiently computed. Its m-th term can be evaluated in $O(n \log n \log m)$ arithmetic operations. For small m it can be computed in $O(mn)$ operations. For each m, $D_m(w)$ can also be represented as a special Toeplitz determinant. When the given input w is sufficiently close to a root, for any $m \geq 2$ the Basic Sequence can be turned into an iterative method of order m when the root is simple. This is done by replacing w with $B_m(w)$, and repeating. There are many computational implication of these results. As an example one possible algorithm for polynomial root-finding is the following: for a given input w continue computing $B_m(w)$ until the gap between $B_m(w)$ and $B_{m+1}(w)$ is sufficiently small. Then for a desirable m replace w with $B_m(w)$ and repeat in order to obtain higher and higher accuracy. It is also possible to switch back and forth between the two schemes.

Another approach, suggested by the properties of the Basic Sequence, is a procedure to compute all the roots of a given polynomial: first obtain a rectangle that contains all the roots (see Chapter 15). Then by selecting a sparse number of gird point we can generate the corresponding Basic Sequence and thereby obtain good approximation to a subset of the roots. Then after deflation the same approach can be applied to find additional roots. This method could provide an excellent alternative, or one complementary to the method of Weyl (1924) (see also Pan (1997)).

In this chapter we have also shown how to estimate the error $|B_m(w) - \theta|$, where θ is a root. The Basic Sequence can even be defined for functions that are not polynomial. In Kalantari (2000b) we have used the pointwise evaluation of the Basic Family (the Basic Sequence in the terminology of this chapter) to give new formulas for approximation of π and e (see Chapter 14). Although the definition of the Basic Sequence is motivated by the properties of the Basic Family of iteration functions, the results have implied deeper properties of the Basic Family. Furthermore, the relationship between the Basic Sequence and the Basic Family has motivated new results on general homogeneous linear recurrence relations that are defined via a single nonzero initial condition.

Finally, we have shown the equivalence of the Basic Sequence and the Bernoulli method, once defined at an arbitrary point w. This connection gives a deeper understanding of the Bernoulli method. In particular, using the determinantal generalization of Taylor's theorem one can actually estimate the error in approximation of the extreme roots via the Bernoulli method. In general the results should find many theoretical and practical applications in approximation of roots of polynomials and more general functions.

In polynomiography, other than the use of individual iteration functions, it is possible to generate many new types of visualizations using the Basic Sequence. We will consider this in Chapters 17 and 18. The connections to homogeneous linear recurrence relations, HLRR, also justify polynomiography associated to HLRR, opening yet even newer possibilities in visualizations.

Problem 1. Given a polynomial $p(z) = a_n z^n + \cdots + a_a z + a_0$ and an arbitrary given point w, can we determine if it is not equidistant to two distinct roots of $p(z)$, hence a point in the Voronoi region of one of the zeros? Given an arbitrary complex number w, possibly equidistant to two roots, can we determine if the corresponding Basic Sequence would be convergent to a root?

Problem 2. Consider the polytope $P_{t,k}$ with respect to the Generalized Fibonacci of order n, where $b^{(k)}$ is a nonnegative integral vector, or even one with only zero-one elements (i.e. the Generalized Fibonacci polytope). Classify the integral vertices. Classify the zero-one vertices.

Chapter 10

Generalization of Taylor's Theorem and Newton's Method and Their Applications

The ordinary Taylor's Theorem expresses a given function of real or complex number as a sum of a Taylor polynomial and a remainder term. The two major applications of the Taylor's Theorem are in approximation of the function or its inverse, in particular its zeros. In this chapter we present a very powerful generalization of Taylor's Theorem where the given function is expressed as a certain rational function and a corresponding remainder term. The generalization leads to formulas that play a significant role in the approximation of the function, or its inverse. On the one hand each ordinary Taylor polynomial unfolds into an infinite spectrum of rational approximants. On the other hand the theorem gives rise to an infinite spectrum of rational inverse approximants as well as a fundamental family of single or multipoint iteration functions which includes the Basic Family of iteration functions. The chapter then exhibits a number of distinct applications of the results.

10.1 Introduction

Taylor's Theorem is one of the most basic and fundamental theorems of mathematics. The classical Taylor's Theorem is the most familiar case. However, the formula can also be written with respect to Newton's interpolating polynomial, corresponding to distinct nodes. These are two extreme cases of the general form of Taylor's Theorem, where the nodes are allowed to be "confluent", i.e. divided into distinct groups of identical copies. It is convenient to view the non-confluent case of distinct nodes also as a special case of confluent nodes, so that with one definition we cover all cases. In case each group contains the same number, s, of identical copies, the corresponding interpolation problem is known as Hermite interpolation, if $s = 2$;

and hyperosculatory interpolation, if $s > 2$. In order to write the general case of the interpolating polynomial, if a group contains k copies of a node, the interpolating polynomial must have available to it the values of the first $(k-1)$ derivatives at that node. In all cases Taylor's Theorem represents a given function as an "interpolating polynomial", and a "remainder term".

The two major applications of Taylor's Theorem are in approximation of a given function, or its inverse. The former approximation is the obvious Taylor polynomial. The latter approximation is achieved, trivially, by rewriting the formula so as to get a linear approximation to the inverse. This in particular gives rise to Newton's method, as well as its two-point variant, the secant method. Using Taylor's Theorem, one can also trivially obtain an iteration function with cubic rate of convergence, discovered independently by Euler, Schröder, and Chebyshev. Moreover, by a relatively simple manipulation of Taylor's Theorem one can derive Halley's iteration function, another third order iteration function that with respect to certain criterion is better than Euler-Schröder-Chebyshev's. More generally, Taylor's Theorem together with two recursive formulas gives rise to infinite families of high order iteration functions that includes a family known as the Euler-Schröder family, as well as the Basic Family and its multipoint versions.

In this chapter we describe a family of "determinantal interpolation formulas" which includes Taylor's formula and gives rise to new schemes for approximation of functions, or their inverses. The formulas make use of Taylor's Theorem and through a sequence of recursive iterations, represent a given function f as a rational function in x and f itself, described in terms of certain determinants. While the determinantal formulas and the recursive formula that generates them are relatively simple, their derivation is indeed tedious and in particular requires the proof of several special determinantal identities. The determinantal formulas play a dual role and can approximate f, when x is known; or approximate x, when $f(x)$ is known. On the one hand a single Taylor polynomial unfolds into an infinite set of rational approximations that includes the Taylor polynomial itself. Each such rational approximation is defined with respect to the exact same information as is available to the Taylor polynomial. On the other hand the determinantal formulas give inverse approximation formulas of arbitrary high order, giving rise to a family of single and multipoint iteration functions. Each iteration function in this family is defined as the ratio of two determinants. In case the confluent vector of nodes consists of multiple copies of a single point, the determinants reduce to special Toeplitz matrix determinants. These iteration functions are ideal for finding real or complex roots. Other applications of the determinantal formulas include: a very novel method for polynomial and general root-finding, as well as new approximation formulas for special numbers such as π, e, roots of numbers. Also, we may derive a new rational expansion formula that gives Halley's method, as well as a Padé approximant.

An overview of the results of the chapter can be summarized as follows: The general form of Taylor's Theorem for a function $f : K \to K$, where K is the real line or the complex plane, gives the formula, $f = P_n + R_n$, where P_n is the Newton interpolating polynomial computed with respect to a confluent vector of nodes, and R_n is the remainder. Whenever $f' \neq 0$, for each $m = 2, \ldots, n+1$, we describe a "determinantal interpolation formula", $f = P_{m,n} + R_{m,n}$, where $P_{m,n}$ is a rational function in x and f itself. These formulas play a dual role in the approximation of f or its inverse. For $m = 2$, the formula is Taylor's and for $m = 3$ is a new expansion formula and a Padé approximant. By applying the formulas to P_n, for each $m \geq 2$, $P_{m,m-1}, \ldots, P_{m,m+n-2}$ is a set of n rational approximations that includes P_n, and may provide a better approximation to f, than P_n. Hence each Taylor polynomial unfolds into an infinite spectrum of rational approximations. The formulas also give an infinite spectrum of rational inverse approximations, as well as a fundamental k-point iteration function $B_m^{(k)}$, for each $k \leq m$, defined as the ratio of two determinants that depend on the first $m - k$ derivatives. Application of our formulas have motivated several new results to be covered in sequel chapters: (i) theoretical analysis of the order of $B_m^{(k)}$, $k = 1, \ldots, m$, proving that it ranges from m to the limiting ratio of generalized Fibonacci numbers of order m; (ii) computational results with the first nine members of $B_m^{(k)}$ indicating that they outperform traditional root-finding methods, e.g., Newton's; (iii) polynomial root-finding method requiring only a single input and the evaluation of the sequence of iteration functions $B_m^{(1)}$ at that input; (iv) a new strategy for general root-finding; (v) new formulas for approximation of π, e, other special numbers, and zeros of general analytic functions.

The following sections of the chapter are organized as follows. First, we formally state the general form of Taylor's Theorem with respect to confluent divided differences (Theorem 10.1) and describe some elementary applications. Next, we give preliminary definitions and describe the determinantal formulas. More specifically, the precise statement of two theorems (Theorems 10.2 and 10.3) and their corollaries (Corollary 10.1 and Corollary 10.2). Then, we give the proof of the two theorems. In the final section we state several significant applications of the theorems to be analyzed in sequel chapters. These applications include: an infinite spectrum of rational approximation formulas; an infinite spectrum of rational inverse approximation formulas; infinite families of single and multipoint iteration functions; determinantal approximation of roots of polynomials; a rational expansion formula. Another derivative of the theorems, the truncated Basic Family and its root-finding application has already been analyzed in Chapter 8.

10.2 Taylor's Theorem with Confluent Divided Differences

In order to state Taylor's Theorem with confluent divided differences, as well as its determinantal generalization, it is convenient to give some definitions. We will let K denote either the real field or the complex field. Let $f(x)$ be a function from K into itself. Thus, x is a real variable when K is the real field, and a complex variable, otherwise. We will denote an open ball in K by I ($I = \{a \in K : |a - a_0| < r\}$, for some $a_0 \in K$ and $r > 0$).

For the sake of convenience and in order to be able to state a single Taylor's Theorem, both for real and complex cases, as well as the determinantal generalization to be established in this chapter, we will make statements such as, "f is $(n+1)$-times continuously differentiable over I". Of course if K is the complex field, it suffices to speak only of differentiability of f over I, since in that case f is analytic and has derivatives of all order.

Definition 10.1. A vector $a = (x_1, \ldots, x_{n+1}) \in K^{n+1}$ (K real or complex field) is said to be an *admissible vector of nodes* if whenever $x_i = x_j$, $i \leq j$, then, $x_i = x_{i+1} = \cdots = x_j$. If the number of distinct x_i's is k, we shall say a is *k-point* admissible. In the special case where $k = 1$, we identify a with the common value, x_1. We say a is *monotonic k-point*, if it is k-point admissible and $a = (x_1, \ldots, x_1, x_2, \ldots, x_k)$, where $x_i \neq x_j$, if $i \neq j$.

Remark 10.1. The index k is to be interpreted as the number of input variables in the formulas that describe Taylor's Theorem with confluent divided differences or the determinantal generalizations to be described in the subsequent section.

Definition 10.2. Assume the function $f : K \to K$ (K real or complex field) and is $(n+1)$-times continuously differentiable on an open ball $I \subset K$. Let $a = (x_1, \ldots, x_{n+1})$ be an admissible vector of nodes with $x_i \in I$. Let

x be in I, $x \neq x_i$, $i = 1, \ldots, n+1$. Set $x_{n+2} = x$. For any pair of indices i, j satisfying $1 \leq i \leq j \leq (n+2)$, inductively define the *confluent divided differences* as

$$f_{ij} = \begin{cases} \frac{f^{(j-i)}(x_i)}{(j-i)!}, & \text{if } x_i = x_j; \\ \frac{f_{i+1,j} \; - \; f_{i,j-1}}{(x_j - x_i)}, & \text{otherwise.} \end{cases}$$

Remark 10.2. In case the interpolating points are distinct, f_{ij} is the usual divided difference $f[x_i, \ldots, x_j]$. We have chosen to index the interpolating points from 1 to $(n+1)$, as opposed to the usual 0 through n. As we shall see f_{ij} will appear as a bona fide ij-entry of a matrix of divided differences.

Theorem 10.1 (Taylor's with Confluent Divided Differences). *Assume that $f : K \to K$ (K real or complex field) and is $(n+1)$-times continuously differentiable on an open ball $I \subset K$. Given $x \in I$, for any admissible vector of nodes $a = (x_1, \ldots, x_{n+1})$ with $x_i \in I$, $x_i \neq x \equiv x_{n+2}$, we have*

$$f(x) = P_n(x, a) + R_n(x, a),$$

$$P_n(x, a) = f(x_1) + \sum_{i=1}^{n} f_{1,i+1} \prod_{l=1}^{i} (x - x_l), \quad R_n(x, a) = f_{1,n+2} \prod_{l=1}^{n+1} (x - x_l).$$

Equivalently, if $f'(x) \neq 0$, there exists an open ball I_x centered at x and contained in I such that for any admissible vector of nodes $a = (x_1, \ldots, x_{n+1})$ with $x_i \in I_x$, $x_i \neq x$, the quantity f_{12} is nonzero, and we have

$$x = f^{-1}(y) = Q_n(x, y, a) + \rho_n(x, y, a),$$

$$Q_n(x, y, a) = x_1 + (y - f(x_1)) \frac{1}{f_{12}} - \sum_{i=2}^{n} \frac{f_{1,i+1}}{f_{12}} \prod_{l=1}^{i} (x - x_l),$$

$$\rho_n(x, y, a) = -\frac{f_{1,n+2}}{f_{12}} \prod_{l=1}^{n+1} (x - x_l).$$

Moreover, for K the real field and the complex field we have, respectively

$$f_{1,n+2} = \frac{f^{(n+1)}(\xi)}{(n+1)!}, \quad f_{1,n+2} = \frac{1}{2\pi i} \int_C \frac{f(\eta) d\eta}{\prod_{l=1}^{n+1} (\eta - x_l)^{n+1} (\eta - x)},$$

where ξ is a number that lies in the smallest interval containing $x_1, \ldots, x_{n+1}$, and x, denoted by $Span(a, x)$; and where C is the circumference of a ball within I, centered at x, enclosing the nodes. □

Remark 10.3. From the historical point of view, what is so widely known as "Newton's method", should also be credited to Raphson and Simpson (see Ypma (1995)). It is interesting to note that it was Halley's work on root-finding that inspired Taylor to state the famous "Taylor's Theorem" (see Ypma (1995), Bailey (1989), Scavo and Thoo (1995)) which in turn gives Newton's iteration function, its rate of convergence, and its asymptotic error constant. On the other hand, the limiting case of Newton's interpolation formula for the case of distinct nodes reduces to Taylor's Theorem. It is perhaps for the latter reason that the general case of Taylor's Theorem, i.e. what we have labeled as, "Taylor's Theorem with confluent divided differences", is a nameless theorem in the literature.

To prove Theorem 10.1, one first considers the case of distinct nodes, and establishes by induction Newton's identity

$$f(x) = f(x_1) + \sum_{i=1}^{n+1} f_{1,i+1} \prod_{l=1}^{i} (x - x_l)$$

(see e.g. Hildebrand (1974)). Then, using the continuity of $f^{(j-i)}(x)$, one establishes the fact that f_{ij} converges to $f^{(j-i)}(x_i)/(j-i)!$ as $x_{i+1}, \ldots, x_j$ converge to x_i. This also justifies the definition of divided differences under complete or partial confluence. For an alternative description of the confluent divided differences see Traub (1964), and Ostrowski (1966), two major books on iteration functions. The treatment of Taylor's Theorem for the case of distinct nodes may be found in standard numerical analysis textbooks such as Hildebrand (1974), Dahlquist and Björck (1974), and Atkinson (1989). When K is the complex field the formula for $f_{1,n+2}$ can be found in Baker and Graves-Morris (1996) (Hermite's formula). This formula is a more general formula for the special case of complete confluence (e.g. Ahlfors (1979)): $f_{1,n+2} = \frac{1}{2\pi i} \int_C f(\eta) d\eta / [(\eta - a)^{n+1}(\eta - x)]$. This formula can be used to estimate the error term in Taylor's polynomial (e.g. Nehari (1961)). In particular, when K is the real field and a is one-point, Theorem 10.1 reduces to the classical Taylor's Theorem.

Convention. As is customary, we shall denote $P_n(x, a)$ by $P_n(x)$, and $R_n(x, a)$ by $R_n(x)$. However, it is in fact the interchange of the role of x and a in the equivalent inverse form in Theorem 10.1 that results in iteration functions such as Newton's, Halley's, and more generally infinite families of iteration functions.

In order to evaluate $P_n(x)$ at a given x_0, we form a table of confluent divided differences analogous to the case of distinct nodes, see e.g. Hildebrand (1974), Dahlquist and Björck (1974), Atkinson (1989). Once the divided differences are computed, efficient evaluation of $P_n(x_0)$ is as follows: Let $\beta_n = f_{1,n+1}$. For $i = n-1, \ldots, 0$, compute

$$\beta_i = f_{1,i+1} + \beta_{i+1}(x_0 - x_{i+1}).$$

Then, $\beta_0 = P_n(x_0)$. In particular in the special case of $a = 0$, the above scheme reduces to Horner's method. A polynomial of degree n is characterized and represented by a given admissible vector $a = (x_1, \ldots, x_{n+1})$. Even when x_i's are selected from a specific set of small cardinality, Theorem 10.1 implies that a given polynomial can have a combinatorially large number of representations.

10.2.1 *Basic Applications*

Two fundamental applications of Theorem 10.1 are in approximation of f and its inverse. The first part of Theorem 10.1 trivially gives rise to the well-known formula for approximation of $f(x)$, given an admissible vector a, and the function values at x_i's, namely:

$$f(x) \quad \approx \quad P_n(x).$$

The equivalent inverse form of Theorem 10.1, in case $f'(x)$ is nonzero, allows the interchange of the role of x and the admissible vector a, and gives a formula for approximation of x, given $y = f(x)$:

$$x \quad \approx \quad \kappa_n(a) \equiv Q_n\Big(Q_1(x, y, a), y, a\Big).$$

Note that $\kappa_1(a) = Q_1(x, y, a)$. The function $\kappa_n(a)$ also gives rise to a fixed-point iterative method for approximation of x: given a k-point monotonic vector $a \in K^{n+1}$, the *fixed-point* iteration is defined as the substitution

$$a = (x_1, \ldots, x_1, x_2, \ldots, x_k) \quad \longleftarrow \quad (\kappa_n(a), \ldots, \kappa_n(a), x_1, \ldots, x_{k-1}) \in K^{n+1}.$$

Assuming that $x = \theta$ is a root of $f(x)$, setting $n = 1$ in κ_n, one obtains

$$\kappa_1(x_1) = x_1 - \frac{f(x_1)}{f'(x_1)}, \qquad \kappa_1(a) = x_1 - \frac{f(x_1)}{f_{12}},$$

i.e., the iteration functions of the classical Newton's method, and secant method. From Theorem 10.1 one can immediately conclude that if $f'(\theta) \neq 0$, then there exists a neighborhood I_θ of θ such that for any $a^{(0)} = (x_1^{(0)}, x_2^{(0)})$ with $x_i^{(0)} \in I_\theta$, the sequence of fixed-point iterates $\{\kappa_1(a^{(t)})\}_{t=0}^{\infty}$ is well-defined, and it converges to θ satisfying

$$\lim_{t\to\infty} \frac{\theta - \kappa_1(a^{(t)})}{(\theta - x_1^{(t)})(\theta - x_2^{(t)})} = -\frac{f''(\theta)}{2f'(\theta)}.$$

In particular, the above implies the quadratic rate of convergence of Newton's method, and also gives the corresponding asymptotic error constant. For the secant method it only gives partial information. Subsequently, this partial information can be used to derive a precise statement on the order of convergence, and the corresponding asymptotic error constant. Setting $n = 2$ in κ_n, we get $\kappa_2(a) = Q_2(\kappa_1(a), 0, a)$. In particular, letting a to be (x_1, x_1, x_1), (x_1, x_1, x_2), (x_1, x_2, x_3), we get one-point, monotonic two-point, and three-point iteration functions, the first two of which can be viewed as "corrected Newton methods", and the third, a "corrected secant method". The one-point case becomes

$$\kappa_2(x_1) = \kappa_1(a) - \frac{f''(x_1)}{2f'(x_1)}(\kappa_1(a) - x_1)^2 = x_1 - f(x_1)\frac{2f'(x_1)^2 + f''(x_1)f(x_1)}{2f'(x_1)^3}.$$

This function apparently was obtained by Euler, Schröder, and Chebyshev (for which he received a silver medal in a student contest), see Traub (1964), and Bailey (1989). If $f'(\theta) \neq 0$, and $a^{(0)}$ is properly chosen, the fixed-point iteration corresponding to $\kappa_2(a^{(t)})$ is locally well-defined and converges to θ. Moreover, from Theorem 10.1 and definitions of κ_1 and κ_2, one can show that

$$\lim_{t \to \infty} \frac{\theta - \kappa_2(a^{(t)})}{(\theta - x_1^{(t)})(\theta - x_2^{(t)})} = 0.$$

The above implies that both the one-point and the monotonic two-point iteration functions corresponding to κ_2 have better rate of convergence than Newton's method. Also the three-point iteration function corresponding to κ_2 has a better rate of convergence than the secant method. In fact $\kappa_2(x_1)$ has cubic rate of convergence. It is the second member of the family of Euler-Schröder iteration functions, $\{E_m\}_{m=2}^{\infty}$, where E_m has order of convergence equal to m, and E_2 is Newton's. For the history of this family, as well as different schemes for its generation, see Traub (1964), Henrici (1974), Householder (1970), Smale (1985). In fact as was shown in Chapter 7 this family can also be generated from Taylor's Theorem, via a simple recursive formula. For instance, using this formula we obtained

$$E_5 = x_1 - (\frac{f}{f'}) - (\frac{f}{f'})^2 \frac{f''}{2f'} + (\frac{f}{f'})^3 (\frac{f'''}{6f'} - \frac{f''^2}{2f'^2})$$

$$-(\frac{f}{f'})^4 (\frac{f^{(4)}}{4!f'} - \frac{5f''f'''}{12f'^2} + \frac{5f''^3}{8f'^3}), \qquad (10.1)$$

where E_i is the sum of multiples of the first $(i-1)$ powers of (f/f').

The derivation of the Euler-Schröder family of iteration functions established here has been via an elementary application of the classical Taylor's Theorem. In the subsequent sections we will see that through a nontrivial determinantal generalization of Taylor's Theorem we can in particular derive the fundamental multipoint Basic Family which in particular includes the previously defined one-point Basic Family. As we saw in Chapter 7 the Basic Family has many optimal properties with respect to several criteria and is in particular advantageous over the Euler-Schröder family.

Via the determinantal generalization of Taylor's Theorem, to be developed in this chapter, we will obtain not only the Basic Family for more general functions, but also its multipoint versions, as well as a precise expansion formula that allows many novel applications described in the final section of the chapter.

10.3 The Determinantal Taylor Theorem

In this section we describe the determinantal Taylor formulas. The description of these formulas require the definition of the *matrix of divided differences* and some of its submatrices. The reader needs to become familiarized with these submatrices.

Definition 10.3. [Matrix of Divided Differences] Let $m \geq 2$ be a given natural number, and $n \geq (m-1)$. Assume that $f : K \to K$ (K real or complex field) and is $(n+1)$-times continuously differentiable on an open ball $I \subset K$. Let $a = (x_1, \ldots, x_{n+1})$ be an admissible vector of nodes with $x_i \in I$. Let x be in I, $x \neq x_i$, $i = 1, \ldots, n+1$. Set $x_{n+2} = x$, $y = f(x)$. We shall refer to the $(m-1) \times (n+2)$ matrix

$$F = \begin{pmatrix} f_{11} - y & f_{12} & f_{13} & \cdots & f_{1,m-1} & f_{1,m} & \cdots & f_{1,n+2} \\ 0 & f_{22} - y & f_{23} & \cdots & f_{2,m-1} & f_{2,m} & \cdots & f_{2,n+2} \\ 0 & 0 & f_{33} - y & \cdots & f_{3,m-1} & f_{3,m} & \cdots & f_{3,n+2} \\ \vdots & \vdots & \vdots & \ddots & \vdots & \cdots & \cdots & \vdots \\ 0 & 0 & 0 & \cdots & f_{m-1,m-1} - y & f_{m-1,m} & \cdots & f_{m-1,n+2} \end{pmatrix}$$

as the *matrix of divided differences at* x. We also denote F by $[u_1, \ldots, u_{n+2}]$, where u_i is the i-th column of F.

Definition 10.4. (Determinantal Components) Define

$$a_i = (x_1, \ldots, x_i), \quad i = 1, \ldots, n+2, \quad a \equiv a_{n+1}.$$

By *determinantal components* of F we shall refer to three classes of submatrix determinants. The superscripts used below will denote the corresponding matrix dimension. For a matrix A, $|A| = det(A)$, its determinant. Let

$$D^{(m-1)}(y, a_m) = |u_2, \ldots, u_m|, \tag{10.2}$$

$$\widehat{D}_i^{(m-1)}(y, a_{i+1}) = |u_3, \ldots, u_m, u_{i+1}|, \quad i = m, \ldots, n+1, \tag{10.3}$$

and

$$N^{(m-2)}(y, a_m) = \begin{vmatrix} f_{23} & \cdots & f_{2,m-1} & f_{2,m} \\ f_{33} - y & \cdots & f_{3,m-1} & f_{3,m} \\ \vdots & \ddots & \vdots & \vdots \\ 0 & \cdots & f_{m-1,m-1} - y & f_{m-1,m} \end{vmatrix}, \quad N^{(0)}(y, a_2) \equiv 1. \tag{10.4}$$

In case a is one-point, we get

$$D^{(m-1)}(y, a) = \begin{vmatrix} f'(a) & \frac{f''(a)}{2!} & \cdots & \frac{f^{(m-2)}(a)}{(m-2)!} & \frac{f^{(m-1)}(a)}{(m-1)!} \\ f(a) - y & f'(a) & \ddots & \ddots & \frac{f^{(m-2)}(a)}{(m-2)!} \\ 0 & f(a) - y & \ddots & \ddots & \vdots \\ \vdots & \vdots & \ddots & \ddots & \frac{f''(a)}{2!} \\ 0 & 0 & \cdots & f(a) - y & f'(a) \end{vmatrix}, \tag{10.5}$$

and

$$\widehat{D}_i^{(m-1)}(y, a) = \begin{vmatrix} \frac{f''(a)}{2!} & \frac{f'''(a)}{3!} & \cdots & \frac{f^{(m-1)}(a)}{(m-1)!} & \frac{f^{(i)}(a)}{i!} \\ f'(a) & \frac{f''(a)}{2!} & \ddots & \frac{f^{(m-2)}(a)}{(m-2)!} & \frac{f^{(i-1)}(a)}{(i-1)!} \\ f(a) - y & f'(a) & \ddots & \vdots & \vdots \\ \vdots & \vdots & \ddots & \frac{f''(a)}{2!} & \frac{f^{(i-m+3)}(a)}{(i-m+3)!} \\ 0 & 0 & \cdots & f'(a) & \frac{f^{(i-m+2)}(a)}{(i-m+2)!} \end{vmatrix}, \tag{10.6}$$

and for all $m \geq 2$, $N^{(m-2)}(y, a) = D^{(m-2)}(y, a)$. In particular, $D^{(m-1)}$ and $\widehat{D}_m^{(m-1)}$ are determinants corresponding to special Toeplitz matrices.

Definition 10.5. (Error Determinant) The *error determinant* refers to the quantity

$$\widehat{D}_{n+1}^{(m-1)}(y, a_{n+2}) = |u_3, \ldots, u_m, u_{n+2}|.$$

Remark 10.4. If $f(x)$ is a polynomial of degree ν, then for any $n > m+\nu-2$, and each $m \geq 2$, $u_{n+2} = 0$ so that the error determinant is zero.

When K is the real field we can represent the error determinant in two different forms. We use the first representation of the error determinant when approximating x, and the second representation when approximating $f(x)$. In the first form we define $\widehat{\delta}_{n+1}^{(m-1)}(y, \widehat{\xi}, a) \equiv \widehat{D}_{n+1}^{(m-1)}(y, a_{n+2})$ but such that the determinant is computed while replacing u_{n+2} with the equivalent vector

$$u_{n+2} = [\frac{f^{(n+1)}(\xi_1)}{(n+1)!}, \frac{f^{(n)}(\xi_2)}{n!}, \ldots, \frac{f^{(n-m+3)}(\xi_{m-1})}{(n-m+3)!}]^T,$$

where $\widehat{\xi} = (\xi_1, \ldots, \xi_{m-1})$, with $\xi_i \in Span(a, x)$. The equivalence of this form is a straightforward application of Theorem 10.1. In the second form of the error determinant we represent u_{n+2} in terms of the divided differences at $x_1, \ldots, x_{n+1}$, and a single unknown ξ, appearing in the error term $R_n(x)$ of Taylor polynomial (Theorem 10.1). Then, we employ a backward recursion using the confluent divided differences

$$f_{i+1,n+2} = f_{i,n+2}(x - x_i) + f_{i,n+1}.$$

We define

$$\widehat{\Delta}_{n+1}^{(m-1)}(x, \xi, a) = \widehat{D}_{n+1}^{(m-1)}(y, a_{n+2}),$$

but such that the determinant is computed after the following substitutions are applied:

$$f_{1,n+2} = \frac{f^{(n+1)}(\xi)}{(n+1)!}, \quad \xi \in Span(a, x),$$

also for $i = 2, \ldots, m-1$,

$$f_{i,n+2} = \frac{f^{(n+1)}(\xi)}{(n+1)!} \prod_{l=1}^{i-1}(x - x_l) + \sum_{j=1}^{i-2} f_{j,n+1} \prod_{l=j+1}^{i-1}(x - x_l) + f_{i-1,n+1},$$

as well as

$$f_{ii} - f(x) = -\sum_{j=i+1}^{n+2} f_{ij} \prod_{l=i}^{j}(x - x_l), \quad i = 3, \ldots, m-1.$$

10.3.1 *Determinantal Interpolation Formulas*

In this section we formally state two equivalent determinantal generalizations of Taylor's Theorem and their one-point corollaries that give determinantal generalizations of the classical Taylor's Theorem, and are of particular interest. The nontrivial proof of the two theorems will be given in the next section.

Theorem 10.2 (Kalantari (2000a)). *Let $m \geq 2$ be a natural number, and $n \geq (m-1)$. Assume that $f : K \to K$ (K real or complex field) and is $(n+1)$-times continuously differentiable on an open ball $I \subset K$. Assume that for a given $x \in I$, $f'(x) \neq 0$. Let $y = f(x)$. Then, there exists an open ball I_x centered at x and contained in I such that for any admissible vector of nodes $a = (x_1, \dots, x_{n+1})$ with $x_i \in I_x$, $x_i \neq x$, the quantities $D^{(m-1)}(y, a_m)$ and $N^{(m-2)}(y, a_m)$ are nonzero, and we have*

$$y = P_{m,n}(x, y, a) + R_{m,n}(x, y, a), \tag{10.7}$$

where

$$P_{m,n}(x, y, a)$$

$$= f(x_1) + \frac{D^{(m-1)}(y, a_m)}{N^{(m-2)}(y, a_m)}(x - x_1) + (-1)^m \sum_{i=m}^{n} \frac{\widehat{D}_i^{(m-1)}(y, a_{i+1})}{N^{(m-2)}(y, a_m)} \prod_{l=1}^{i}(x - x_l),$$

$$R_{m,n}(x, y, a) = (-1)^m \frac{\widehat{D}_{n+1}^{(m-1)}(y, a)}{N^{(m-2)}(y, a_m)} \prod_{l=1}^{n+1}(x - x_l).$$

In particular, if $f(x)$ is a polynomial of degree ν, then for any $j \geq m+\nu-2$, we have

$$f(x) = P_{m,j}(x, f(x), a).$$

Equivalently,

$$x = f^{-1}(y) = Q_{m,n}(x, y, a) + \rho_{m,n}(x, y, a), \tag{10.8}$$

where

$$Q_{m,n}(x, y, a) =$$

$$x_1 + (y - f(x_1))\frac{N^{(m-2)}(y, a_m)}{D^{(m-1)}(y, a_m)} + (-1)^{m-1} \sum_{i=m}^{n} \frac{\widehat{D}_i^{(m-1)}(y, a_{i+1})}{D^{(m-1)}(y, a_m)} \prod_{l=1}^{i}(x - x_l),$$

$$\rho_{m,n}(x, y, a) = (-1)^{m-1} \frac{\widehat{D}_{n+1}^{(m-1)}(y, a)}{D^{(m-1)}(y, a_m)} \prod_{l=1}^{n+1}(x - x_l).$$

In particular, if $f(x)$ is a polynomial of degree ν, then for any $j \geq m+\nu-2$, we have

$$x = Q_{m,j}(x, f(x), a).$$

Remark 10.5. The equivalence of (10.7) and (10.8) in Theorem 10.2 is straightforward: From (10.7) we have

$$\frac{D^{(m-1)}(y, a_m)}{N^{(m-2)}(y, a_m)}(x - x_1)$$

$$= y - f(x_1) + (-1)^{m-1} \sum_{i=m}^{n} \frac{\widehat{D}_i^{(m-1)}(y, a_{i+1})}{N^{(m-2)}(y, a_m)} \prod_{l=1}^{i}(x - x_l) - R_{m,n}(x, y, a).$$

Now by multiplying the above by $N^{(m-2)}(y, a_m)/D^{(m-1)}(y, a_m)$ and by moving $-x_1$ to the right-hand side we get (*ii*). Conversely, (10.8) implies (10.7).

Remark 10.6. Note that $P_{2,n}(x, y, a) = P_n(x)$, so that Theorem 10.2 implies Theorem 10.1, given that $f'(x) \neq 0$. If $f(x)$ is a polynomial of degree n, since it is infinitely differentiable, for each $m \geq 2$, Theorem 10.2, associates n determinantal formulas $P_{m,j}(x, P_n(x), a)$, $j = m-1, \ldots, m+n-2$. Hence an infinite sequence of determinantal formulas can be associated to a single polynomial. The following is an immediate one-point version of Theorem 10.2 (i.e. the case with $x_1 = x_2 = \cdots = x_{n+1}$). It generalizes the classical Taylor's Theorem.

Corollary 10.1. *Let $m \geq 2$ be a natural number, and $n \geq (m-1)$. Assume that $f : K \to K$ (K real or complex field) and is $(n+1)$-times continuously differentiable on an open ball $I \subset K$. Assume that for a given $x \in I$, $f'(x) \neq 0$. Let $y = f(x)$. Then, there exists an open ball I_x centered at x and contained in I such that for any $a \in I_x$, $a \neq x$, the quantities $D^{(m-1)}(y, a)$ and $D^{(m-2)}(y, a)$ are nonzero, and we have*

$$y = P_{m,n}(x, y, a) + R_{m,n}(x, y, a),$$

$$P_{m,n}(x, y, a)$$

$$= f(a) + \frac{D^{(m-1)}(y, a)}{D^{(m-2)}(y, a)}(x - a) + (-1)^m \sum_{i=m}^{n} \frac{\widehat{D}_i^{(m-1)}(y, a)}{D^{(m-2)}(y, a)}(x - a)^i,$$

$$R_{m,n}(x, y, a) = (-1)^m \frac{\widehat{D}_{n+1}^{(m-1)}(y, a)}{D^{(m-2)}(y, a)}(x - a)^{n+1}.$$

In particular, if $f(x)$ is a polynomial of degree ν, then for any $j \geq m+\nu-2$, we have

$$f(x) = P_{m,j}(x, f(x), a).$$

Equivalently, we have

$$x = Q_{m,n}(x,y,a) + \rho_{m,n}(x,y,a),$$

$$Q_{m,n}(x,y,a)$$

$$= a + (y - f(a))\frac{D^{(m-2)}(y,a)}{D^{(m-1)}(y,a)} + (-1)^{m-1}\sum_{i=m}^{n}\frac{\widehat{D}_i^{(m-1)}(y,a)}{D^{(m-1)}(y,a)}(x-a)^i,$$

$$\rho_{m,n}(x,y,a) = (-1)^{m-1}\frac{\widehat{D}_{n+1}^{(m-1)}(y,a)}{D^{(m-1)}(y,a)}(x-a)^{n+1}.$$

In particular, if $f(x)$ is a polynomial of degree ν, then for any $j \geq m+\nu-2$, we have

$$x = Q_{m,j}(x, f(x), a).$$

□

The following theorem is an equivalent formulation of Theorem 10.2 and gives rise to an infinite family of single and multipoint iteration functions for root-finding.

Theorem 10.3 (Kalantari (2000a)). *Let $m \geq 2$ be a natural number, and $n \geq (m-1)$. Assume that $f : K \to K$ (K real or complex field) and is $(n+1)$-times continuously differentiable on an open ball $I \subset K$ containing a simple root θ. Then, there exists an open ball I_θ centered at θ and contained in I so that for any admissible vector of nodes $a = (x_1, \ldots, x_{n+1})$ with $x_i \in I_\theta$, $x_i \neq \theta \equiv x_{n+2}$, $i = 1, \ldots, n+1$, the quantity $D^{(m-1)}(0, a_m)$ is nonzero, and we have*

$$B_m^{(k)}(a_m) \equiv x_1 - f(x_1)\ \frac{N^{(m-2)}(0,a_m)}{D^{(m-1)}(0,a_m)}$$

$$= \theta + \sum_{i=m}^{n}\gamma_i^{(m)}(a_{i+1})\prod_{l=1}^{i}(\theta - x_l) + \gamma_{n+1}^{(m)}(a,\theta)\prod_{l=1}^{n+1}(\theta - x_l),$$

where $a_{i+1} = (x_1, \ldots, x_{i+1})$, a_m is k-point, and

$$\gamma_i^{(m)}(a_{i+1}) = (-1)^m\frac{\widehat{D}_i^{(m-1)}(0,a_{i+1})}{D^{(m-1)}(0,a_m)}, \quad i = m, \ldots, n+1.$$

In particular, if $f(x)$ is a polynomial of degree ν, for any $i \geq m+\nu-2$, $\gamma_{i+1}^{(m)} \equiv 0$.

Remark 10.7. The relevance of the index k becomes evident when Theorem 10.2 is viewed in the context of iteration functions. Note that $B_m^{(k)}$ can be viewed as a function of k indeterminates. The following is an immediate consequence of Theorem 10.2.

Corollary 10.2. *Let $m \geq 2$ be a natural number, and $n \geq (m-1)$. Assume that $f : K \to K$ (K real or complex field) and is $(n+1)$-times continuously differentiable on an open ball $I \subset K$ containing a simple root θ. Then, there exists an open ball I_θ centered at θ and contained in I such that for any $x_1 \in I_\theta$, $x_1 \neq \theta$, $D^{(m-1)}(0, x_1)$ is nonzero, and we have*

$$B_m^{(1)}(x_1) \equiv x_1 - f(x_1)\ \frac{D^{(m-2)}(0, x_1)}{D^{(m-1)}(0, x_1)} =$$

$$\theta + \sum_{i=m}^{n} \gamma_i^{(m)}(x_1)(\theta - x_1)^i + \gamma_{n+1}^{(m)}(x_1, \theta)(\theta - x_1)^{n+1},$$

$$\gamma_i^{(m)}(x_1) = (-1)^m \frac{\widehat{D}_i^{(m-1)}(0, x_1)}{D^{(m-1)}(0, x_1)}, \qquad i = m, \dots, n,$$

$$\gamma_{n+1}^{(m)}(x_1, \theta) = (-1)^m \frac{\widehat{D}_{n+1}^{(m-1)}(0, x_1, \theta)}{D^{(m-1)}(0, x_1)}.$$

In particular, if $f(x)$ is a polynomial of degree ν, for any $i \geq m + \nu - 2$, $\gamma_{i+1}^{(m)} \equiv 0$. □

Theorem 10.3 and Corollary 10.2 are fundamental from several points of view. It describes a more general development of the Basic Family, and its multipoint version. Even for $k = 1$, the expansion formula for $B_m^{(k)}$ in Theorem 10.3 becomes a more precise formula than the one in Chapter 7. For example, the formula in Theorem 10.3 allows the definition of a *corrected Basic Family* of k-point iteration functions,

$$\widehat{B}_m^{(k)}(a_m) \equiv B_m^{(k)}(a_m) - \gamma_m^{(m)}(a_{m+1}) \prod_{l=1}^{m} (B_m^{(k)}(a_m) - x_l).$$

This is derived by moving the term $\gamma_m^{(m)}(a_{m+1}) \prod_{l=1}^{m} (\theta - x_l)$ from the right-hand side of the equation of $B_m^{(k)}(a_m)$ (see Theorem 10.3) to the left-hand side, while also replacing θ by an approximate value, namely $B_m^{(k)}(a_m)$. In particular, one can define a corrected Halley method with super-cubic rate of convergence. There are indeed numerous results that have been motivated by Theorem 10.3 and its Corollary 10.2 in sequel chapters some of which are summarized in the final section.

10.4 Proof of Determinantal Taylor Theorem and Equivalent Form

In this section we prove Theorems 10.2 and 10.3. We first need to define some determinants. The corresponding matrices are all submatrices of the matrix F (see Definition 3), corresponding to two consecutive m's, with $y = f(x) = 0$. The explicit description of these matrices facilitates the long proof. Let $m \geq 2$ be a natural number, and $n \geq (m-1)$. Assume that f is $(n+1)$-times continuously differentiable in an open ball I containing a simple root θ. Let $a = (x_1, \ldots, x_{n+1})$ be an admissible vector of nodes with $x_i \in I$, $x_i \neq \theta$, $i = 1, \ldots, n+1$. Set $x_{n+2} = \theta$. For each $i = 2, \ldots, n+2$, and each $j = i, \ldots, n+2$, let f_{ij} be the confluent divided difference (see Definition 2). The superscript used below will denote the corresponding matrix dimensions.

For each $i = m, \ldots, n+1$ define $D_i^{(m)}$ as

$$D_i^{(m)} = \begin{vmatrix} f_{12} & f_{13} & \cdots & f_{1,m-1} & f_{1,m} & f_{1,i+1} \\ f_{22} & f_{23} & \cdots & f_{2,m-1} & f_{2,m} & f_{2,i+1} \\ 0 & f_{33} & \cdots & f_{3,m-1} & f_{3,m} & f_{3,i+1} \\ \vdots & \vdots & \ddots & \vdots & \vdots & \vdots \\ 0 & 0 & \cdots & f_{m-1,m-1} & f_{m-1,m} & f_{m-1,i+1} \\ 0 & 0 & \cdots & 0 & f_{m,m} & f_{m,i+1} \end{vmatrix}. \tag{10.9}$$

Also define,

$$D_i^{(1)} = f_{1,i+1}, \quad i = 1, \ldots, n+1. \tag{10.10}$$

For each $i = m+1, \ldots, n+1$ define $\widehat{D}_i^{(m)}$ as

$$\widehat{D}_i^{(m)} = \begin{vmatrix} f_{13} & \cdots & f_{1,m-1} & f_{1,m} & f_{1,m+1} & f_{1,i+1} \\ f_{23} & \cdots & f_{2,m-1} & f_{2,m} & f_{2,m+1} & f_{2,i+1} \\ f_{33} & \cdots & f_{3,m-1} & f_{3,m} & f_{3,m+1} & f_{3,i+1} \\ \vdots & \ddots & \vdots & \vdots & \vdots & \vdots \\ 0 & \cdots & f_{m-1,m-1} & f_{m-1,m} & f_{m-1,m+1} & f_{m-1,i+1} \\ 0 & \cdots & 0 & f_{m,m} & f_{m,m+1} & f_{m,i+1} \end{vmatrix}. \tag{10.11}$$

Also define,

$$\widehat{D}_i^{(1)} = f_{1,i+1}, \quad i = 2, \ldots, n+1. \tag{10.12}$$

For $i = m+1, \ldots, n+1$ define

$$\bar{D}_i^{(m-1)} = \begin{vmatrix} f_{13} & \cdots & f_{1,m-1} & f_{1,m+1} & f_{1,i+1} \\ f_{23} & \cdots & f_{2,m-1} & f_{2,m+1} & f_{2,i+1} \\ f_{33} & \cdots & f_{3,m-1} & f_{3,m+1} & f_{3,i+1} \\ \vdots & \cdots & \vdots & \vdots & \vdots \\ 0 & \cdots & f_{m-1,m-1} & f_{m-1,m+1} & f_{m-1,i+1} \end{vmatrix}. \tag{10.13}$$

For $j = m-2$ and $j = m-1$ define

$$N_j^{(m-2)} = N_j^{(m-2)} = \begin{vmatrix} f_{23} & \cdots & f_{2,m-1} & f_{2,j+2} \\ f_{33} & \cdots & f_{3,m-1} & f_{3,j+2} \\ \vdots & \ddots & \vdots & \vdots \\ 0 & \cdots & f_{m-1,m-1} & f_{m-1,j+2} \end{vmatrix}. \tag{10.14}$$

Note that $D_{m-1}^{(m-1)}$, $N_{m-2}^{(m-2)}$, and $\widehat{D}_i^{(m-1)}$ coincide with previously defined $D^{(m-1)}(y, a_m)$, $N^{(m-2)}(y, a_m)$, and $\widehat{D}_i^{(m-1)}(y, a_m)$, respectively, evaluated at $y = 0$ (see (3c)-(3e)). For simplicity of notation we have suppressed the argument to these functions. We have chosen to represent these determinants with more detail in order to facilitate the proofs.

Lemma 10.1. *Let $m \geq 2$ be a natural number, and $n \geq (m-1)$. Assume that $f : K \to K$ (K real or complex field) and is $(n+1)$-times continuously differentiable on an open ball $I \subset K$ containing a simple root θ. Then there exists an open ball I_θ containing θ so that for any admissible vector of nodes $a = (x_1, \ldots, x_{n+1})$ with $x_i \in I_\theta$, $x_i \neq \theta$, the corresponding $D_{m-1}^{(m-1)}$ and $\gamma_m^{(m)} \equiv (-1)^m \widehat{D}_m^{(m-1)} / D_{m-1}^{(m-1)}$, are nonzero. In particular, $\widehat{D}_m^{(m-1)}$ is nonzero in this neighborhood.*

Proof. If $x_1 = \cdots = x_{n+1} = \theta$, then the corresponding $D_{m-1}^{(m-1)}$ is nonzero, since it reduces to an upper triangular matrix whose diagonal entries are $f'(\theta) \neq 0$. From this and continuity, there exists an open ball I_θ containing θ such that for any admissible a with coordinates contained in I_θ, the corresponding $D_{m-1}^{(m-1)}$ is nonzero. Assuming that $f(x)$ is a polynomial, it is shown in Chapter 7 that the corresponding one-point $\gamma_m^{(m)}$ is not identically zero. Since $\gamma_m^{(m)}$ is not identically zero for polynomials, it follows that it is not identically zero for the general case of functions considered here. From this and continuity it follows that for the case of an arbitrary admissible vector of nodes a, the corresponding $\gamma_m^{(m)}$ is not identically zero, provided that its coordinates lie within an open ball containing θ. □

Lemma 10.2. *Theorem 10.2 and Theorem 10.3 are equivalent.*

Proof. First we prove that Theorem 10.2 implies Theorem 10.3. Assume Theorem 10.2 is valid. From the equivalence of (i) and (ii) of Theorem 10.2 (see Remark 10.5) it suffices to note that since $y = f(\theta) = 0$, we have

$$Q_{m,n}(x, y, a) = B_m^{(k)}(a_m).$$

Hence Theorem 10.3 is valid. Next assume that Theorem 10.3 is valid. We will show that Theorem 10.2 must hold. Let x_0 be a point in I such that $f'(x_0) \neq 0$. Now consider the function

$$g(w) = f(w) - f(x_0), \quad w \in I.$$

Then, $g(x_0) = 0$, and $g^{(i)}(w) = f^{(i)}(w)$, for all $i \geq 1$. Since $g'(x_0) \neq 0$, Theorem 10.3 applies to $g(w)$ at $\theta = x_0$. In particular, the corresponding $D^{(m-1)}(0, a_m) \neq 0$. But $D^{(m-1)}(0, a_m)$, written with respect to $g(w)$ at x_0, coincides with $D^{(m-1)}(y, a_m)$, written with respect to $f(x)$ at x_0. This also applies to $N^{(m-2)}$ and $\widehat{D}_i^{(m-1)}$. This implies that $Q_{m,n}(x_0, y, a)$, written with respect to $f(x)$ at x_0, is identical with $B_m^{(k)}(a_m)$, written with respect to $g(w)$ at $\theta = x_0$. Thus, the expansion given by part (ii) of Theorem 10.2 is valid. To show that part (i) is also valid we need to show that $N^{(m-2)}(y, a_m)$ is not identically zero in a neighborhood of x_0. But this follows from Lemma 10.1 and the definition of $N^{(m-2)}(y, a_m)$ (see (3e)).$\square$

What follows from here on is toward the proof of Theorem 10.3. However, we will need to prove four lemmas before the proof of Theorem 10.3 can begin. These lemmas consist of some determinantal identities. The first one is a determinantal identity reminiscent of Sylvester's theorem (see Baker and Graves-Morris (1996)), but not equivalent to that theorem.

Lemma 10.3. *Let $a = (a_1, a_2, 0, \ldots, 0)^T$ be a column vector in K^k, $k \geq 2$. Let c and d, and e be arbitrary row vectors in K^k. Let $A = (a_{ij})$ be an $k \times (k-2)$ matrix having the property that $a_{ij} = 0$, for all i and j satisfying $i - j \geq 3$, and $a_{i,i-2} \neq 0$, for all $i = 3, \ldots, k$, i.e.*

$$A = \begin{pmatrix} a_{11} & a_{12} & \cdots & a_{1,k-2} \\ a_{21} & a_{22} & \cdots & a_{2,k-2} \\ a_{31} & a_{32} & \cdots & a_{3,k-2} \\ 0 & a_{42} & \cdots & a_{4,k-2} \\ \vdots & \vdots & \ddots & \vdots \\ 0 & 0 & \cdots & a_{k,k-2} \end{pmatrix}.$$

Then for $k = 2$

$$\big|a\ d\big|\ \big|c\ e\big| - \big|a\ e\big|\ \big|c\ d\big| = \big|a\ c\big|\ \big|d\ e\big|\,, \tag{10.15}$$

and for $k \geq 3$

$$\big|a\ A\ d\big|\ \big|A\ c\ e\big| - \big|a\ A\ e\big|\ \big|A\ c\ d\big| = \big|a\ A\ c\big|\ \big|A\ d\ e\big|\,. \tag{10.16}$$

Proof. We prove the lemma by induction on k. For $k = 2$ the identity (10.15) can easily be verified by direct computation of the determinants. Assume that $k \geq 3$, and (10.16) is true for $k-1$. We prove that (10.16) is true for k. Using last column of A as the pivot column and adding scalar multiples of it to other columns of the matrices appearing in the left-hand side of (10.16), we can reduce each matrix in (10.16) into a matrix whose last row has a single nonzero, namely $a_{k,k-2}$. Then by expanding the determinants along the last row, the left-hand side of (10.16) reduces to

$$-a_{k,k-2}^2\big(\big|a'\ A'\ d'\big|\ \big|A'\ c'\ e'\big| - \big|a'\ A'\ e'\big|\ \big|A'\ c'\ d'\big|\big), \tag{10.17}$$

where a', c', d', e' are vectors in K^{k-1}, and $a' = (a'_1, a'_2, 0, \ldots, 0)^T$, and $A' = (a'_{ij})$ is a $(k-1)\times(k-3)$ matrix satisfying $a'_{ij} = 0$, for all i and j satisfying $i - j \geq 3$, and $a'_{i,i-2} \neq 0$, for all $i = 3, \ldots, k-1$. Similarly the right-hand side of (ii) reduces to

$$-a_{k,k-2}^2\big(\big|a'\ A'\ c'\big|\ \big|A'\ d'\ e'\big|\big). \tag{10.18}$$

Since $a_{k,k-2}$ is nonzero, by induction hypothesis (i') and (ii') are identical, hence the proof. □

Lemma 10.4. *For each* $m \geq 2$, $i \geq m+1$, *we have*

$$D_m^{(m-1)}\widehat{D}_i^{(m-1)} - D_i^{(m-1)}\widehat{D}_m^{(m-1)} = D_{m-1}^{(m-1)}\bar{D}_i^{(m-1)}.$$

Proof. In Lemma 10.3 choose $k = m-1$. Then, let a be the first column of the matrix corresponding to $D_m^{(m-1)}$, let A be the submatrix corresponding to columns 2 through $m-2$, and let d be the last column of this matrix. Thus, $D_m^{(m-1)} = \big|a\ A\ d\big|$. Also denote the last two columns of the corresponding matrix of $\widehat{D}_i^{(m-1)}$ by c and e, respectively. Thus $\widehat{D}_i^{(m-1)} = \big|A\ c\ e\big|$. Similarly, the next four determinants appearing in the claimed identity, represent corresponding determinants stated in Lemma 10.3. □

Lemma 10.5. *For each* $m \geq 2$, *and* $i = m+1, \ldots, n$, *we have*

$$D_{m-1}^{(m-1)}\widehat{D}_i^{(m-2)} + D_{m-2}^{(m-2)}\widehat{D}_i^{(m-1)} = D_i^{(m-1)}\widehat{D}_{m-1}^{(m-2)}, \tag{10.19}$$

$$D_m^{(m)}\widehat{D}_i^{(m-1)} - D_i^{(m)}\widehat{D}_m^{(m-1)} = -D_{m-1}^{(m-1)}\widehat{D}_i^{(m)}. \tag{10.20}$$

Proof. We will prove (10.20). The equation in (10.19) is simply (10.20) written for $m-1$ with the terms rearranged. By expanding the determinant of the matrices corresponding to $D_m^{(m)}$, and $D_i^{(m)}$ about their last row, we get

$$D_m^{(m)} = -f_{m,m}D_m^{(m-1)} + f_{m,m+1}D_{m-1}^{(m-1)},$$

$$D_i^{(m)} = -f_{m,m}D_i^{(m-1)} + f_{m,i+1}D_{m-1}^{(m-1)}.$$

Substituting these into the left-hand side of (10.20), it becomes

$$-f_{m,m}D_m^{(m-1)}\widehat{D}_i^{(m-1)} + f_{m,m+1}D_{m-1}^{(m-1)}\widehat{D}_i^{(m-1)} + f_{m,m}D_i^{(m-1)}\widehat{D}_m^{(m-1)}$$

$$-f_{m,i+1}D_{m-1}^{(m-1)}\widehat{D}_m^{(m-1)} = -f_{m,m}\left(D_m^{(m-1)}\widehat{D}_i^{(m-1)} - D_i^{(m-1)}\widehat{D}_m^{(m-1)}\right)$$

$$+f_{m,m+1}\left(D_{m-1}^{(m-1)}\widehat{D}_i^{(m-1)}\right) - f_{m,i+1}\left(D_{m-1}^{(m-1)}\widehat{D}_m^{(m-1)}\right).$$

On the other hand, by expanding the determinant of the matrix corresponding to $\widehat{D}_i^{(m)}$ about its last row, the right-hand side of (10.20) reduces to

$$-f_{m,m}\left(\bar{D}_i^{(m-1)}D_{m-1}^{(m-1)}\right) + f_{m,m+1}\left(\widehat{D}_i^{(m-1)}D_{m-1}^{(m-1)}\right)$$

$$-f_{m,i+1}\left(\widehat{D}_m^{(m-1)}D_{m-1}^{(m-1)}\right).$$

Thus to prove the lemma it suffices to show that the coefficients of $f_{m,m}$ are identical. But this follows from Lemma 10.4. □

Lemma 10.6. *For each $m \geq 3$, we have*

$$N_{m-2}^{(m-2)}D_m^{(m-1)} - N_{m-1}^{(m-2)}D_{m-1}^{(m-1)} = f_{22}\cdots f_{m-1,m-1}\widehat{D}_m^{(m-1)}.$$

Proof. The proof of the lemma uses the special format of the matrices and the fact that $f_{ii} \neq 0$, as opposed to the use of specific entries. We will prove it by induction. For $m = 3$, from (10.13) the right-hand side of the claimed identity is

$$f_{22}\widehat{D}_3^{(2)} = f_{22}\begin{vmatrix} f_{13} & f_{14} \\ f_{23} & f_{24} \end{vmatrix} = f_{22}(f_{13}f_{24} - f_{23}f_{14}).$$

On the other hand from (10.9) and (10.14) we have

$$N_1^{(1)}D_3^{(2)} - N_2^{(1)}D_2^{(2)} = f_{23}\begin{vmatrix} f_{12} & f_{14} \\ f_{22} & f_{24} \end{vmatrix} - f_{24}\begin{vmatrix} f_{12} & f_{13} \\ f_{22} & f_{23} \end{vmatrix}.$$

It is now easy to check that the equation of the lemma is true for $m = 3$. Let us denote the matrices corresponding to $N_{m-2}^{(m-2)}$, $D_m^{(m-1)}$, $N_{m-1}^{(m-2)}$, $D_{m-1}^{(m-1)}$, and $\widehat{D}_m^{(m-1)}$, by A_1, A_2, A_3, A_4, and A_5, respectively. Note that for $i = 1, \ldots, 5$, the first nonzero entry of the last row of A_i is $f_{m-1,m-1}$ (see (10.9)-(10.14)). Using the last row in A_i as the pivot row, $i = 1, \ldots, 5$, we can reduce A_i into a matrix A_i', where in the column corresponding to $f_{m-1,m-1}$ all other entries are zero. Now by expanding the determinant of these matrices along the column corresponding to $f_{m-1,m-1}$, the left-hand side of the claimed identity reduces to

$$f_{m-1,m-1}^2 |\bar{A}_1||\bar{A}_2| - |\bar{A}_3||\bar{A}_4|, \tag{10.21}$$

and the right-hand side reduces to

$$f_{22} \cdots f_{m-2,m-2} f_{m-1,m-1}^2 |\bar{A}_5|, \tag{10.22}$$

where $\bar{A}_i$'s preserve the format of A_i's. We will demonstrate this for $m = 4$

$$N_2^{(2)} D_4^{(3)} - N_3^{(2)} D_3^{(3)} = \begin{vmatrix} f_{23} & f_{24} \\ f_{33} & f_{34} \end{vmatrix} \begin{vmatrix} f_{12} & f_{13} & f_{15} \\ f_{22} & f_{23} & f_{25} \\ 0 & f_{33} & f_{35} \end{vmatrix} - \begin{vmatrix} f_{23} & f_{25} \\ f_{33} & f_{35} \end{vmatrix} \begin{vmatrix} f_{12} & f_{13} & f_{14} \\ f_{22} & f_{23} & f_{24} \\ 0 & f_{33} & f_{34} \end{vmatrix}$$

$$= \begin{vmatrix} 0 & f_{24}' \\ f_{33} & f_{34} \end{vmatrix} \begin{vmatrix} f_{12} & 0 & f_{15}' \\ f_{22} & 0 & f_{25}' \\ 0 & f_{33} & f_{35} \end{vmatrix} - \begin{vmatrix} 0 & f_{25}' \\ f_{33} & f_{35} \end{vmatrix} \begin{vmatrix} f_{12} & 0 & f_{14}' \\ f_{22} & 0 & f_{24}' \\ 0 & f_{33} & f_{34} \end{vmatrix}$$

$$= f_{33}^2 \left(f_{24}' \begin{vmatrix} f_{12} & f_{15}' \\ f_{22} & f_{25}' \end{vmatrix} - f_{25}' \begin{vmatrix} f_{12} & f_{14}' \\ f_{22} & f_{24}' \end{vmatrix} \right), \tag{10.23}$$

where f_{ij}''s denote the modified entries. On the other hand

$$f_{22} f_{33} \widehat{D}_4^{(3)} = f_{22} f_{33} \begin{vmatrix} f_{13} & f_{14} & f_{15} \\ f_{23} & f_{24} & f_{25} \\ f_{33} & f_{34} & f_{35} \end{vmatrix} = f_{22} f_{33} \begin{vmatrix} 0 & f_{14}' & f_{15}' \\ 0 & f_{24}' & f_{25}' \\ f_{33} & f_{34} & f_{35} \end{vmatrix} = f_{22} f_{33}^2 \begin{vmatrix} f_{14}' & f_{15}' \\ f_{24}' & f_{25}' \end{vmatrix}. \tag{10.24}$$

Dividing (10.23) and (10.24) by f_{33}^2, we are back to the previous case of $m = 2$ as applied to analogous matrices. For the general case of m, we divide both sides of (10.21) and (10.22) by $f_{m-1,m-1}^2$, and use the induction hypothesis to conclude that they are identical. □

Lemma 10.7. *For each $m \geq 2$, we have,*

$$N_{m-2}^{(m-2)} D_m^{(m)} - N_{m-1}^{(m-1)} D_{m-1}^{(m-1)} = -f_{22} \cdots f_{m,m} \widehat{D}_m^{(m-1)}.$$

Proof. For $m = 2$ the formula is valid since from (10.9)-(10.14) we have

$$N_0^{(0)} D_2^{(2)} - N_1^{(1)} D_1^{(1)} = 1. \begin{vmatrix} f_{12} & f_{13} \\ f_{22} & f_{23} \end{vmatrix} - f_{23} f_{12} = -f_{22} f_{13} = -f_{22} \widehat{D}_2^{(1)}.$$

Assume that $m \geq 3$. By expanding the determinant for the matrices corresponding to $D_m^{(m)}$ and $N_{m-1}^{(m-1)}$ about their last row we get

$$N_{m-2}^{(m-2)} D_m^{(m)} - N_{m-1}^{(m-1)} D_{m-1}^{(m-1)} = N_{m-2}^{(m-2)} (-f_{m,m} D_m^{(m-1)} + f_{m,m+1} D_{m-1}^{(m-1)})$$

$$-(-f_{m,m} N_{m-1}^{(m-2)} + f_{m,m+1} N_{m-2}^{(m-2)}) D_{m-1}^{(m-1)}$$

$$= -f_{m,m} (N_{m-2}^{(m-2)} D_m^{(m-1)} - N_{m-1}^{(m-2)} D_{m-1}^{(m-1)}).$$

From Lemma 10.6, the above is equivalent to $-f_{22} \cdots f_{m,m} \widehat{D}_m^{(m-1)}$. □

Proof of Theorem 10.3. Let I_θ be an open ball containing θ such that for any admissible vector of nodes $a = (x_1, \ldots, x_{n+1}) \in K^{n+1}$, with $x_i \in I_\theta$, $x_i \neq \theta$, $i = 1, \ldots, n+1$, we have $D_j^{(j)} \not\equiv 0$, $j = 1, \ldots, m-1$. Set $x_{n+2} = \theta$. We wish to derive the formula for $B_m^{(k)}(x_1, \ldots, x_m)$. Since the superscript k is irrelevant as far as the proof is concerned, we shall omit it throughout the rest of the section. With the interpolating nodes $x_1, \ldots, x_{n+1}$, and $x = \theta = x_{n+2}$, from Theorem 10.1 we have

$$0 = f_{11} + \sum_{i=1}^{n+1} f_{1,i+1} \prod_{l=1}^{i} (\theta - x_l). \tag{10.25}$$

Adding $x_1 - f_{11} = [\theta - (\theta - x_1) - f_{11}]$ to (10.25) we get

$$B_1(x_1) \equiv x_1 - f_{11} = \theta + \sum_{i=1}^{n+1} \gamma_i^{(1)} \prod_{l=1}^{i} (\theta - x_l), \tag{10.26}$$

where

$$\gamma_1^{(1)} = f_{12} - 1, \quad \gamma_i^{(1)} = f_{1,i+1}, \quad i = 2, \ldots, n+1.$$

Also from (10.25) we easily obtain

$$B_2(x_1, x_2) \equiv x_1 - \frac{f_{11}}{f_{12}} = \theta + \sum_{i=2}^{n+1} \gamma_i^{(2)} \prod_{l=1}^{i} (\theta - x_l), \tag{10.27}$$

where

$$\gamma_i^{(2)} = \frac{f_{1,i+1}}{f_{12}} = \frac{\widehat{D}_i^{(1)}}{D_1^{(1)}}, \quad i = 2, \ldots, n+1,$$

and where the second equality above follows from the definitions of $D_1^{(1)}$ and $\widehat{D}_i^{(1)}$ (see (10.10) and (10.11)). Thus Theorem 10.3 is true for $m = 2$. Next we show that the theorem is valid for $m = 3$. Subtracting (10.27) from (10.26) we get

$$B_1(x_1) - B_2(x_1, x_2)$$

$$= -f_{11} + \frac{f_{11}}{f_{12}} = \gamma_1^{(1)}(\theta - x_1) + \sum_{i=2}^{n+1} \left(\gamma_i^{(1)} - \gamma_i^{(2)}\right) \prod_{l=1}^{i} (\theta - x_l). \tag{10.28}$$

Next using the points $x_2, \ldots, x_{n+1}$ as the interpolating nodes and $x = \theta = x_{n+2}$, from Theorem 10.1 we get

$$0 = f_{22} + \sum_{i=2}^{n+1} f_{2,i+1} \prod_{l=2}^{i} (\theta - x_l). \tag{10.29}$$

Multiplying (10.29) by $-\gamma_1^{(1)}(\theta - x_1)$ and (10.28) by f_{22} and then adding the results we get

$$f_{22}(B_1(x_1) - B_2(x_1, x_2)) = f_{11} f_{22} \left(\frac{1 - f_{12}}{f_{12}}\right) = \sum_{i=2}^{n+1} u_i^{(2)} \prod_{l=1}^{i} (\theta - x_l), \tag{10.30}$$

where

$$u_i^{(2)} = f_{22}\left(\gamma_i^{(1)} - \gamma_i^{(2)}\right) - \gamma_1^{(1)} f_{2,i+1}, \quad i = 2, \ldots, n+1.$$

We can easily verify that

$$u_i^{(2)} = \frac{(1 - f_{12}) D_i^{(2)}}{f_{12}}, \quad i = 2, \ldots, n+1.$$

Multiplying (10.30) by

$$-\frac{\gamma_2^{(2)}}{u_2^{(2)}} = -\frac{f_{13}}{(1 - f_{12}) D_2^{(2)}},$$

and adding the result to the equation of $B_2(x_1, x_2)$, and simplifying we obtain a new iteration function defined as

$$B_3(x_1, x_2, x_3) \equiv x_1 - f_{11} \frac{f_{23}}{D_2^{(2)}} = \theta + \sum_{i=3}^{n+1} \gamma_i^{(3)} \prod_{l=1}^{i} (\theta - x_l), \tag{10.31}$$

where

$$\gamma_i^{(3)} = \gamma_i^{(2)} - \frac{\gamma_2^{(2)}}{u_2^{(2)}} u_i^{(2)} = \frac{\widehat{D}_i^{(1)}}{D_1^{(1)}} - \frac{f_{13} D_i^{(2)}}{f_{12} D_2^{(2)}}, \qquad i = 3, \ldots, n+1.$$

It can be verified that

$$\gamma_i^{(3)} = -\frac{\widehat{D}_i^{(2)}}{D_2^{(2)}}, \qquad i = 3, \ldots, n+1.$$

Thus Theorem 10.3 is valid for $m = 3$.

The inductive step: Assume that the theorem is true for $m-1$ and m. Thus for $r = m-1$ and $r = m$ we have

$$B_r(x_1, \ldots, x_r) = x_1 - f_{11} \frac{N_{r-2}^{(r-2)}}{D_{r-1}^{(r-1)}} = \theta + \sum_{i=r}^{n+1} \gamma_i^{(r)} \prod_{l=1}^{i} (\theta - x_l), \tag{10.32}$$

where

$$\gamma_i^{(r)} = (-1)^r \frac{\widehat{D}_i^{(r-1)}}{D_{r-1}^{(r-1)}}, \qquad i = r, \ldots, n+1. \tag{10.33}$$

We shall obtain $B_{m+1}(x_1, \ldots, x_{m+1})$. Subtracting $B_m(x_1, \ldots, x_m)$ from $B_{m-1}(x_1, \ldots, x_{m-1})$, using (10.32), and suppressing their arguments for simplicity we get

$$B_{m-1} - B_m = \gamma_{m-1}^{(m-1)} \prod_{l=1}^{m-1} (\theta - x_l) + \sum_{i=m}^{n+1} (\gamma_i^{(m-1)} - \gamma_i^{(m)}) \prod_{l=1}^{i} (\theta - x_l). \tag{10.34}$$

Using the interpolating nodes $x_m, \ldots, x_{n+1}$, and letting $x = \theta = x_{n+2}$, from Theorem 10.1 we get

$$0 = f_{m,m} + \sum_{i=m}^{n+1} f_{m,i+1} \prod_{l=m}^{i} (\theta - x_l). \tag{10.35}$$

Multiplying (10.35) by $-\gamma_{m-1}^{(m-1)} \prod_{l=1}^{m-1} (\theta - x_l)$ and (10.34) by $f_{m,m}$ and then adding the results we get

$$f_{m,m}(B_{m-1} - B_m) = \sum_{i=m}^{n+1} u_i^{(m)} \prod_{l=1}^{i} (\theta - x_l), \tag{10.36}$$

where

$$u_i^{(m)} = f_{m,m}(\gamma_i^{(m-1)} - \gamma_i^{(m)}) - \gamma_{m-1}^{(m-1)} f_{m,i+1}, \qquad i = m, \ldots, n+1. \tag{10.37}$$

We first need to obtain a simplified form for $u_i^{(m)}$. From (10.37) and the induction hypothesis we have

$$u_i^{(m)} = f_{m,m}\left((-1)^{m-1}\frac{\widehat{D}_i^{(m-2)}}{D_{m-2}^{(m-2)}} - (-1)^m\frac{\widehat{D}_i^{(m-1)}}{D_{m-1}^{(m-1)}}\right)$$

$$-(-1)^{m-1}\frac{\widehat{D}_{m-1}^{(m-2)}}{D_{m-2}^{(m-2)}}f_{m,i+1} = (-1)^{m-1}\frac{M}{D_{m-2}^{(m-2)}D_{m-1}^{(m-1)}}, \qquad (10.38)$$

where

$$M = f_{m,m}\left(D_{m-1}^{(m-1)}\widehat{D}_i^{(m-2)} + D_{m-2}^{(m-2)}\widehat{D}_i^{(m-1)}\right) - f_{m,i+1}D_{m-1}^{(m-1)}\widehat{D}_{m-1}^{(m-2)}.$$

From (10.19) of Lemma 10.5 we get

$$M = f_{m,m}D_i^{(m-1)}\widehat{D}_{m-1}^{(m-2)} - f_{m,i+1}D_{m-1}^{(m-1)}\widehat{D}_{m-1}^{(m-2)}.$$

By expanding the determinant corresponding to $D_i^{(m)}$, see (10.9), along its last row we have

$$D_i^{(m)} = -f_{m,m}D_i^{(m-1)} + f_{m,i+1}D_{m-1}^{(m-1)}.$$

From the above the equation of M can further be simplified as

$$M = -\widehat{D}_{m-1}^{(m-2)}D_i^{(m)}.$$

This implies that

$$u_i^{(m)} = (-1)^m\frac{\widehat{D}_{m-1}^{(m-2)}D_i^{(m)}}{D_{m-2}^{(m-2)}D_{m-1}^{(m-1)}}. \quad i = m, \ldots, n+1. \qquad (10.39)$$

From Lemma 10.1 it follows that $u_m^{(m)}$ is not identically zero. Again from Lemma 10.1 the quantity $-\gamma_m^{(m)}/u_m^{(m)}$ is not identically zero. Multiplying (10.36) by $-\gamma_m^{(m)}/u_m^{(m)}$ and adding the result to the equation of B_m from (i), we get

$$B_{m+1} \equiv B_m - f_{m,m}\frac{\gamma_m^{(m)}}{u_m^{(m)}}(B_{m-1} - B_m)$$

$$= \theta + \sum_{i=m+1}^{n+1}\gamma_i^{(m+1)}(\theta - x_1)\cdots(\theta - x_i), \qquad (10.40)$$

where

$$\gamma_i^{(m+1)} = \gamma_i^{(m)} - \frac{\gamma_m^{(m)}}{u_m^{(m)}} u_i^{(m)}, \quad i = m+1, \ldots, n+1. \tag{10.41}$$

Next we show that

$$\gamma_i^{(m+1)} = (-1)^{m+1} \frac{\widehat{D}_i^{(m)}}{D_m^{(m)}}, \quad i = m+1, \ldots, n+1. \tag{10.42}$$

From (10.39), and induction hypothesis, we have

$$-\frac{\gamma_m^{(m)}}{u_m^{(m)}} u_i^{(m)} = (-1)^{m-1} \frac{\widehat{D}_m^{(m-1)} D_i^{(m)}}{D_m^{(m)} D_{m-1}^{(m-1)}}. \tag{10.43}$$

Substituting (10.33) and (10.43) into (10.41) gives

$$\begin{aligned} \gamma_i^{(m+1)} &= (-1)^m \frac{\widehat{D}_i^{(m-1)}}{D_{m-1}^{(m-1)}} + (-1)^{m-1} \frac{\widehat{D}_m^{(m-1)} D_i^{(m)}}{D_m^{(m)} D_{m-1}^{(m-1)}} \\ &= (-1)^m \frac{\widehat{D}_i^{(m-1)} D_m^{(m)} - \widehat{D}_m^{(m-1)} D_i^{(m)}}{D_m^{(m)} D_{m-1}^{(m-1)}}. \end{aligned} \tag{10.44}$$

But from (10.20) of Lemma 10.5, the numerator of the above fraction is $(-1)^{m+1} \widehat{D}_i^{(m)} D_{m-1}^{(m-1)}$, and hence we get the desired formula for $\gamma_i^{(m+1)}$, claimed in (10.42).

To complete the proof of the theorem we need to derive the desired formula for B_{m+1}. Since

$$u_m^{(m)} = (-1)^m \frac{\widehat{D}_{m-1}^{(m-2)} D_m^{(m)}}{D_{m-2}^{(m-2)} D_{m-1}^{(m-1)}}, \quad \gamma_m^{(m)} = (-1)^m \frac{\widehat{D}_m^{(m-1)}}{D_{m-1}^{(m-1)}}, \tag{10.45}$$

we get

$$\frac{\gamma_m^{(m)}}{u_m^{(m)}} = \frac{\widehat{D}_m^{(m-1)} D_{m-2}^{(m-2)}}{\widehat{D}_{m-1}^{(m-2)} D_m^{(m)}}. \tag{10.46}$$

From (10.40), (10.46), and the inductive hypothesis we have

$$B_{m+1} = B_m - f_{m,m} \frac{\gamma_m^{(m)}}{u_m^{(m)}} (B_{m-1} - B_m)$$

$$= x_1 - f_{11} \; \frac{N_{m-2}^{(m-2)}}{D_{m-1}^{(m-1)}} + f_{11} f_{m,m} \frac{\widehat{D}_m^{(m-1)} D_{m-2}^{(m-2)}}{\widehat{D}_{m-1}^{(m-2)} D_m^{(m)}} \left(\frac{N_{m-3}^{(m-3)}}{D_{m-2}^{(m-2)}} - \frac{N_{m-2}^{(m-2)}}{D_{m-1}^{(m-1)}} \right)$$

$$= x_1 - f_{11} \; \frac{N_{m-2}^{(m-2)}}{D_{m-1}^{(m-1)}}$$

$$+ f_{11} f_{m,m} \frac{\widehat{D}_m^{(m-1)}}{\widehat{D}_{m-1}^{(m-2)} D_m^{(m)} D_{m-1}^{(m-1)}} \left(N_{m-3}^{(m-3)} D_{m-1}^{(m-1)} - N_{m-2}^{(m-2)} D_{m-2}^{(m-2)} \right). \tag{10.47}$$

From Lemma 10.7 we have

$$N_{m-3}^{(m-3)} D_{m-1}^{(m-1)} - N_{m-2}^{(m-2)} D_{m-2}^{(m-2)} = -f_{22} \cdots f_{m-1,m-1} \widehat{D}_{m-1}^{(m-2)}. \tag{10.48}$$

From (10.47) and (10.48) we get

$$B_{m+1} = x_1 - f_{11} \; \frac{N_{m-2}^{(m-2)}}{D_{m-1}^{(m-1)}}$$

$$- f_{11} \cdots f_{m,m} \frac{\widehat{D}_m^{(m-1)}}{D_m^{(m)} D_{m-1}^{(m-1)}}. \tag{10.49}$$

Applying Lemma 10.7 once more to (10.49) we get

$$B_{m+1} = x_1 - f_{11} \; \frac{N_{m-2}^{(m-2)}}{D_{m-1}^{(m-1)}} + f_{11} \frac{\left(N_{m-2}^{(m-2)} D_m^{(m)} - N_{m-1}^{(m-1)} D_{m-1}^{(m-1)} \right)}{D_m^{(m)} D_{m-1}^{(m-1)}}. \tag{10.50}$$

But this trivially simplifies into the desired formula

$$B_{m+1} = x_1 - f_{11} \; \frac{N_{m-1}^{(m-1)}}{D_m^{(m)}}.$$

□

10.5 Applications of Determinantal Formulas

In this section we will give an overview of some of the many applications of the determinantal formulas. Some of these results will be dealt with in detail in subsequent chapters.

10.5.1 *Infinite Spectrum of Rational Approximation Formulas*

Motivated by Theorem 10.2, once we have computed $P_n(x)$, the Taylor polynomial corresponding to a given function $f(x)$, for each $m \geq 2$, we can associate a spectrum of n rational approximations, i.e.

$$f(x) \quad \approx \quad \pi_{n,j}^{(m)}(x) \equiv P_{m,j}(x, P_n(x), a), \quad j = m-1, \ldots, (m+n-2).$$

These approximations give rise to a table with an infinite number or rows and n columns whose last column entries coincide with $P_n(x)$. We shall refer to this table as the P_n-table of rational approximations, see Table 10.1.

Table 10.1 The P_n-Table of Rational Approximations $\pi_{n,j}^{(m)}(x)$

$m \setminus j$	$m-1$	m	$m+1$	$\ldots$	$m+n-2$
2	$\pi_{n,1}^{(2)}(x)$	$\pi_{n,2}^{(2)}(x)$	$\pi_{n,3}^{(2)}(x)$	$\ldots$	$\pi_{n,n}^{(2)}(x)$
3	$\pi_{n,2}^{(3)}(x)$	$\pi_{n,3}^{(3)}(x)$	$\pi_{n,4}^{(3)}(x)$	$\ldots$	$\pi_{n,n+1}^{(3)}(x)$
$\vdots$	$\vdots$	$\vdots$	$\vdots$	$\ldots$	$\vdots$
m	$\pi_{n,m-1}^{(m)}(x)$	$\pi_{n,m}^{(m)}(x)$	$\pi_{n,m+1}^{(m)}(x)$	$\ldots$	$\pi_{n,m+n-2}^{(m)}(x)$
$\vdots$	$\vdots$	$\vdots$	$\vdots$	$\ldots$	$\vdots$

From Theorem 10.1, for each $j = m-1, \ldots, m+n-2$, we obtain the error formula

$$R_{m,j}(x, P_n(x), a) \equiv f(x) - \pi_{n,j}^{(m)}(x) = R_n(x) + P_n(x) - \pi_{n,j}^{(m)}(x).$$

In some cases (see example below) even the error of the special functions

$$\pi_{m-1,m-1}^{(m)}(x) = f(x_1) + \frac{D^{(m-1)}(P_{m-1}(x), a_m)}{N^{(m-2)}(P_{m-1}(x), a_m)}(x - x_1),$$

can be better than the error of using $P_{m-1}(x)$, i.e.

$$|R_{m,m-1}(x, P_{m-1}(x), a)| \leq |R_{m-1}(x)|. \tag{10.51}$$

This is surprising in the sense that $\pi_{m-1,m-1}^{(m)}(x)$ makes use of the same information as available to $P_{m-1}(x)$. Its computation only requires the evaluation of the two additional determinants, $D^{(m-1)}(P_{m-1}(x), a_m)$, and $N^{(m-1)}(P_{m-1}(x), a_m)$. In general, given an arbitrary function f, by considering the error inequality, (10.51), we may be able to determine if for a given x, $\pi_{n,j}^{(m)}(x)$ provides a better approximation than $P_n(x)$.

Example 10.1. Consider the case where $f(x) = e^x$, $a = 0$. Then $P_{m-1}(x) = 1 + x + x^2/2! + \cdots + x^{m-1}/(m-1)!$. The matrix corresponding to $D^{(m-1)}(P_{m-1}(x), 0)$ is the Toeplitz matrix whose subdiagonal entries are $1 - P_{m-1}(x)$, the remaining lower triangular part is filled with zeros, and for each row the entries starting with the diagonal entry are filled with the numbers $1, 1/2!, -1/3!, 1/4!$, etc. For instance

$$D^{(5)}(P_5(x), 0) = \begin{vmatrix} 1 & \frac{1}{2!} & \frac{1}{3!} & \frac{1}{4!} & \frac{1}{5!} \\ 1 - P_5(x) & 1 & \frac{1}{2!} & \frac{1}{3!} & \frac{1}{4!} \\ 0 & 1 - P_5(x) & 1 & \frac{1}{2!} & \frac{1}{3!} \\ 0 & 0 & 1 - P_5(x) & 1 & \frac{1}{2!} \\ 0 & 0 & 0 & 1 - P_5(x) & 1 \end{vmatrix}.$$

Note that $\pi^{(m)}_{m-1,m-1}(1) = 1 + D^{(m-1)}(P_{m-1}(1), 0)/D^{(m-2)}(P_{m-1}(1), 0)$. Simple calculation gives Table 10.2.

Table 10.2 Approximation of $e = 2.7182\cdots$

m	2	3	4	5	6	7
$P_{m-1}(1)$	2.0000	2.5000	2.6667	2.7083	2.7167	2.7181
$\pi^{(m)}_{m-1,m-1}(1)$	2.0000	2.7500	2.7071	2.7130	2.7174	2.7182

We see that $\pi^{(m)}_{m-1,m-1}(1)$, $m = 3 - 7$, consistently gives a better approximation to e, than $P_{m-1}(1)$.

The verification of this fact for all m is nontrivial and will not be done here. However, using Theorem 10.2 and a determinantal lower bound we will prove that for each $x \in [0, 1/8]$, $\pi^{(m)}_{m-1,m-1}(x)$ converges to e^x. Thus, e itself can be approximated by $\pi^{(m)}_{m-1,m-1}(1/8)^8$. Before formal statement we will first need to state an auxiliary result.

Theorem 10.4 (Determinantal Lower Bound, Kalantari (1997)). *Let A be an $n \times n$ real or complex matrix. Assume we are given positive numbers l and u such that if λ is an eigenvalue of A, then $l \leq |\lambda| \leq u$, where $|\lambda|$ is the modulus of λ. Let*

$$\kappa = \kappa(l, u) \equiv \lceil \frac{nu - |trace(A)|}{u - l} \rceil.$$

If $|trace(A)| \geq nl$, then we have

$$l^{\kappa} u^{n-\kappa} \leq |det(A)|.$$

□

Now we prove the following

Theorem 10.5 (Approximation of e^x, Kalantari (2000a)). *Let $f(x) = e^x$, $a = 0$. Then for each $x \in I = [0, 1/8]$, the sequence of functions*

$$P_{m,m-1}(x, e^x, 0) = 1 + x\,\frac{D^{(m-1)}(e^x, 0)}{N^{(m-2)}(e^x, 0)}, \qquad m = 2, 3, \cdots,$$

converge to e^x satisfying

$$|e^x - P_{m,m-1}(x, e^x, 0)| \le (\sqrt[4]{200x})^m.$$

Moreover, the sequence

$$\pi^{(m)}_{m-1,m-1}(x) = 1 + x\,\frac{D^{(m-1)}(P_{m-1}(x), 0)}{N^{(m-2)}(P_{m-1}(x), 0)}, \qquad m = 2, 3, \cdots,$$

satisfies

$$|P_{m-1}(x) - \pi^{(m)}_{m-1,m-1}(x)| \le (\sqrt[4]{200x})^m.$$

In particular, since $P_{m-1}(x)$ converges to e^x, so does $\pi^{(m)}_{m-1,m-1}(x)$.

Proof. From Corollary 10.1, we need to bound the error term $|R_{m,m-1}(x, e^x, 0)| = |\widehat{D}^{(m-1)}_m(e^x, 0)|/\ |D^{(m-2)}_m(e^x, 0)|$. The matrix corresponding to $D^{(m-2)}(e^x, 0)$ is the Toeplitz matrix whose subdiagonal entries are $1 - e^x$, the remaining lower triangular part is filled with zeros, and for each row the entries starting with the diagonal entry are filled with the numbers $1, 1/2!, -1/3!, 1/5!$, etc. For instance

$$D^{(5)}(e^x, 0) = \begin{vmatrix} 1 & \frac{1}{2!} & \frac{1}{3!} & \frac{1}{4!} & \frac{1}{5!} \\ 1 - e^x & 1 & \frac{1}{2!} & \frac{1}{3!} & \frac{1}{4!} \\ 0 & 1 - e^x & 1 & \frac{1}{2!} & \frac{1}{3!} \\ 0 & 0 & 1 - e^x & 1 & \frac{1}{2!} \\ 0 & 0 & 0 & 1 - e^x & 1 \end{vmatrix}.$$

For $x \in I$, the matrix corresponding to $D^{(m-2)}(e^x, 0)$ is diagonally dominant, so that we can apply Theorem 10.4. We will use Hadamard's inequality to find an upper bound on the error determinant $\widehat{D}^{(m-1)}_m(e^x, 0)$. We use the fact that it is equivalent to $\widehat{\delta}^{(m-1)}_m(e^x, \widehat{\xi}, 0)$, where $\widehat{\xi} \in \Re^{m-1}$, with each component in I. For $x \in I$, we have $e^x - 1 \le 1/2$. Thus each column of the matrix of $\widehat{\delta}^{(m-1)}_m$, except possibly for its last column, is a vector whose Euclidean norm is bounded above by

$$\left(1 + \frac{1}{2} + \frac{1}{2^2} + \frac{1}{(3!)^2} + \frac{1}{(5!)^2} + \cdots\right)^{1/2} \le \left(1 + \frac{1}{2} + \frac{1}{2^2} + \frac{1}{2^3} + \frac{1}{2^4} + \cdots\right)^{1/2} = \sqrt{2}.$$

In fact we claim that the norm of the last column, u_{m+1}, is also bounded by the same number. Note that by using the same bounding approach,

$$\|u_{m+1}\| \le \sqrt{e}(\frac{1}{2^2} + \frac{1}{(3!)^2} + \frac{1}{(5!)^2} + \cdots)^{1/2} \le \sqrt{e/2}.$$

Hence from Hadamard's inequality we have

$$|\widehat{\delta}_m^{(m-1)}| \le \sqrt{2}^{\frac{m-1}{2}}.$$

To find a lower bound on $|D^{(m-2)}(1/2, 0)|$, we apply Theorem 6 with $n = m-2$, $trace = m-2$. We must have $l = 1-\alpha$, $u = 1+\alpha$, for some α. Thus $\kappa = \lceil (m-2)/2 \rceil$. For simplicity assume m is even. Note that

$$\alpha = (e^x - 1) + \frac{1}{2!} + \frac{1}{3!} + \frac{1}{4!} + \cdots = e^x + e - 3 \le 0.9.$$

Thus

$$|D^{(m-1)}(1/2, 0)| \ge (1-\alpha^2)^{\frac{m-2}{2}} \ge (0.1)^{\frac{m-2}{2}}.$$

Therefore we get the following error bound

$$|R_{m,m-1}(x, e^x, 0)| \le (10\sqrt{2})^{\frac{m-1}{2}}\ x^m \le (\sqrt[4]{200}x)^m.$$

This completes the first part of the proof, since for $x \in I$, $\sqrt[4]{200}x < 1$. By applying Corollary 10.1 to $f(x) = P_{m-1}(x)$, then using similar techniques we can obtain the claimed bound on $|P_{m-1}(x) - \pi_{m-1,m-1}^{(m)}(x)|$. The convergence of $\pi_{m-1,m-1}^{(m)}(x)$ to e^x now follows from the bound on $|P_{m-1}(x) - \pi_{m-1,m-1}^{(m)}(x)|$ and since $P_{m-1}(x)$ converges to e^x. □

10.5.2 *Infinite Spectrum of Rational Inverse Approximation Formulas*

Using Theorem 10.2, we also define formulas that approximates x, given $y = f(x)$. Note that

$$Q_{m,m-1}(x, y, a) = x_1 + (y - f(x_1))\frac{N^{(m-2)}(y, a_m)}{D^{(m-1)}(y, a_m)},$$

does not depend on x. By itself it gives rise to an iterative method. It also gives rise to a spectrum of inverse approximations. For each given pair (m, n), $m \ge 2$ we associate a spectrum of n approximations, i.e.

$$x \approx \kappa_{n,j}^{(m)}(y, a) \equiv Q_{m,j}\Big(Q_{m,m-1}(x, y, a), y, a\Big), \quad j = m-1, \ldots, (m+n-2).$$

These approximations give rise to a table with an infinite number of rows and n columns whose first column entries are $Q_{m,m-1}$, see Table 10.3.

Table 10.3 The $Q_{m,m-1}$-Table of Inverse Rational Approximations $\kappa_{n,j}^{(m)}(x)$

$m \setminus j$	$m-1$	m	$m+1$	$\dots$	$m+n-2$
2	$\kappa_{n,1}^{(2)}(x)$	$\kappa_{n,2}^{(2)}(x)$	$\kappa_{n,3}^{(2)}(x)$	$\dots$	$\kappa_{n,n}^{(2)}(x)$
3	$\kappa_{n,2}^{(3)}(x)$	$\kappa_{n,3}^{(3)}(x)$	$\kappa_{n,4}^{(3)}(x)$	$\dots$	$\kappa_{n,n+1}^{(3)}(x)$
$\vdots$	$\vdots$	$\vdots$	$\vdots$	$\dots$	$\vdots$
m	$\kappa_{n,m-1}^{(m)}(x)$	$\kappa_{n,m}^{(m)}(x)$	$\kappa_{n,m+1}^{(m)}(x)$	$\dots$	$\kappa_{n,m+n-2}^{(m)}(x)$
$\vdots$	$\vdots$	$\vdots$	$\vdots$	$\dots$	$\vdots$

In fact one can also define recursive inverse approximations, as follows:

$$\widehat{\kappa}_{n,m-1}^{(m)}(y,a) \equiv Q_{m,m-1}(x,y,a),$$

$$\widehat{\kappa}_{n,j}^{(m)}(y,a) \equiv Q_{m,j}\left(Q_{m,j-1}(x,y,a),y,a\right), \quad j = m,\dots,n.$$

However, even the family $Q_{m,m-1}$, $m = 2,3,\dots$, is a significant family by itself. For instance when $y = f(x) = 0$, and a is 1-point, then $Q_{m,m-1} = B_m^{(1)}$ gives:

Theorem 10.6 (Determinantal approximation of square-roots). *Let $\nu > 1$ be given. Let $f(x) = x^2 - \nu$. Assume that a satisfies the inequalities $2a - (a^2 - \nu + 1) \geq 1$, and $0 \leq a - \sqrt{\nu} < 1$. Then,*

$$|\sqrt{\nu} - B_m^{(1)}(a)| \leq \frac{|\sqrt{\nu} - a|^m}{(2a+1)^{\lfloor \frac{(m-1)}{2} \rfloor}}, \quad B_m^{(1)}(a) = a - f(a)\frac{D^{(m-2)}(0,a)}{D^{(m-1)}(0,a)}.$$

Note that $D^{(m-1)}(0,a)$ is a tridiagonal Toeplitz matrix whose upper diagonal, diagonal, and subdiagonal entries are filled with the numbers 1, $2a$, and $a^2 - \nu$, respectively. For instance for $m = 5$ we get the following general case and the special case of $\nu = 2$, $a = 2$

$$D^{(4)}(0,a) = \begin{vmatrix} 2a & 1 & 0 & 0 \\ a^2-\nu & 2a & 1 & 0 \\ 0 & a^2-\nu & 2a & 1 \\ 0 & 0 & a^2-\nu & 2a \end{vmatrix}.$$

Theorem 10.7 (Approximation of π, Kalantari (2000b)). *Let $f(x) = \sin x - 0.5$, $a = 0$. Then,*

$$|\frac{\pi}{6} - B_m^{(1)}(0)| \leq (\frac{\sqrt{3}\pi}{6})^{m-1}, \quad B_m^{(1)}(0) = \frac{1}{2}\frac{D^{(m-2)}(0,0)}{D^{(m-1)}(0,0)}.$$

In this case the matrix $D^{(m-1)}(0,0)$ is the Toeplitz matrix whose subdiagonal entries are $-1/2$, the remaining lower triangular part is filled with zeros, and for each row the entries starting with the diagonal entry are filled with the numbers $1, 0, -1/3!, 0, 1/5!, 0, -1/7!$, etc. For instance

$$D^{(5)}(0,0) = \begin{vmatrix} 1 & 0 & \frac{-1}{3!} & 0 & \frac{1}{5!} \\ \frac{-1}{2} & 1 & 0 & \frac{-1}{3!} & 0 \\ 0 & \frac{-1}{2} & 1 & 0 & \frac{-1}{3!} \\ 0 & 0 & \frac{-1}{2} & 1 & 0 \\ 0 & 0 & 0 & \frac{-1}{2} & 1 \end{vmatrix}.$$

Remark 10.8. Many other convergent sequences to π are possible. Also, it is possible to construct high order iterative methods of arbitrary order, starting from the initial point (Kalantari (2000b)). These give novel methods for approximation of π, very different from the existing formulas, see Chapter 14.

10.5.3 *Infinite Families of Single and Multipoint Iteration Functions*

Consider the function defined in Theorem 10.3:

$$B_m^{(k)}(a_m) = x_1 - f(x_1)\,\frac{N^{(m-2)}(0, a_m)}{D^{(m-1)}(0, a_m)},$$

where a_m is a monotonic k-point admissible vector in K^m. This function defines a k-point iteration function. The case of monotonic k-point is the most meaningful of the k-points. For each $i = 1, \dots, m-1$, locally, it is natural to take x_i as a more accurate approximation of the root of f being approximated, than x_{i+1}. More precisely, given an initial monotonic k-point vector

$$a_m = (x_1, \dots, x_1, x_2, \dots, x_k),$$

the *fixed-point* iteration is the substitution

$$a_m \quad \longleftarrow \quad \left(B_m^{(k)}(a_m), \dots, B_m^{(k)}(a_m), x_1, \dots, x_{k-1}\right) \in K^m.$$

For $k = 1, 2$, $B_2^{(k)}(a_2)$ gives Newton's method and the secant method, respectively. For $k = 1$, $B_3^{(k)}(a_3)$ gives the ordinary Halley's method, and for $k = 2$ and $k = 3$, it gives two- and three-point variations of that method. There are four versions of $B_4^{(k)}(a_4)$.

Theorem 10.8 (Order of Convergence of Multipoint Family, Kalantari (1999)). *The sequence of fixed-point iterates* $\{x_1^{(t)}\}_{t=0}^{\infty}$ *satisfy*

$$\lim_{t\to\infty} \frac{(\theta - x_1^{(t+1)})}{(\theta - x_1^{(t)})^p} = c_m(\theta)^{\left(\frac{p-1}{m-1}\right)}, \quad c_m(\theta) = \frac{(-1)^{m-1}}{p'(\theta)^{m-1}} \widehat{D}_m^{(m-1)}(0,\theta),$$

p is the unique positive root of $P(z) = z^k - (m-k+1)z^{k-1} - \sum_{j=0}^{k-2} z^j$. *Moreover, for $k > 1$, $m-k+1 < p < m-k+2$; for fixed k it is monotonically increasing in m; and for fixed m it is monotonically decreasing in k, ranging from m to the limiting ratio of generalized Fibonacci numbers of order m.*

In Kalantari and Park (2001), computational results with the first few members of the basic family for random polynomials, and using the first nine members of the family $B_m^{(k)}$, for $m = 2, 3, 4$, and $k \leq m$ reveals the following: As the degree of the polynomial increases, the new iteration functions become more and more efficient than the traditional methods. For a given polynomial, all nine iteration functions were applied with the same seed and its total arithmetic operations for approximation of the same root was compared with that of Newton's. The results indicate that Newton's method, the secant method, and Halley's method are less efficient than the other six iteration functions. For larger degree polynomials, the iteration function

$$B_4^{(4)}(a_4) = x_1 - f_{11} \frac{\begin{vmatrix} f_{23} & f_{24} \\ f_{33} & f_{34} \end{vmatrix}}{\begin{vmatrix} f_{12} & f_{13} & f_{14} \\ f_{22} & f_{23} & f_{24} \\ 0 & f_{33} & f_{34} \end{vmatrix}},$$

which is derivative-free and has almost quadratic order of convergence is the most efficient iteration function. Thus these iteration functions should find practical applications in root-finding. These will be detailed in Chapter 12.

10.5.4 *Determinantal Approximation of Roots of Polynomials*

Here we describe yet another application of the Basic Family and the determinantal Taylor's Theorem. We can approximate polynomial roots by evaluation of the sequence of iteration functions $\{B_m^{(1)}\}_{m=2}^{\infty}$, all at the same input. The following surprising result holds:

Theorem 10.9 (Kalantari (1998a)). *Let $f(x)$ be a polynomial with complex coefficients. Let θ be a simple root of $f(x)$. There exists $r^* \in (0,1)$ such that given any $x_0 \in N_{r^*}(\theta) = \{z : |z-\theta| \leq r^*\}$, we have*

$$\theta = \lim_{m\to\infty} B_m^{(1)}(x_0) = \lim_{m\to\infty} x_0 - f(x_0)\frac{D^{(m-2)}(0,x_0)}{D^{(m-1)}(0,x_0)}.$$

More precisely, let

$$w(x) = \frac{1}{f'(x)}\sqrt{\sum_{i=0}^{n}\left(\frac{|f^{(i)}(x)|}{i!}\right)^2}.$$

There exists $r^* \in (0,1)$ such that given any $x_0 \in N_{r^*}(\theta)$, for all $m \geq 2$, we have

$$|B_m^{(1)}(x_0)-\theta| \leq \frac{|2w(x_0)(x_0-\theta)|^m}{1-|x_0-\theta|} \leq \frac{|4w(\theta)(x_0-\theta)|^m}{1-|x_0-\theta|}, \quad |4w(\theta)(x_0-\theta)| < 1.$$

It is important to mention that the determinant $D^{(m)}(x)$ satisfies the recursive property:

Proposition 10.1 (Kalantari (1998a)). *For $j < 0$, set $D^{(j)}(x) = 0$. Then, for all $m \geq 1$ we have,*

$$D^{(m)}(x) = \sum_{i=1}^{n}(-1)^{i-1}\frac{(f(x))^{i-1}f^{(i)}(x)}{i!}D^{(m-i)}(x).$$

The proof which is straightforward is in fact based on a recursive formula for computing the determinant of an arbitrary Toeplitz matrix that is also upper Hessenberg. The formula becomes apparent once the determinant of such a matrix is expanded along its first column.

Proposition 10.1 implies that $D^{(m)}(x)$ can be computed efficiently. It should also be mentioned that the entries of $D^{(m)}$ can be evaluated efficiently since given $f(x)$, a polynomial of degree n, and an input x_0, the evaluation of $f^{(j)}(x_0)/j!$, $j = 0,\ldots,n$, called the *normalized derivatives*, can be established in $O(n\log^2 n)$ arithmetic operations, see Kung (1974).

10.5.5 *A Rational Expansion Formula and Connection to Padé Approximant*

Consider the determinantal formula $f = P_{m,n} + R_{m,n}$ for f real-valued. For $m = 2$, this formula gives Taylor's Theorem (with confluent divided differences). Also, for $m = 3$, we can convert the formula into a formula purely in terms of $f(x)$, i.e. $f(x) = H_n(x,a) + E_n(x,a)$, where $H_n(x,a)$ is a rational function, and $E_n(x,a)$ is the error term. This expansion gives an alternate approximation to Taylor expansion, and its inverse form $x = Q_{3,2}(x,0,a) + \rho_{3,2}(x,0,a)$, gives rise to the famous Halley's iteration function. Also, $H_2(x,a)$ gives a Padé approximant. To derive $H_n(x,a)$, let us consider the case where a is one-point. Note that

$$D^{(2)}(y,a) = \begin{vmatrix} f'(a) & \frac{f''(a)}{2} \\ f(a)-f(x) & f'(a) \end{vmatrix},$$

$$\widehat{D}_i^{(2)}(y,a) = \begin{vmatrix} \frac{f''(a)}{2} & \frac{f^{(i)}(a)}{i!} \\ f'(a) & \frac{f^{(i-1)}(a)}{(i-1)!} \end{vmatrix}, \quad i = 3,\ldots,n.$$

Now substituting in the formula $f(x) = P_{3,n}(x, f(x), a) + R_{3,n}(x, f(x), a)$, and solving for $f(x)$, with the assumption that $g(x) = 2f'(a) - (x-a)f''(a)$ is nonzero, which is valid locally if $f'(a) \neq 0$, we get $f(x) = H_n(x,a) + E_n(x,a)$, where

$$H_n(x,a) = f(a) + \frac{2(x-a)f'(a)^2}{g(x)} - 2\sum_{i=3}^{n} \frac{\widehat{D}_i^{(2)}(y,a)}{g(x)} f'(a)(x-a)^i,$$

$$E_n(x,a) = -\frac{2f'(a)\widehat{\Delta}_{n+1}^{(2)}(x,\xi,a)}{g(x)}(x-a)^{n+1}, \quad \xi \in Span(a,x),$$

$$\widehat{\Delta}_{n+1}^{(2)}(x,\xi,a) = \begin{vmatrix} \frac{f''(a)}{2} & f_{1,n+2} \\ f'(a) & f_{2,n+2} \end{vmatrix}$$

$$= \frac{f^{(n+1}(\xi)}{(n+1)!}\left(\frac{f''(a)}{2}(x-a) - f'(a)\right) + \frac{f^{(n)}(a)}{n!}\frac{f''(a)}{2}.$$

If for each $x \in I$, $f^{(j)}(x)$ exists for all j, and $f^{(j)}(x)/j!$ converges to zero, as j approaches infinity, then the finite summation can be replaced with an infinite expansion. We consider an example next.

We claim that $H_2(x,a)$ is the Padé approximant $[1/1]$. In general the $[L/M]$ Padé approximant to a function $f(z) = \sum_{i=0}^{\infty} c_i(z-a)^i$, where $c_i = \frac{f^{(i)}(a)}{i!}$, normally considered for the Maclaurian expansion, $a = 0$, and the case of analytic functions, refers to a rational approximation $[L/M] \equiv p_L(z)/g_M(z)$ of f, where the polynomials $p_L(z) = \sum_{i=0}^{L} a_i(z-a)^i$, and $q_M(z) = \sum_{i=0}^{M} b_i(z-a)^i$ are to be chosen so that the expansion of $[L/M]$ at a agrees with that of f, as far as possible. Writing $q_M(z)f(z) =$

$p_L(z) + O((z-a)^{L+M+1})$, and by equating the coefficients of $(z-a)^i$, $i = L, \ldots, L+M$, we obtain a set of linear equations,

$$b_M c_{L-M+j} + b_{M-1} c_{L-M+j+1} + \cdots + b_0 c_{L+j} = 0, \quad j = 1, \ldots, M,$$

where $c_k = 0$, if $k < 0$. Without loss of generality we can take $b_0 = 1$. The above system can then be solved for b_i, $i = 0, \ldots, M$. Given the b_i's, we can compute the coefficients a_j, $j = 0, \ldots, M$, by equating the coefficients of z_k, $k = 1, \ldots, L+M$. In particular for $L = M = 1$ this gives,

$$b_0 = 1, \quad c_1 b_1 + c_2 = 0, \quad a_0 = c_0, \quad a_1 = c_1 + b_1 c_0.$$

Alternatively, p_L and q_M can be described in a determinantal form, where their evaluation does not require their coefficients explicitly. For the vast theory of Padé approximants which usually takes place over the complex plane, see Baker and Graves-Morris (1996).

Upon solving the above equations we see that the corresponding Padé approximant, [1/1], is in fact $H_2(x,a)$. When viewed in the context of Padé approximants, it is interesting that the corresponding error, i.e. $E_2(x,a)$, becomes available thorough our formulas. For instance, the following example shows that the [1/1] approximant to the exponential function gives a better approximation than the quadratic Taylor polynomial when $x \in [0, 0.5]$. Moreover, it gives an infinite expansion of e^x at zero, different from the ordinary Taylor series. In general, $H_n(x,a)$ offers an approximation to Padé approximants $[n,1]$, $n \geq 3$.

Example 10.2. Consider the case where $f(x) = e^x$, $a = 0$ and any $x \neq 2$. Then we have

$$H_n(x,a) = \frac{2+x}{2-x} + \frac{1}{2-x} \sum_{i=3}^{n} \frac{i-2}{i!} x^i,$$

$$E_n(x,0) = \left(\frac{e^\xi}{(n+1)!} - \frac{1}{n!(2-x)} \right) x^{n+1},$$

where ξ lies between 0 and x. In particular we get the following expansion

$$e^x = 1 + \frac{2x}{2-x} + \frac{1}{2-x} \sum_{i=3}^{\infty} \frac{i-2}{i!} x^i,$$

and for $x = 1$ we get

$$e = 3 - \sum_{i=3}^{\infty} \frac{i-2}{i!}.$$

Let us compare $H_n(x,0)$ with the ordinary Taylor polynomial, $P_n(x) = 1 + x + \cdots + x^n/n!$, $R_n(x) = [e^{\xi}/(n+1)!]x^{n+1}$. It is easy to verify that for $x \in [0, 0.5]$, $H_2(x,0) = (2+x)/(2-x)$ is a better approximation than $P_2(x)$, i.e. $|E_2(x,0)| \leq |R_2(x)|$. Also, for $n \geq 3$, for all $x \in [0, 2-(n+1)/(n+2)]$, we have $|E_n(x,0)| \leq |P_{n-1}(x)|$.

More generally, for $m \geq 3$, to solve the determinantal interpolation formula purely in $f(x)$, requires the computation of roots of a polynomial of degree $(m-2)$. Thus the case of $m=4$ can be solved explicitly, giving two solutions. The cases of $m = 5, 6$ can also be solved explicitly using well-known but more difficult formulas. Aside from the unsolvability of polynomial equations of degree greater than four (via closed formulas in radicals), for $m \geq 4$, one does not obtain a rational approximation to $f(x)$. Rather than trying to express the formulas purely in terms of $f(x)$, our spectrum of rational approximation formulas discussed earlier makes it possible to give direct rational approximations using the ordinary Taylor approximation, $P_n(x)$.

10.5.6 *Algebraic Approximation Formulas*

We close this section by pointing out that there is yet another type of approximation. For a given x, we can take as an approximation of $f(x)$, any solution Φ of the equation

$$\Phi = P_{m,n}(x, \Phi, a).$$

From Theorem 10.2 it follows that if $n = m + \nu - 2$, then the Taylor polynomial $P_\nu(x, a)$ is an algebraic solution. Analogously, given $f(x)$, we can take as an approximation to x any solution of the equation

$$X = Q_{m,n}(X, f(x), a).$$

Algebraic approximations are polynomial root-finding problem. One possible way to approximate the desired roots is to start with a good initial approximation Φ_k to $f(x)$, and iteratively compute the next iterate as

$$\Phi_{k+1} = P_{m,n}(x, \Phi_k, a).$$

Analogously, given an initial approximation X_k to $f^{-1}(x)$, set $X_{k+1} = Q_{m,n}(X_k, f(x), a)$.

Example 10.3. Consider e^x, $a = 1$, $x = 1$, $n = m - 1$. Then for $m = 3$ the equation that need to be solved is

$$\Phi - 1 = \begin{vmatrix} 1 & \frac{1}{2!} \\ 1-\Phi & 1 \end{vmatrix},$$

having the solution $\Phi_2 = 3$, as an approximation to e. For $m = 4$, we get the equation

$$\Phi - 1 = \frac{\begin{vmatrix} 1 & \frac{1}{2!} & \frac{1}{3!} \\ 1-\Phi & 1 & \frac{1}{2!} \\ 0 & 1-\Phi & 1 \end{vmatrix}}{\begin{vmatrix} 1 & \frac{1}{2!} \\ 1-\Phi & 1 \end{vmatrix}}.$$

The best root of the above equation is $\Phi_3 = 1+\sqrt{3} \approx 2.732$. As we proved in this section for $x \in [0, 1/8]$, the sequence $\{P_{m,m-1}(x, e^x, 0)\}_{m=2}^{\infty}$ converges to e^x. This implies that for such x, the algebraic solutions should also converge to e^x.

10.6 Concluding Remarks

In this chapter we have developed determinantal formulas, $f = P_{m,n}+R_{m,n}$, and $x = Q_{m,n}+\rho_{m,n}$, generalizing Taylor's Theorem with confluent divided differences over the real or complex field. We used these formulas to describe several schemes for the approximation of functions, or their inverses.

On the one hand, while the Taylor polynomial provides a single approximation, $P_n(x)$ to a given function, via the determinantal formulas for each $m \geq 2$, $P_n(x)$ unfolds into a set of n approximations $\pi_{n,j}^{(m)}(x) \equiv P_{m,j}(x, P_n(x), a)$, $j = m-1, \ldots, (m+n-2)$. This set includes $P_n(x)$ itself.

On the other hand, even the particular family, $\pi_{m-1,m-1}^{(m)}$, viewed as a sequence in m, provides a determinantal approximation to functions (see e.g. the approximation of e^x, Theorem 10.5). For the case of distinct nodes, the evaluation of $\pi_{m-1,m-1}^{(m)}$ (or $Q_{m,m-1}$) can be done in $O(m^2)$ arithmetic operations, which is the same as the number of operations needed to compute Newton's interpolating polynomial P_{m-1}.

In case P_{m-1} is the classic one-point Taylor polynomial, the determinants in the corresponding $\pi_{m-1,m-1}^{(m)}$ are Toeplitz determinants and can be computed efficiently (see Golub and Loan (1996), Chan and Ng (1996), and Bini and Pan (1994)). In fact the determinants in this case are special correspond to matrices that are both Toeplitz and upper Hessenberg and can be computed via a linear recurrence relation (see Chapter 9).

Another interesting family is $\pi_{m-1,m}^{(m)}$. Yet a third family is $\pi_{1,m-1}^{(m)}$, where for $m \geq 4$, it is the quotient of two polynomials of degree $m-1$, and $m-3$, respectively. As rational approximants, the latter family provides an alternative approximation to the Padé approximants $[(m-1)/(m-3)]$.

A very important application of the determinantal formulas is in terms of inverse approximation, as direct formulas, or as iteration functions. Firstly, the one-point version of $Q_{m,m-1}$, when viewed as a sequence in m, gives a determinantal approximation of the inverse of f. Secondly, $Q_{m,m-1}$ gives rise to an infinite table of inverse approximations. Thirdly, for fixed m, $Q_{m,m-1}$ gives rise to m different k-point iteration functions, $B_m^{(k)}$.

Using $Q_{m,m-1}$ as a sequence, we can derive determinantal sequences converging to a root of an arbitrary polynomial, based on the evaluation of $B_m^{(1)}$, all at a single input (see Chapter 9). Such technique can be proven to hold for more general functions. For instance, we can approximate π using very novel formulas (see Kalantari (2000b), Chapter 14).

The determinantal family, $B_m^{(k)}$, is a fundamental family of iteration functions for root-finding with many ideal features. These features are in contrast to iteration functions based on the use of Padé approximants (see Baker and Graves-Morris (1996), Traub (1964)), the Euler-Schröder family (see Traub (1964), Henrici (1974), Householder (1970), Smale (1985), also Chapter 7 for a simple recursive formula that generates this family), or the methods based on continued fractions (see Jones and Thron (1980), Brezinski (1990), and Cuyt *et al.* (2008) for the general theory and history of continued fraction).

Despite the fact that there are good practical and theoretical methods for finding all roots of polynomials, e.g. see Pan (1994), Pan (1997), there is no reason to believe that these methods cannot improve upon the existing algorithms. In particular, it is natural to expect practical applications of the family $B_m^{(k)}$, even in polynomial root-finding. Indeed the computational results in Kalantari and Park (2001) (see Chapter 12) confirm the practical significance of this family.

Many interesting research problems remain to be investigated. The results in the present chapter and its many derivatives of them, some of which were summarized and will be analyzed in subsequent chapters offer sufficient evidence for the importance of the formulas. Polynomiography is yet another application of the iteration functions, offering infinitely many methods for the visualization of even a single polynomial.

Chapter 11

The Multipoint Basic Family and its Order of Convergence *

In this chapter we will analyze the order of convergence of the multipoint Basic Family for functions of real variable. For each natural number m greater than one, and each natural number k less than or equal to m, there exists a multipoint version of the Basic Family, defined as the ratio of two determinants that depend on the first $m-k$ derivatives of the given function. For fixed m, as k increases, the order decreases from m to the positive root of the characteristic polynomial of Generalized Fibonacci numbers of order m. For fixed k, the order increases in m. Newton and Halley methods and their multipoint versions are members of the multipoint family.

11.1 Introduction

In the multipoint version of the Basic Family for each given $m \geq 2$, B_m blossoms into m iteration functions $B_m^{(1)}, \ldots, B_m^{(m)}$. For each $k \leq m$, $B_m^{(k)}$ is defined in terms of the ratio of two determinants that depend on the first $m - k$ derivatives. Thus for each fixed k, we have a family of k-point iteration functions $\{B_m^{(k)}\}_{m=k}^{\infty}$. The single-point Basic Family member B_m coincides with $B_m^{(1)}$.

The functions $B_2^{(1)}$ and $B_3^{(1)}$ are Newton's and Halley's iteration functions, respectively. For $k = 2$, the first two members of the family are two-point versions of Newton's method (i.e. the secant method), and Halley's method, respectively. The orders of convergence of $B_2^{(2)}$ and $B_3^{(2)}$ are the positive roots of the characteristic polynomial of Fibonacci sequence (1.618), and the Pell sequence (2.414), respectively. For $k = 3$, the first member of the family is a three-point variation of Halley's method having

*Part of this chapter has been reprinted from On the Order of Convergence of a Determinantal Family of Root-Finding Methods, *BIT*, Vol. 39 (1999) 96–109, B. Kalantari. With kind permission of Springer Science and Business Media.

order of convergence equal to the largest root of the characteristic polynomial of Fibonacci sequence of order 3 (1.839).

In this chapter we derive the order of convergence of the general member, $B_m^{(k)}$. We show that for each fixed m, the order of convergence of $B_m^{(k)}$ is monotonically decreasing in k, from m to the largest root of the characteristic polynomial of Generalized Fibonacci numbers of order m. We also derive other monotonicity properties, as well as the asymptotic error constant of $B_m^{(k)}$.

11.2 The Multipoint Basic Family

We first state the necessary ingredients, some already defined in the previous chapter.

Definition 11.1. A vector $a = (x_1, \dots, x_{n+1}) \in \mathbb{R}^{n+1}$ is said to be an *admissible vector of nodes* if whenever $x_i = x_j$, $i \leq j$, then, $x_i = x_{i+1} = \cdots = x_j$. If the number of distinct x_i's is k, we shall say a is *k-point* admissible. In the special case where $k = 1$, we identify a with the common value, x_1. We say a is *monotonic k-point*, if it is k-point admissible and $a = (x_1, \dots, x_1, x_2, \dots, x_k)$, where $x_i \neq x_j$, if $i \neq j$.

Definition 11.2. Assume the function $f : \mathbb{R} \to \mathbb{R}$ and is $(n+1)$-times continuously differentiable on an interval I. Let $a = (x_1, \dots, x_{n+1})$ be an admissible vector of nodes with $x_i \in I$. Let x be in I, $x \neq x_i$, $i = 1, \dots, n+1$. Set $x_{n+2} = x$. For any pair of indices i, j satisfying $1 \leq i \leq j \leq (n+2)$, inductively define the *confluent divided differences* as

$$f_{ij} = \begin{cases} \frac{f^{(j-i)}(x_i)}{(j-i)!}, & \text{if } x_i = x_j; \\ \frac{f_{i+1,j} \;-\; f_{i,j-1}}{(x_j - x_i)}, & \text{otherwise.} \end{cases}$$

Let $m \geq 2$ be a given natural number, and $n \geq (m-1)$. For $i > j$, let $f_{ij} = 0$. Let $F = (f_{ij})$ be the $(m-1) \times (n+2)$ matrix

$$F = \begin{pmatrix} f_{11} & f_{12} & f_{13} & \cdots & f_{1,m-1} & f_{1,m} & \cdots & f_{1,n+1} & f_{1,n+2} \\ 0 & f_{22} & f_{23} & \cdots & f_{2,m-1} & f_{2,m} & \cdots & f_{2,n+1} & f_{2,n+2} \\ 0 & 0 & f_{33} & \cdots & f_{3,m-1} & f_{3,m} & \cdots & f_{3,n+2} & f_{3,n+2} \\ \vdots & \vdots & \vdots & \ddots & \vdots & \dots & \dots & \vdots & \vdots \\ 0 & 0 & 0 & \dots & f_{m-1,m-1} & f_{m-1,m} & \cdots & f_{m-1,n+1} & f_{m-1,n+2} \end{pmatrix}.$$

Let u_i be the ith column of F, $i = 1, \ldots, n+2$. Set $a_i = (x_1, \ldots, x_i)$, $i = 1, \ldots, n+2$. Thus $a = a_{n+1}$. Set

$$D^{(m-1)}(a_m) = |u_2, \ldots, u_m|,$$

$$\widehat{D}_i^{(m-1)}(a_{i+1}) = |u_3, \ldots, u_m, u_{i+1}|, \quad i = m, \ldots, n+1.$$

Set $N^{(0)}(a_2) = 1$, and for $m \geq 3$, define

$$N^{(m-2)}(a_m) = \begin{vmatrix} f_{23} & \cdots & f_{2,m-1} & f_{2,m} \\ f_{33} & \cdots & f_{3,m-1} & f_{3,m} \\ \vdots & \ddots & \vdots & \vdots \\ 0 & \cdots & f_{m-1,m-1} & f_{m-1,m} \end{vmatrix}.$$

If $x_1 = \cdots = x_{n+1}$, then $N^{(m-2)}(a) = D^{(m-2)}(a)$.

Theorem 11.1 (Theorem 10.3, Chap. 10). *Let $m \geq 2$ be a natural number, and $n \geq (m-1)$. Assume that $f : K \to K$ (K real or complex field) and is $(n+1)$-times continuously differentiable on an open ball $I \subset K$ containing a simple root θ. Then, there exists an open ball I_θ centered at θ and contained in I so that for any admissible vector of nodes $a = (x_1, \ldots, x_{n+1})$ with $x_i \in I_\theta$, $x_i \neq \theta \equiv x_{n+2}$, $i = 1, \ldots, n+1$, the quantity $D^{(m-1)}(a_m)$ is nonzero, and we have*

$$B_m^{(k)}(a_m) \equiv x_1 - f(x_1) \frac{N^{(m-2)}(a_m)}{D^{(m-1)}(a_m)}$$

$$= \theta + \sum_{i=m}^{n} \gamma_i^{(m)}(a_{i+1}) \prod_{l=1}^{i} (\theta - x_l) + \gamma_{n+1}^{(m)}(a, \theta) \prod_{l=1}^{n+1} (\theta - x_l),$$

where $a_{i+1} = (x_1, \ldots, x_{i+1})$, a_m is k-point, and

$$\gamma_i^{(m)}(a_{i+1}) = (-1)^m \frac{\widehat{D}_i^{(m-1)}(a_{i+1})}{D^{(m-1)}(a_m)}, \quad i = m, \ldots, n+1.$$

In particular, if $f(x)$ is a polynomial of degree ν, for any $i \geq m + \nu - 2$, $\gamma_{i+1}^{(m)} \equiv 0$. □

The index k becomes relevant when $B_m^{(k)}$ is viewed as an iteration function. As a k-point admissible vector, the most meaningful case of a_m is when it is monotonic (Definition 11.1). For each monotonic k-point admissible vector a_m, $k = 1, \ldots, m$, $B_m^{(k)}(a_m)$ defines a k-point iteration function for the approximation of roots of $f(x)$, via the *fixed-point* iteration

$$a_m \quad \longleftarrow \quad \left(B_m^{(k)}(a_m), \ldots, B_m^{(k)}(a_m), x_1, \ldots, x_{k-1} \right) \in K^m.$$

For $m = 2, 3, 4$, and $k \leq m$, we get the following iteration functions

$$B_2^{(k)}(a_2) = x_1 - f_{11}\frac{1}{f_{12}},$$

$$B_3^{(k)}(a_3) = x_1 - f_{11}\frac{f_{23}}{\begin{vmatrix} f_{12} & f_{13} \\ f_{22} & f_{23} \end{vmatrix}},$$

$$B_4^{(k)}(a_4) = x_1 - f_{11}\frac{\begin{vmatrix} f_{23} & f_{24} \\ f_{33} & f_{34} \end{vmatrix}}{\begin{vmatrix} f_{12} & f_{13} & f_{14} \\ f_{22} & f_{23} & f_{24} \\ 0 & f_{33} & f_{34} \end{vmatrix}}.$$

For $k = 1$ and $k = 2$, $B_2^{(k)}$ gives Newton's method and the secant method, respectively. For $k = 1$, $B_3^{(k)}(a_3)$ is Halley's function, and for $k = 2$ and $k = 3$, it gives two- and three-point variation of Halley's.

The set of iteration functions, $B_m^{(k)}$, forms the multipoint Basic Family, also applicable for the approximation of complex roots of analytic functions. The formula for $B_m^{(k)}$ allows the definition of a "corrected" family of k-point iteration functions,

$$\widehat{B}_m^{(k)}(a_m) \equiv B_m^{(k)}(a_m) - \gamma_m^{(m)}(a_{m+1})\prod_{l=1}^{m}(B_m^{(k)}(a_m) - x_l).$$

In particular, one can define a corrected Halley method with super-cubic rate of convergence.

For general k, once we have computed the divided differences, each fixed-point iteration corresponding to $B_m^{(k)}$ takes $O(m^2)$ arithmetic operations. This is essentially the complexity of Gaussian elimination, taking into account the special form of the matrices involved. For $k = 1$, $B_m^{(1)}$ is essentially the ratio of two Toeplitz determinants. A Toeplitz determinant of order $O(m)$ can be computed in $O(m \log^2 m)$ arithmetic operations, and in some cases even faster, see Golub and Loan (1996), Chan and Ng (1996).

11.3 Description of the Order of Convergence

Theorem 11.2 below describes the order of convergence of the general monotonic k-point iteration functions. In Theorem 11.3 the result will be turned into a meaningful statement on the order of convergence.

Theorem 11.2 (Kalantari (1999)). *Let f be m-times continuously differentiable at a simple root θ, $m \geq 2$. For each $k = 1, \ldots, m$, there exists a neighborhood I_θ of θ such that given any initial k-point monotonic vector $a_m^{(0)} = (x_1^{(0)}, \ldots, x_1^{(0)}, x_2^{(0)}, \ldots, x_k^{(0)}) \in \mathbb{R}^m$ with $x_i^{(0)} \in I_\theta$, the fixed-point iteration that for each $r \geq 0$ replaces $a_m^{(r)}$ with the monotonic k-point vector*

$$a_m^{(r+1)} = \left(B_m^{(k)}(a_m^{(r)}), \ldots, B_m^{(k)}(a_m^{(r)}), x_1^{(r)}, \ldots, x_{k-1}^{(r)}\right) \in \mathbb{R}^m,$$

is well-defined, and $B_m^{(k)}(a_m^{(r)}) \in I_\theta$. Moreover, the sequence of points $\{x_1^{(r)}\}_{r=0}^{\infty}$ converges to θ satisfying

$$\lim_{r\to\infty} \frac{(\theta - x_1^{(r+1)})}{(\theta - x_1^{(r)})^{(m-k+1)}(\theta - x_2^{(r)}) \cdots (\theta - x_k^{(r)})}$$

$$= -\gamma_m^{(m)}(\theta) = \frac{(-1)^{m-1}}{p'(\theta)^{m-1}} \widehat{D}_m^{(m-1)}(\theta) \equiv c_m(\theta),$$

where

$$\widehat{D}_m^{(m-1)}(\theta) = \begin{vmatrix} \frac{f''(\theta)}{2!} & \frac{f'''(\theta)}{3!} & \cdots & \frac{f^{(m-1)}(\theta)}{(m-1)!} & \frac{f^{(m)}(\theta)}{m!} \\ f'(\theta) & \frac{f''(\theta)}{2!} & \ddots & \frac{f^{(m-2)}(\theta)}{(m-2)!} & \frac{f^{(m-1)}(\theta)}{(m-1)!} \\ 0 & f'(\theta) & \ddots & \vdots & \vdots \\ \vdots & \vdots & \ddots & \frac{f''(\theta)}{2!} & \frac{f'''(\theta)}{3!} \\ 0 & 0 & \cdots & f'(\theta) & \frac{f''(\theta)}{2!} \end{vmatrix}.$$

In particular, if $k = 1$, $\lim_{r\to\infty} (\theta - x_1^{(r+1)})/(\theta - x_1^{(r)})^m = c_m(\theta)$, i.e. the order of convergence is m.

Proof. Consider Theorem 11.1 for $n = m - 1$. Since θ is a simple root, $D^{(m-1)}(\theta) = p'(\theta)^{m-1} \neq 0$. This implies that there exists a neighborhood of θ so that for any admissible vector of nodes $a_m = (x_1, \ldots, x_m)$ with x_i in this neighborhood, $|\gamma_m^{(m)}(a_{m+1})|$ is bounded by a constant, say M. This implies that there exists a neighborhood so that

$$|\theta - x_1^{(r+1)}| = |\theta - B_m^{(k)}(a_m^{(r)})| \leq M \prod_{i=1}^{m} |\theta - x_i^{(r)}| < |\theta - x_1^{(r)}|/2,$$

i.e. the fixed-point iteration is a contraction mapping. This implies convergence. To prove the claimed statement on the limit, we need to use continuity, the fact that $f_{1,n+2}$ can be replaced by $f^{(n+1)}(\xi)/(n+1)!$ (Theorem 10.1, (*ii*) - Taylor's Theorem with confluent divided difference), and that f_{ij} converges to $f^{(j-i)}/(j-i)!(x_i)$, as $x_{i+1}, \ldots, x_j$ converge to x_i. $\square$

The quantity $c_m(\theta)$ is the asymptotic error constant of the one-point iteration function $B_m^{(1)}$. It can be computed conveniently, and it depends only on the first m derivatives of f. For example

$$c_2(\theta) = -\frac{f''(\theta)}{2f'(\theta)}$$

is the asymptotic error constant of Newton's method, and

$$c_3(\theta) = \frac{2f'(\theta)f'''(\theta) - 3f''(\theta)^2}{12f'(\theta)^2}$$

is the asymptotic error constant of Halley's method. For general k the computation of the order of convergence of $B_m^{(k)}$ is nontrivial. One needs to convert the limit statement of Theorem 11.2, into a statement on the order of convergence. Even the rigorous analysis of the rate of convergence of the secant method, i.e. the case of $m = k = 2$, is nontrivial and is often omitted in numerical analysis books. It can be shown that the order of convergence of the secant method is the positive root of the characteristic polynomial of the Fibonacci numbers. The proof of this fact requires results from the theory of difference equations. Two major books that contain complete proof are Ostrowski (1966), and Traub (1964). Ostrowski's proof makes use of the specific way in which the iterates are defined, and the mean-valued theorem (see Ostrowski (1966), p. 95). A theorem of Traub (see Traub (1964), Theorem 3-3, p. 56) substantially simplifies the analysis, removes dependence on the mean value theorem, and is applicable to more general cases of multipoint iteration methods. In particular, using this theorem one can conclude that the rate of convergence of $B_m^{(m)}$ is the unique positive root p of the polynomial $P(t) = t^k - \sum_{j=0}^{k-1} t^j$. This polynomial is the characteristic polynomial of Fibonacci numbers of order m. If $k = 2$, $p = 1.618$, also the limiting ratio of Fibonacci numbers, and if $k = 3$, $p = 1.839$, also the limiting ratio of Fibonacci numbers of order 3, sometimes referred as Tribonacci numbers, see Miles (1960), Feinberg (1963) and Bergum *et al.* (1992). However, in general in order to determine the order of convergence of $B_m^{(k)}$ for k not equal to 1 or m, new results are needed. We shall prove the following theorem that in particular generalizes the above mentioned theorem of Traub, and implies the desired order for various m and k. Note that the theorem is independent of iteration function.

Theorem 11.3. *Let $\{e_n\}_{n=0}^{\infty}$ be a sequence of numbers converging to zero. Let $k \geq 2$, and μ be natural numbers. The polynomial*

$$P_{k,\mu}(t) = t^k - \mu t^{k-1} - \sum_{j=0}^{k-2} t^j$$

has only simple roots, and a unique positive root $p \equiv p_{k,\mu}$, *satisfying*

$$\mu < p_{k,\mu} < \mu + 1. \tag{i}$$

All other roots have moduli less than unity. Moreover, the following inequalities are satisfied

$$p_{k,\mu} < p_{k+1,\mu}, \tag{ii}$$

$$p_{k,\mu} < p_{k,\mu+1}, \tag{iii}$$

$$p_{k+1,\mu} < p_{k,\mu+1}. \tag{iv}$$

Suppose that

$$\lim_{n\to\infty} \frac{e_{n+1}}{e_n^{\mu} e_{n-1} \cdots e_{n-k+1}} = c,$$

where c *is a nonzero constant. Then,*

$$\lim_{n\to\infty} \frac{e_{n+1}}{e_n^{p}} = c^{\left(\frac{p-1}{\mu+k-2}\right)}. \tag{v}$$

Before proving Theorem 11.3 we state the following theorem that completely characterizes the behavior of the multipoint Basic Family.

Theorem 11.4 (Order of Convergence of Multipoint Family). *Let* f *be* m *times continuously differentiable at a simple root* θ, $m \geq 2$. *Given* $k \in \{1, \ldots, m\}$, *set* $\mu = m - k + 1$. *Then there exists a neighborhood* I_θ *of* θ *such that given any initial* k*-point monotonic vector* $a_m^{(0)}$ *with* $x_i^{(0)} \in I_\theta$, *the fixed-point iteration corresponding to* $B_m^{(k)}$ *is well-defined, the sequence of points* $\{x_1^{(r)}\}_{r=0}^{\infty}$ *converge to* θ *satisfying*

$$\lim_{r\to\infty} \frac{(\theta - x_1^{(r+1)})}{(\theta - x_1^{(r)})^{p_{k,\mu}}} = c_m(\theta)^{\left(\frac{p_{k,\mu}-1}{m-1}\right)},$$

where $p_{k,\mu}$ *is the unique positive root of* $P(t) = t^k - \mu t^{k-1} - \sum_{j=0}^{k-2} t^j$. *Moreover, for* $k > 1$, $p_{k,\mu}$ *satisfies* $\mu < p_{k,\mu} < \mu + 1$; *for fixed* k, $p_{k,\mu}$ *is monotonically increasing in* m; *and for fixed* m, *it is monotonically decreasing in* k.

Proof. We need to apply Theorem 11.2, Theorem 11.3, and make the observation that after a few iterations we will have

$$a_m^{(r+1)} = (x_1^{(r)}, \ldots, x_1^{(r)}, x_1^{(r-1)}, \ldots, x_1^{(r-k+1)}),$$

and set $e_r = (\theta - x_1^{(r)})$. □

The following diagram represents the ascending order of convergence of $B_m^{(k)}$, and the corresponding orders, $p_{k,\mu}$ for a partial table of iteration functions:

$$
\begin{array}{ccccccccccc}
B_2^{(1)} & \leftarrow & B_2^{(2)} & & & & & & & & \\
\downarrow & & \downarrow & \searrow & & & & & & & \\
B_3^{(1)} & \leftarrow & B_3^{(2)} & \leftarrow & B_3^{(3)} & & & & & & \\
\downarrow & & \downarrow & & \downarrow & \searrow & & & & & \\
B_4^{(1)} & \leftarrow & B_4^{(2)} & \leftarrow & B_4^{(3)} & \leftarrow & B_4^{(4)} & & & & \\
\downarrow & & \downarrow & & \downarrow & & \downarrow & \searrow & & & \\
B_5^{(1)} & \leftarrow & B_5^{(2)} & \leftarrow & B_5^{(3)} & \leftarrow & B_5^{(4)} & \leftarrow & B_5^{(5)} & & \\
\downarrow & & \downarrow & & \downarrow & & \downarrow & \cdots & & \searrow &
\end{array}
$$

Fig. 11.1 Partial list of actual orders.

$$
\begin{array}{ccccccccccc}
2 & \leftarrow & 1.618 & & & & & & & & \\
\downarrow & & \downarrow & \searrow & & & & & & & \\
3 & \leftarrow & 2.414 & \leftarrow & 1.839 & & & & & & \\
\downarrow & & \downarrow & & \downarrow & \searrow & & & & & \\
4 & \leftarrow & 3.302 & \leftarrow & 2.546 & \leftarrow & 1.927 & & & & \\
\downarrow & & \downarrow & & \downarrow & & \downarrow & \searrow & & & \\
5 & \leftarrow & 4.236 & \leftarrow & 3.383 & \leftarrow & 2.592 & \leftarrow & 1.965 & & \\
\downarrow & & \downarrow & & \downarrow & & \downarrow & \cdots & & \searrow &
\end{array}
$$

Fig. 11.2 The ascending order of the Multipoint Basic Family.

11.4 Proof of the Order of Convergence

To prove Theorem 11.3 we need some preliminary results. Let σ_n and w_n be sequences of numbers defined by the difference equation

$$\sum_{j=0}^{k} K_j \sigma_{n+j} = w_n, \quad n = 0, 1, \ldots,$$

where K_j's are constants, $K_k = 1$, and $\sigma_0, \ldots, \sigma_k$, a given set of initial conditions. The characteristic polynomial of the difference equation is defined to be

$$P(t) = \sum_{j=0}^{k} K_j t^j.$$

Lemma 11.1 (Traub (1964), Theorem 3-1 and corollary, p. 44). *Assume that the roots of $P(t)$ are simple and all have moduli less than one. If ω_n converges to ω, then*

$$\lim_{n\to\infty} \sigma_n = \frac{\omega}{\sum_{j=0}^{k} K_j}.$$

Corollary 11.1. *Assume that the roots of $P(t)$ are all simple, and there is a unique root p of modulus larger than unity, and all other roots have moduli less than unity. Then*

$$\lim_{n\to\infty} (\sigma_n - p\sigma_{n-1}) = \omega \frac{1-p}{P(1)}.$$

Proof. Let $P(t) = (t-p)Q(t)$, where $Q(t) = q_{k-1}t^{k-1} + \cdots + q_1 t + q_0$. Define $D_n = \sigma_n - p\sigma_{n-1}$. Then it is easy to see that the characteristic polynomial of D_n is $Q(t)$. Clearly, the condition of the previous lemma now applies to the new sequence. Also since $\sum_{j=0}^{k} K_j$ in the lemma is equal to $P(1)$, the proof follows. □

From Corollary 11.1 it follows that to prove part (v) of the theorem it suffices to take $\sigma_n = \ln e_n$, and show that all roots of $P_{k,\mu}(t)$ are simple, and there exists a unique root with modulus larger than one. We first need an auxiliary results (see Ostrowski (1966) and Traub (1964)):

Lemma 11.2. *Let $g(x) = a_0 + a_1 x + \cdots + a_n x^n$ be a polynomial with each $a_i > 0$, then any root ξ of g satisfies*

$$|\xi| \le \max\{\frac{a_0}{a_1}, \frac{a_1}{a_2}, \ldots, \frac{a_{n-1}}{a_n}\}.$$

We now have the necessary ingredients to prove Theorem 11.3. First we prove that there exists a root satisfying (i). Clearly, $P_{k,\mu}(\mu) < 0$. We claim that $P_{k,\mu}(\mu+1) > 0$. Dividing by $(\mu+1)^k$ and from geometric series we get:

$$\frac{P_{k,\mu}(\mu+1)}{(\mu+1)^k} = \frac{1}{(\mu+1)} - \sum_{i=2}^{k} \frac{1}{(\mu+1)^i}$$

$$> \frac{1}{(\mu+1)} - \sum_{i=2}^{\infty} \frac{1}{(\mu+1)^i} \ge \frac{1}{(\mu+1)} - \frac{1}{\mu(\mu+1)} \ge 0.$$

This proves that a positive root p exists, satisfying (i). To prove its uniqueness we first show that

$$P_{k,\mu}(t) = (t-p)Q(t),$$

where

$$Q(t) = \frac{1}{p} + (\frac{1}{p} + \frac{1}{p^2})t + \cdots + (\frac{1}{p} + \frac{1}{p^2} + \cdots + \frac{1}{p^k})t^{k-1}.$$

To prove the above, from synthetic division we have

$$Q(t) = t^{k-1} + (p-\mu)t^{k-2} + \cdots + (p^{k-2} - p^{k-3}\mu$$

$$-p^{k-4} - \cdots - 1)t + (p^{k-1} - p^{k-2}\mu - p^{k-3} - \cdots - 1).$$

Since p is a root of $P_{k,\mu}(t)$,

$$p^k - p^{k-1}\mu - p^{k-2} - \cdots - 1 = 0.$$

Adding 1 to both sides of the above and dividing by p implies

$$p^{k-1} - p^{k-2}\mu - p^{k-3} - \cdots - 1 = \frac{1}{p}.$$

Adding 1 to both sides of the above and dividing by p implies that the coefficient of t in $Q(t)$ is $1/p + 1/p^2$. By repeating the above we obtain the desired description of the coefficients of $Q(t)$. Since for $t > 0$, $Q(t) > 0$, it implies that p is a unique positive simple root of $P_{k,\mu}$. Applying Lemma 11.2 to $Q(t)$, it follows that the modulus of each root of $P_{k,\mu}$ is less than one. It thus suffices to prove that the roots of $P_{k,\mu}$ are simple. To this end consider $G_{k,\mu}(t) = (t-1)P_{k,\mu}(t)$. We will prove that $G_{k,\mu}$ has simple roots. It is easy to show that $G_{k,\mu}(t) = t^{k+1} - (\mu+1)t^k + (\mu-1)t^{k-1} + 1$. From this we get

$$G'_{k,\mu}(t) = t^{k-2}[(k+1)t^2 - k(\mu+1)t + (k-1)(\mu-1)].$$

Thus the roots of $G'_{k,\mu}(t)$ are $t = 0$ and

$$t = \frac{1}{2(k+1)}(k(\mu+1) \pm \sqrt{\Delta}),$$

where $\Delta = k^2(\mu+1)^2 - 4(k^2-1)(\mu-1)$. We prove by induction on μ that Δ is nonnegative. This implies that the latter two roots are nonnegative. Since $P_{k,\mu}$ has a simple positive root, the simplicity of all roots follows. The statement is true for $\mu = 1$. Assume true for μ. We have

$$[k(\mu+2)]^2 = [k(\mu+1)+k]^2 = k^2(\mu+1)^2 + 2k^2(\mu+1) + k^2 \geq$$

$$4(k^2-1)(\mu-1) + k^2(2\mu+3) \geq 4(k^2-1)\mu + k^2(2\mu+3) - 4k^2 =$$

$$4(k^2 - 1)\mu + k^2[(2\mu + 3) - 4] \geq 4(k^2 - 1)\mu.$$

Thus, we have proved the initial desired properties of $P_{k,\mu}(t)$. To prove (ii) note that

$$P_{k+1,\mu}(t) = tP_{k,\mu}(t) - 1.$$

Thus, $P_{k+1,\mu}(p_{k,\mu}) = -1 < 0$. Since from (i) we also have $\mu < p_{k+1,\mu} < \mu + 1$, (ii) must be true since otherwise this contradicts that $P_{k+1,\mu}$ has a unique positive root. The proof (iii) is immediate from (i). To prove (iv) let $p = p_{k+1,\mu}$. On the one hand we have

$$P_{k,\mu+1}(p) = p^k - (\mu + 1)p^{k-1} - p^{k-2} - \cdots - 1.$$

On the other hand we have

$$P_{k+1,\mu}(p) = p^{k+1} - \mu p^k - p^{k-1} - \cdots - 1 = 0.$$

Adding $-p^k + 1$ to both sides of the above and dividing by p we see that

$$P_{k,\mu+1}(p) = \frac{-p^k + 1}{p},$$

and hence negative since $p > 1$. From this, and since from what has been proved earlier it follows that $P_{k,\mu+1}(t) > 0$, for $t > p_{k,\mu+1}$, it follows that (iv) is also satisfied. The proof of Theorem 11.3 is now complete. □

Chapter 12

A Computational Study of the Multipoint Basic Family

In this chapter we give a computational study of the first nine members of the multipoint Basic Family. These include Newton, secant, and Halley methods. Our computational results with polynomials of degree up to 30 reveal that for small degree polynomials $B_m^{(k-1)}$ is more efficient than $B_m^{(k)}$, but as the degree increases, $B_m^{(k)}$ becomes more efficient than $B_m^{(k-1)}$. The most efficient of the nine methods is $B_4^{(4)}$, having theoretical order of convergence equal to 1.927. Newton's method which is often viewed as the method of choice is in fact the least efficient of the nine tested.

12.1 Introduction

In the previous chapter we have defined and analyzed the multipoint Basic Family member, $B_m^{(k)}$, $m \geq k$, dependent on the first $m-k$ derivatives of the underlying function. Here we are interested in the computational comparison of the first nine members where the diagrams below represent the ascending order of convergence of these members and their corresponding theoretical order of convergence:

$$
\begin{array}{ccccccc}
B_2^{(1)} & \leftarrow & B_2^{(2)} & & & & \\
\downarrow & & \downarrow & \searrow & & & \\
B_3^{(1)} & \leftarrow & B_3^{(2)} & \leftarrow & B_3^{(3)} & & \\
\downarrow & & \downarrow & & \downarrow & \searrow & \\
B_4^{(1)} & \leftarrow & B_4^{(2)} & \leftarrow & B_4^{(3)} & \leftarrow & B_4^{(4)}
\end{array}
\qquad
\begin{array}{ccccccc}
2 & \leftarrow & 1.618 & & & & \\
\downarrow & & \downarrow & \searrow & & & \\
3 & \leftarrow & 2.414 & \leftarrow & 1.839 & & \\
\downarrow & & \downarrow & & \downarrow & \searrow & \\
4 & \leftarrow & 3.302 & \leftarrow & 2.546 & \leftarrow & 1.927
\end{array}
$$

Fig. 12.1 The first nine members and their orders.

In this chapter we will make a computational comparison of the first nine members in finding roots of the polynomial

$$f(x) = c_t x^t + c_{t-1} x^{t-1} + \cdots + c_0,$$

where the c_i's are reals. The iteration functions can also be applied in the complex plane. However, in this chapter we will restrict ourselves to experimentation with finding real roots of polynomials.

In the following sections of the chapter we give the explicit form of $B_m^{(k)}(a_m)$, their iteration complexity, our experimental results, and our conclusions.

12.2 The Iteration Functions

The monotonic vector of interest for $m = 2, 3$, and 4 are respectively:

$$a_2 : \quad (x_1, x_1), \quad (x_1, x_2),$$

$$a_3 : \quad (x_1, x_1, x_1), \quad (x_1, x_1, x_2), \quad (z_1, x_2, x_3),$$

$$a_4 : \quad (x_1, x_1, x_1, x_1), \quad (x_1, x_1, x_1, x_2), \quad (x_1, x_1, x_2, x_3), \quad (x_1, x_2, x_3, x_4),$$

where $i < j$ implies $x_i < x_j$.

The matrices needed to compute $B_m^{(k)}$ for our cases of interest are all submatrices of the matrix

$$F = \begin{pmatrix} f_{11} & f_{12} & f_{13} & f_{14} \\ 0 & f_{22} & f_{23} & f_{24} \\ 0 & 0 & f_{33} & f_{34} \end{pmatrix},$$

where for any pair of indices i, j satisfying $1 \le i \le j \le 4$, the corresponding confluent divided difference as defined in the previous chapter is

$$f_{ij} = \begin{cases} \frac{f^{(j-i)}(x_i)}{(j-i)!}, & \text{if } x_i = x_j; \\ \frac{f_{i+1,j} \; - \; f_{i,j-1}}{(x_j - x_i)}, & \text{otherwise.} \end{cases}$$

The iteration functions for $m = 2, 3, 4$, and $k \le m$ are:

$$B_2^{(k)}(a_2) = x_1 - f_{11} \frac{1}{f_{12}},$$

$$B_3^{(k)}(a_3) = x_1 - f_{11} \; \frac{f_{23}}{\begin{vmatrix} f_{12} & f_{13} \\ f_{22} & f_{23} \end{vmatrix}}$$

$$B_4^{(k)}(a_4) = x_1 - f_{11} \frac{\begin{vmatrix} f_{23} & f_{24} \\ f_{33} & f_{34} \end{vmatrix}}{\begin{vmatrix} f_{12} & f_{13} & f_{14} \\ f_{22} & f_{23} & f_{24} \\ 0 & f_{33} & f_{34} \end{vmatrix}}.$$

Given m, $k \leq m$, and an initial k-point monotonic vector

$$a_m^{(0)} = (x_1^{(0)}, \ldots, x_1^{(0)}, x_2^{(0)}, \ldots, x_k^{(0)}) \in \mathbb{R}^m,$$

the fixed-point iteration that for each $r \geq 0$ replaces $a_m^{(r)}$ with the monotonic k-point vector

$$a_m^{(r+1)} = \left(B_m^{(k)}(a_m^{(r)}), \ldots, B_m^{(k)}(a_m^{(r)}), x_1^{(r)}, \ldots, x_{k-1}^{(r)} \right) \in \mathbb{R}^m.$$

12.3 The Iteration Complexity

Here we will analyze the iteration complexity of the nine root-finding methods of interest for the case of polynomials. More precisely, we are interested in the number of arithmetic operation, $+,-,\times,\div$ within a given number of iterations of $B_m^{(k)}$, as applied to a given polynomial.

Let $i_m^{(k)}$ be the number of iterations of $B_m^{(k)}$ as applied to a given polynomial. If $k = 1$, in order to go from one iteration to the next, we would have to recompute new function-derivative values. However, for $k > 1$, to go from one iteration to the next it is possible to make use of some precomputed values. Thus, after the initial iteration, the complexity of evaluation of $B_m^{(k)}$ could improve. Let $N_{m,j}^{(k)}$ denote the number of arithmetic operation needed to compute $B_m^{(k)}$ in the j-th iteration. Clearly, we have

$$N_{m,1}^{(k)} \geq N_{m,2}^{(k)}, \quad N_{m,2}^{(k)} = N_{m,j}^{(k)},$$

for all $j > 1$. Let $N_m^{(k)}$ be the number of arithmetic operations performed after $i_m^{(k)}$ iterations of $B_m^{(k)}$. Then, we have

$$N_m^{(k)} = N_{m,1}^{(k)} + (i_m^{(k)} - 1) N_{m,2}^{(k)}.$$

The above equation can be written in terms of $i_m^{(k)}$, and the quantity $T(f^{(j)})$, defined as the number of arithmetic operations needed to compute the j-th derivative of the given polynomial.

As an example, consider $B_4^{(4)}(a_4)$, where $a_4 = (x_1, x_2, x_3, x_4)$, $x_i \neq x_j$. We have

$$B_4^{(4)}(a_4) = x_1 - \frac{f_{11}(f_{23}f_{34} - f_{33}f_{24})}{f_{12}(f_{23}f_{34} - f_{33}f_{24}) - f_{22}(f_{13}f_{34} - f_{33}f_{14})}.$$

It is easy to see that having computed the divided differences, we need 12 arithmetic operations to compute $B_4^{(4)}(a_4)$. The computation of the divided differences can be represented by the following diagram:

$$\begin{array}{ccccccc}
f(x_1) & & & & & & \\
| & \searrow & & & & & \\
f(x_2) & \rightarrow & f_{12} & & & & \\
| & \searrow & & \searrow & & & \\
f(x_3) & \rightarrow & f_{23} & \rightarrow & f_{13} & & \\
| & \searrow & & \searrow & & \searrow & \\
f(x_4) & \rightarrow & f_{34} & \rightarrow & f_{24} & \rightarrow & f_{14}
\end{array}$$

Fig. 12.2 The divided differences needed for the first iteration of $B_4^{(4)}$.

Since x_i's are distinct, the computation of each f_{ij} requires three arithmetic operations. Thus, $N_{4,1}^{(4)} = 4T(f) + 6 \times 3 + 12 = 4T(f) + 30$. In the next iteration the input (x_1, x_2, x_3, x_4) is replaced with $(B_4^{(4)}(a_4), x_1, x_2, x_3)$. Thus, we need to compute the new divided differences:

$$\begin{array}{ccccccc}
f(B_4^{(4)}(a_4)) & & & & & & \\
| & \searrow & & & & & \\
f(x_1) & \rightarrow & f_{12} & & & & \\
| & \searrow & & \searrow & & & \\
f(x_2) & \rightarrow & f_{23} & \rightarrow & f_{13} & & \\
| & \searrow & & \searrow & & \searrow & \\
f(x_3) & \rightarrow & f_{34} & \rightarrow & f_{24} & \rightarrow & f_{14}
\end{array}$$

Fig. 12.3 The divided differences needed for the second iteration of $B_4^{(4)}$.

However, we note that the only new calculation is the computation of $f(B_4^{(4)}(a_4))$, f_{12}, f_{13}, and f_{14}. Thus, $N_{4,2}^{(4)} = T(f) + 3 \times 3 + 12 = T(f) + 21$. Therefore,

$$N_4^{(4)} = 4T(f) + 30 + (i_4^{(4)} - 1)(T(f) + 21).$$

Likewise, we can compute $N_m^{(k)}$ for all the nine iteration functions. This would require the computational complexity of the first and second iterations, taking into account the computation of the corresponding confluent divided differences. It can be shown that this gives:

$$N_2^{(k)} = \begin{cases} i_2^{(1)}[T(f) + T(f') + 2], & \text{if } k = 1; \\ i_2^{(2)}[T(f) + 5] + T(f), & \text{if } k = 2. \end{cases}$$

$$N_3^{(k)} = \begin{cases} i_3^{(1)}[T(f) + T(f') + T(f'') + 7], & \text{if } k = 1; \\ i_3^{(2)}[T(f) + T(f') + 11] + T(f), & \text{if } k = 2; \\ i_3^{(3)}[T(f) + 12] + 2T(f) + 3, & \text{if } k = 3. \end{cases}$$

$$N_4^{(k)} = \begin{cases} i_4^{(1)}[T(f) + T(f') + T(f'') + T(f''') + 14], & \text{if } k = 1; \\ i_4^{(2)}[T(f) + T(f') + T(f'') + 20] + T(f), & \text{if } k = 2; \\ i_4^{(3)}[T(f) + T(f') + 22] + 2T(f) + 3, & \text{if } k = 3; \\ i_4^{(4)}[T(f) + 21] + 3T(f) + 9, & \text{if } k = 4. \end{cases}$$

Fig. 12.4 The iteration complexity of $B_m^{(k)}$ after $i_m^{(k)}$ iterations.

12.4 The Experiment

In this section we describe our experimentation with $B_m^{(k)}$ in finding real roots of polynomials. Specifically, the following guidelines were used.

For each degree n in the range $[2, 30]$, we generated random polynomials in the following fashion:

The coefficient of the highest degree was a randomly chosen integer in the interval $[1, 10]$, and the constant term a randomly chosen integer in $[-100, -1]$. Thus, each generated polynomial had a positive root. All other coefficients were random integers in the interval $[-10, 10]$. All random numbers were generated using uniform distribution. Then we generated a random seed. Then we applied all the nine iteration functions. In the case of multipoint iteration functions, say $k = 4$, we would augment each input, say x_1, to a vector of inputs, (x_1, x_2, x_3, x_4), where $x_i - x_{i-1}$ was chosen to be the same random integer between 2 and 6. The evaluation of a given polynomial and its derivatives were carried out using Horner's method.

We define a *trial* to be the case of applying the nine iteration functions to a given polynomial, for a given seed. If all the iteration functions converged to the same root, then we could consider that a *Case I* trial. If all the iterates of all the nine methods converged, but to different roots, then we would call that a *Case II* trial. If any of the nine methods did not converge we would call that a *Case III* trial. In fact Case III occurred in less than tenth of one percent of the time. Thus, all the nine iteration functions are robust in the sense that they converged to a root. Although not always to same root (see Table 12.1).

For a given polynomial we would try sufficient number of seeds until 15 instances of Case I would occur. If we would witness 30 occurrences of Case II trials, then we would discard that polynomial.

For a given polynomial we applied each of the nine iteration functions until the following stopping criterion was satisfied,

$$|B_m^{(k)}(a_m^{(r)}) - B_m^{(k)}(a_m^{(r-1)})| < 10^{-15}.$$

For each of the nine iteration functions we kept track of the total number of arithmetic operations needed to approximate the same root, i.e., the number $N_m^{(k)}$, until the above stopping criterion was satisfied. The comparison of different methods to a trial of Case I ensured that we would have a fair comparison of the nine iteration functions.

For each given polynomial, we computed the ratio

$$R_m^{(k)} = \frac{N_m^{(k)}}{N_2^{(1)}},$$

for each of the 15 inputs that resulted in trials of Case I. The reciprocal of this ratio can be viewed as the relative efficiency of $B_m^{(k)}$ with respect to Newton's method, i.e., $B_2^{(1)}$. The quantity $R_m^{(k)}$ was then averaged over the 15 random Case I inputs. Finally, this average was averaged over the 10 randomly generated polynomials of a given degree. This final ratio gives the average of $R_m^{(k)}$.

For instance for the polynomial with coefficients

$$(c_{27}, \ldots, c_0) = (1, -4, 0, 0, -5, +5, 5, -4, 1, 5,$$

$$-5, 5, 5, -9, -5, 5, 0, -10, 6, 6, 0, 6, -4, -5, -10, 5, -4, -90),$$

the initial input vector was $a_m = (160, 162, 164, 166)$ and we obtained the quantities shown in Figures 12.5-12.7.

$$\begin{array}{ccccccc}
107 & \leftarrow & 153 & & & & \\
\downarrow & & \downarrow & \searrow & & & \\
56 & \leftarrow & 69 & \leftarrow & 100 & & \\
\downarrow & & \downarrow & & \downarrow & \searrow & \\
39 & \leftarrow & 45 & \leftarrow & 57 & \leftarrow & 81
\end{array}$$

Fig. 12.5 $i_m^{(k)}$: The number of iterations of $B_m^{(k)}$.

$$\begin{array}{ccccccc}
10593 & \leftarrow & 8311 & & & & \\
\downarrow & & \downarrow & \searrow & & & \\
8400 & \leftarrow & 7501 & \leftarrow & 6201 & & \\
\downarrow & & \downarrow & & \downarrow & \searrow & \\
7839 & \leftarrow & 7384 & \leftarrow & 6884 & \leftarrow & 5826
\end{array}$$

Fig. 12.6 $N_m^{(k)}$: The number of arithmetic operations after $i_m^{(k)}$ iterations of $B_m^{(k)}$.

$$R_m^{(k)} : \begin{array}{ccccccc}
1.00 & \leftarrow & 0.78 & & & & \\
\downarrow & & \downarrow & \searrow & & & \\
0.79 & \leftarrow & 0.71 & \leftarrow & 0.59 & & \\
\downarrow & & \downarrow & & \downarrow & \searrow & \\
0.74 & \leftarrow & 0.70 & \leftarrow & 0.65 & \leftarrow & 0.55
\end{array}$$

Fig. 12.7 $R_m^{(k)}$: The ratio of the number of operations of $B_m^{(k)}$ to those of Newton's.

The four subsequent figures graph the average ratio of the average of $R_m^{(k)}$ for the nine iteration functions as a function of degree (computed over trials of Case I). In order to get a better picture, we have enlarged this graph at degrees 9, 16, and 23.

Each trial may be considered as nine *subtrial*, i.e. the case of applying a particular $B_m^{(k)}$ to a given polynomial, with a given seed. We call a subtrial corresponding to trials of Case I or Case II to be within *category A*, if the approximate root obtained by the subtrial corresponds to the root whose function value is closest to zero, among all the approximate roots produced by the nine iteration functions. Otherwise, we call the subtrial to be of category B. Thus, a subtrial of Case I is always of category A. Now it may be of interest to compute for a given $B_m^{(k)}$, the percentage of the subtrials that resulted in category A subtrials. This is tabulated in Table 12.1. As we see this number is almost independent of the method, as well as the polynomial degree. We can thus conclude that all the nine methods have essentially the same performance with this regard.

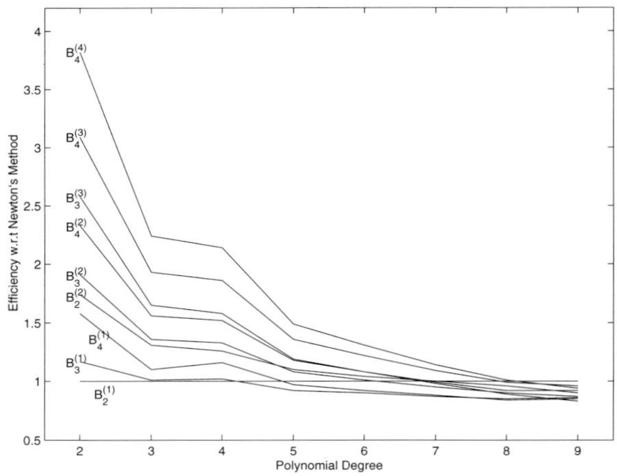

Fig. 12.8 Reciprocal of average efficiency ratio of $B_m^{(k)}$ versus Newton's, $t \leq 8$.

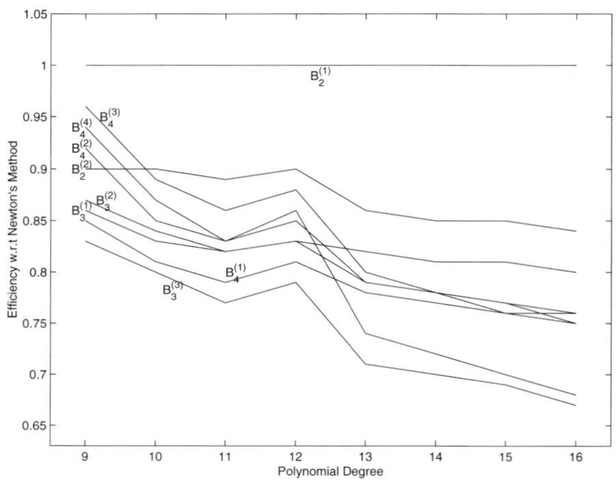

Fig. 12.9 Reciprocal of average efficiency ratio versus Newton's, $8 \leq t \leq 16$.

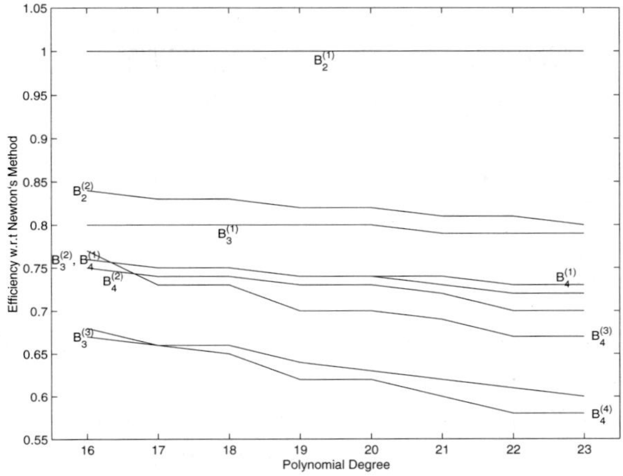

Fig. 12.10 Reciprocal of average efficiency ratio versus Newton's, $16 \leq t \leq 25$.

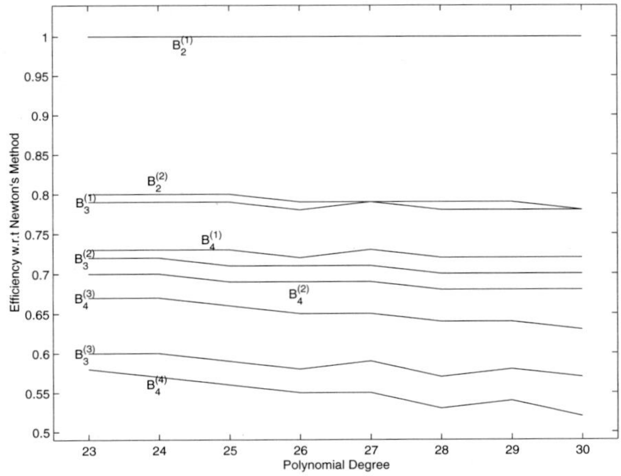

Fig. 12.11 Reciprocal of average efficiency ratio versus Newton's, $25 \leq t \leq 30$.

Table 12.1 Percentage of subtrials of category A.

n	$B_2^{(1)}$	$B_2^{(2)}$	$B_3^{(1)}$	$B_3^{(2)}$	$B_3^{(3)}$	$B_4^{(1)}$	$B_4^{(2)}$	$B_4^{(3)}$	$B_4^{(4)}$
5	83	80	95	82	85	87	83	85	85
10	89	81	88	78	92	77	90	78	93
15	87	78	79	82	82	83	79	84	83
20	86	83	89	86	88	87	85	86	88
25	80	84	80	82	84	77	76	81	80
30	86	84	77	84	89	84	81	85	81

12.5 Conclusions

Based on our experimentation we see that for very small degrees (≤ 4) Newton's method is the best. But, as the degree of polynomial increases, in comparison with other methods, Newton's method becomes less and less effective. Also, this experiment shows that, for very small degrees, $B_m^{(k-1)}$ is better than $B_m^{(k)}$, but, as the degree increases, $B_m^{(k)}$ becomes more efficient than $B_m^{(k-1)}$. As the degree increases, $B_4^{(4)}$ becomes the most efficient among the nine methods. This method which requires no derivative evaluations, has theoretical order of convergence very close to Newton's quadratic order. We expect that for values of m larger than 4, and for larger degree polynomials, one would observe the same behavior. These iteration functions are robust. Finally, one would expect that these experimental results, first demonstrated in Kalantari and Park (2001), would apply to the computation of complex roots of polynomials.

Chapter 13

A General Determinantal Lower Bound and Specific Applications in Root-Finding

There are known upper bounds for the modulus of the determinant of a matrix having real or complex entries, but useful lower bounds are very rare. In this chapter we derive a general lower bound. We then show how it can be used in deriving approximation to roots of complex polynomials. Other applications will be considered in the subsequent chapter in deriving formulas for the approximation of π, e^x, and roots of analytic functions.

13.1 Introduction

Given an $n \times n$ matrix $A = [a_{ij}]$ with real or complex entries we present nontrivial lower bounds for $|\det(A)|$ that depend upon $trace(A)$ and upon upper and lower bound estimates for the moduli of the eigenvalues of A. The result is not restricted to special classes or types of matrices. It will be useful in cases where an estimate for $|\det(A)|$ is needed, or in applications where a lower bound is required for the determinants of a sequence of matrices with growing dimensions such as the cases to be considered in the chapter.

Hadamard inequality (see e.g. Marcus and Minc (1964)) is the best known upper bound for the determinant:

$$|\det(A)| \leq \left(\prod_{i=1}^{n} \sum_{j=1}^{n} |a_{ij}|^2 \right)^{1/2}.$$

We present a lower bound for $|\det(\cdot)|$ that works for any complex matrix provided that eigenvalue bounds are available and a trace inequality is satisfied. The motivation behind the theorem was in fact for the purpose of approximation of roots of polynomials where the following version was established:

Theorem 13.1 (Determinantal Lower Bound, Kalantari (1997)). *Let A be an $n \times n$ real or complex invertible matrix with eigenvalues $\lambda_1, \lambda_2, \dots, \lambda_n$, and let ℓ and u be distinct positive numbers such that $\ell \le |\lambda_i| \le u$ for each $i = 1, \dots, n$. If $|trace(A)| \ge n\ell$, and*

$$\kappa = \lceil \frac{(nu - |trace(A)|)}{(u - \ell)} \rceil,$$

then

$$\ell^{\kappa} u^{n-\kappa} \le |\det(A)|.$$

□

We next state and prove an improved version given below. Throughout by I_n we mean $\{1, 2, \dots, n\}$.

Theorem 13.2 (Kalantari and Pate (2001)). *Let A be an $n \times n$ real or complex invertible matrix with eigenvalues $\lambda_1, \lambda_2, \dots, \lambda_n$, and let ℓ and u be distinct positive numbers such that $\ell \le |\lambda_i| \le u$ for each $i \in I_n$. If $|trace(A)| \ge n\ell$, and*

$$\kappa = \frac{(nu - |trace(A)|)}{(u - \ell)},$$

then

$$\ell^{\kappa} u^{n-\kappa} \le |\det(A)|$$

with equality if and only if there exists $w \in \mathbb{C}$ with $|w| = 1$ such that each of $\lambda_1, \lambda_2, \dots, \lambda_n$ is either $w\ell$ or wu.

We first sketch a proof of the lower bound where the number κ in the inequality is replaced by $\lceil \kappa \rceil$, where $\lceil w \rceil$ denotes $\min\{q \in \mathbb{Z} : q \ge w\}$ for each $w \in \mathbb{R}$. Since $\kappa \le \lceil \kappa \rceil$ the resulting inequality is slightly weaker than the inequality of Theorem 13.2. Nevertheless, the proof is interesting and gives a motivation for the statement of that theorem.

Consider the optimization problem

$$\min \left\{ \prod_{i=1}^{n} x_i \; : \; \sum_{i=1}^{n} x_i \ge |trace(A)|, \quad \ell \le x_i \le u, \quad i = 1, 2, \dots, n \right\}.$$

Since

$$|trace(A)| = |\sum_{i=1}^{n} \lambda_i| \le \sum_{i=1}^{n} |\lambda_i|;$$

it follows that $x = (|\lambda_1|, \ldots, |\lambda_n|)$ is a feasible solution. Thus, to obtain a lower bound for $|\det(A)|$ it suffices to solve the optimization problem. Replacing the objective function with $\ln(\prod_{i=1}^n x_i)$, we obtain an equivalent optimization problem which consists of the minimization of a concave function over a polytope. It is well-known that the optimal solution of such concave programming problem is attained at a vertex of the underlying polytope. It is also well-known that a feasible vertex must have n linearly independent active constraints. It follows that if $x^* = (x_1^*, \ldots, x_n^*)$ is an optimal vertex of the above optimization problem then at most one of its components is not equal to ℓ or u. Since $|trace(A)| \geq n\ell$, the constraint involving the trace is active at x^*, that is,

$$\sum_{i=1}^{n} x_i^* = |trace(A)|.$$

Let $\kappa = (nu - |trace(A)|)/(u - \ell)$, and let m be the number of components of x^* that are equal to ℓ. If $m = n$, then $\kappa = \lceil\kappa\rceil = n$, and $\ell^\kappa u^{n-\kappa} = \ell^{\lceil\kappa\rceil} u^{n-\lceil\kappa\rceil} = \ell^n$; consequently, if $m = n$, then it is trivial that the inequality of Theorem 13.2 holds with κ replaced by $\lceil\kappa\rceil$. Assume that $m < n$. Then at least $(n - m - 1)$ of the components of x^* are set at u, and the last component is set at a value r satisfying $\ell < r \leq u$. Substituting these in the trace equation we get

$$m\ell + (n - m - 1)u + r = |trace(A)|.$$

Thus

$$m(u - \ell) = nu - |trace(A)| - (u - r) \leq nu - |trace(A)|.$$

But, $(u - r)/(u - \ell) < 1$; hence, $m = \lfloor\kappa\rfloor = \max\{q \in \mathbb{Z} : q \leq \kappa\}$, and we have

$$\prod_{i=1}^{n} x_i \geq \ell^m u^{n-m} r \geq \ell^{m+1} u^{n-m-1}$$

for all admissible $x = (x_1, x_2, \ldots, x_n)$. But, $m+1 = \lceil\kappa\rceil$; thus the lower bound given in Theorem 13.2 is true provided that κ is replaced with $\lceil\kappa\rceil$.

We now proceed to give a different proof of the theorem which improves the lower bound by removing the ceiling function. Since the determinant of a matrix is the product of its eigenvalues, any inequality involving determinants is an inequality involving n-fold products of complex numbers. In the present case we are concerned with a sequence $\lambda_1, \lambda_2, \ldots, \lambda_n$ of complex numbers contained in an annulus $\mathcal{A}_{\ell,u} = \{z \in \mathbb{C} : \ell \leq |z| \leq u\}$ in the complex plane. Theorem 13.2 is obviously equivalent to

Theorem 13.3. *If* $0 < \ell < u$, *and* $\lambda_1, \lambda_2, \ldots, \lambda_n$ *are members in* $\mathcal{A}_{\ell,u}$ *such that* $|\sum_{i=1}^n \lambda_i| \geq n\ell$, *then*

$$|\prod_{i=1}^{n} \lambda_i| \geq \ell^{\kappa} u^{n-\kappa}$$

where

$$\kappa = \frac{(nu - |\sum_{i=1}^n \lambda_i|)}{(u-\ell)}$$

with equality if and only if there exists $w \in \mathbb{C}$ *with* $|w| = 1$ *such that each of* $\lambda_1, \lambda_2, \ldots, \lambda_n$ *is either* $w\ell$ *or* wu.

The proof of Theorem 13.3 depends upon the following theorem.

Theorem 13.4. *Suppose* $0 < \ell < u$. *If* $x_1, x_2, \ldots, x_n$ *is a sequence of real numbers such that* $\ell \leq x_i \leq u$ *for each* $i \in I_n$, *and* $\omega_1, \omega_2, \ldots, \omega_n$ *is a sequence of positive numbers such that* $\sum_{i=1}^n \omega_i = n$, *then*

$$\prod_{i=1}^{n} x_i^{\omega_i} \geq \ell^{\kappa} u^{n-\kappa} \qquad \text{where} \qquad \kappa = \frac{(nu - \sum_{i=1}^n \omega_i x_i)}{(u-\ell)} \tag{13.1}$$

with equality if and only if $x_i \in \{\ell, u\}$ *for each* $i \in I_n$.

Proof. Define $\varphi : [0, n] \to \mathbb{R}$ by $\varphi(t) = \ell^t u^{n-t}$. Note that $\varphi'(t) = \ln(\ell/u)\ell^t u^{n-t}$; hence, $\varphi(\cdot)$ is strictly decreasing on $[0, n]$, and the range of $\varphi(\cdot)$ is $[\ell^n, u^n]$. A simple calculation reveals that

$$\varphi^{-1}(y) = \frac{n\ln(u) - \ln(y)}{\ln(u) - \ln(\ell)}$$

for each y such that $\ell^n \leq y \leq u^n$. Let $S = \sum_{i=1}^n \omega_i x_i$, $P = \prod_{i=1}^n x_i^{\omega_i}$, $\kappa = (nu - S)/(u-\ell)$, and $\xi = \varphi^{-1}(P)$. Clearly $0 \leq \kappa \leq n$, and $\ell^n \leq P \leq u^n$; hence to complete the proof it is sufficient to show that $\xi \leq \kappa$. But this is equivalent to showing that

$$\frac{n\ln(u) - \ln(P)}{\ln(u) - \ln(\ell)} \leq \frac{nu - S}{u - \ell}. \tag{13.2}$$

If $nu = S$, then (13.2) is true because both sides reduce to zero. If $nu \neq S$, then hypotheses imply that $nu > S$ in which case (13.2) is equivalent to

$$\frac{n\ln(u) - \ln(P)}{nu - S} \leq \frac{\ln(u) - \ln(\ell)}{u - \ell}. \tag{13.3}$$

Define $\alpha : [\ell, u] \to \mathbb{R}$ by $\alpha(t) = (\ln(u) - \ln(t))/(u - t)$ when $\ell \leq t < u$, and let $\alpha(u) = 1/u$. The inequality $\ln(x) < x - 1$ holds when $x > 0$ and $x \neq 1$;

hence, $\alpha(\cdot)$ is a continuous function whose derivative is negative on $[\ell, u)$. Letting $K_{\ell,u}$ denote $(\ln(u) - \ln(\ell))/(u - \ell)$, we therefore have

$$\ln(u) - \ln(t) \leq K_{\ell,u}(u - t) \quad \text{for all} \quad t \in [\ell, u]. \tag{13.4}$$

Thus, in particular we have $\ln(u) - \ln(x_i) \leq K_{\ell,u}(u - x_i)$ for each $i \in I_n$. Multiplying by ω_i and summing from 1 to n we obtain

$$n\ln(u) - \sum_{i=1}^{n} \omega_i \ln(x_i) \leq K_{\ell,u}(nu - \sum_{i=1}^{n} \omega_i x_i)$$

which is easily seen to be equivalent to (13.3). If $x_i \in \{\ell, u\}$ for each $i \in I_n$, then (13.1) obviously reduces to equality. Suppose conversely that we have equality in (13.1). Since $\varphi(\cdot)$ is strictly decreasing this implies that $\xi = \kappa$, which in turn implies that we have equality in (13.2). There are now two cases. If $nu - S = 0$, then $\sum_{i=1}^{n} \omega_i(u - x_i) = 0$; so, we must have $x_i = u$ for each $i \in I_n$. If $nu - S > 0$, then equality in (13.2) implies equality in (13.3) which implies that

$$\sum_{i=1}^{n} \omega_i(\ln(u) - \ln(x_i)) = K_{\ell,u} \sum_{i=1}^{n} \omega_i(u - x_i). \tag{13.5}$$

But, for each i we have $\ln(u) - \ln(x_i) \leq K_{\ell,u}(u - x_i)$ with equality if and only if $x_i = \ell$ or $x_i = u$; hence, (13.5) cannot hold unless each of $x_1, x_2, \ldots, x_n$ is either ℓ or u. This completes the proof. □

We will now use Theorem 13.4 to prove Theorem 13.3. Suppose $0 < \ell < u$ and let $\lambda_1, \lambda_2, \ldots, \lambda_n$ be complex numbers such that $\ell \leq |\lambda_i| \leq u$ for each i, and $|\sum_{i=1}^{n} \lambda_i| \geq n\ell$. For each i let $x_i = |\lambda_i|$, let $\omega_i = 1$, and note that the hypotheses of Theorem 13.4 are satisfied with respect to ℓ, u, $x_1, x_2, \ldots, x_n$, and $\omega_1, \omega_2, \ldots, \omega_n$. Thus, if $\kappa = (nu - \sum_{i=1}^{n} x_i)/(u - \ell)$, then

$$|\prod_{i=1}^{n} \lambda_i| = \prod_{i=1}^{n} x_i \geq \ell^{\kappa} u^{n-\kappa}. \tag{13.6}$$

Let $\xi = (nu - |\sum_{i=1}^{n} \lambda_i|)/(u - \ell)$. Since $|\sum_{i=1}^{n} \lambda_i| \geq n\ell$, the triangle inequality implies that $0 \leq \kappa \leq \xi \leq n$. But the function $\varphi(t) = \ell^t u^{n-t}$ is strictly decreasing on $[0, n]$; so, $\varphi(\kappa) \geq \varphi(\xi)$. Combining this with (13.6) we immediately obtain

$$|\prod_{i=1}^{n} \lambda_i| \geq \varphi(\xi) = \ell^{\xi} u^{n-\xi} \tag{13.7}$$

which is the inequality of Theorem 13.3. If there exists $w \in \mathbb{C}$ with $|w| = 1$ such that each λ_i is either $w\ell$ or wu, then (13.7) obviously reduces to equality; so, assume conversely that $|\prod_{i=1}^n \lambda_i| = \ell^\xi u^{n-\xi}$ where $\kappa = (nu - \sum_{i=1}^n x_i)/(u-\ell)$. Then, $\kappa = \xi$, and $\prod_{i=1}^n x_i = \varphi(\kappa)$; therefore, according to Theorem 13.4 each x_i is either ℓ or u. But, the equality $\kappa = \xi$ implies that $|\sum_{i=1}^n \lambda_i| = \sum_{i=1}^n |\lambda_i|$, which implies that there exists $w \in \mathbb{C}$ with $|w| = 1$ such that $\lambda_i = w|\lambda_i| = wx_i$ for each $i \in I_n$. Since each x_i is either ℓ or u, each of $\lambda_1, \lambda_2, \ldots, \lambda_n$ must be either $w\ell$ or wu. The proofs of Theorems 13.2 and 13.3 are now complete.

We will now state some corollaries of Theorem 13.2.

If A is an $n \times n$ complex matrix then the optimum choice for u is $\rho(A)$, the spectral radius of A, while the optimum choice for ℓ is $(\rho(A^{-1}))^{-1}$. If λ_+ and λ_- denote the eigenvalues of A of maximum and minimum modulus, respectively, then $\rho(A) = |\lambda_+|$ and $(\rho(A^{-1}))^{-1} = |\lambda_-|$. Therefore, we have

Corollary 13.1. *Suppose A is an $n \times n$ real or complex matrix such that $|\lambda_+| > |\lambda_-|$. If $|trace(A)| \geq n|\lambda_-|$, and*

$$\kappa = \frac{(n|\lambda_+| - trace(A))}{(|\lambda_+| - |\lambda_-|)},$$

then

$$|\lambda_-|^\kappa |\lambda_+|^{n-\kappa} \leq |\det(A)|.$$

□

If A is positive definite and Hermitian, then $\rho(A) = \lambda_+$ and $(\rho(A^{-1}))^{-1} = \lambda_-$. Moreover, the condition $|trace(A)| \geq n\ell$ is automatically satisfied; thus, we have

Corollary 13.2. *If A is an $n \times n$ positive definite Hermitian matrix such that $\lambda_+ > \lambda_-$, and*

$$\kappa = \frac{(n\lambda_+ - trace(A))}{(\lambda_+ - \lambda_-)},$$

then

$$\lambda_-^\kappa \lambda_+^{n-\kappa} \leq \det(A).$$

□

Corollary 13.3. *Suppose A is an $n \times n$ diagonally dominant real or complex matrix. For each $i \in I_n$ let $r_i = \sum_{j \neq i}^n |a_{ij}|$, let $\ell = \min\{|a_{ii}| - r_i : i \in I_n\}$, and let $u = \max\{|a_{ii}| + r_i : i \in I_n\}$. If $|trace(A)| \geq n\ell$, then*

$$\ell^\kappa u^{n-\kappa} \leq |\det(A)|,$$

where $\kappa = (nu - |trace(A)|)/(u - \ell)$.

Proof. From Gerschgorin's theorem each eigenvalue lies in one of the circles $C_i = \{z : |z - a_{ii}| \leq r_i\}$. This and diagonal dominance imply that if λ is an eigenvalue of A, then we must have $0 < l \leq |a_{ii}| - r_i \leq |\lambda| \leq |a_{ii}| + r_i \leq u$. We may now invoke Theorem 13.2 to complete the proof. $\square$

Corollary 13.4. *Let A be an $n \times n$ diagonally dominant real matrix with positive diagonal entries. If l, u, and κ are defined as in Corollary 13.3, then*

$$\ell^{\kappa} u^{n-\kappa} \leq |\det(A)|.$$

Proof. In this case $\ell = \min\{a_{ii} - r_i : i \in I_n\}$; so, $\ell \leq a_{jj}$ for each $j \in I_n$. Therefore, $n\ell \leq \sum_{j=1}^{n} a_{jj} = |\sum_{j=1}^{n} a_{jj}|$; so Corollary 13.4 follows from Corollary 13.3. $\square$

Finally, we will use Theorem 13.2 to obtain an upper bound for $|\det(A)|$.

Corollary 13.5. *Let A be an $n \times n$ real or complex invertible matrix with eigenvalues $\lambda_1, \lambda_2, \ldots, \lambda_n$, and let ℓ and u be distinct positive numbers such that $\ell \leq |\lambda_i| \leq u$ for each $i \in I_n$. If $|trace(A^{-1})| \geq nu^{-1}$, and*

$$\kappa = \frac{(nu - \ell u|trace(A^{-1})|)}{(u - \ell)},$$

then

$$|\det(A)| \leq \ell^{n-\kappa} u^{\kappa} \leq u^n.$$

Proof. Let ξ_i denote λ_i^{-1} for each i, and note that $u^{-1} \leq |\xi_i| \leq \ell^{-1}$. In other words each of $\xi_1, \xi_2, \ldots, \xi_n$ is in $\mathcal{A}_{u^{-1},\ell^{-1}}$. Since the ξ_i's are the eigenvalues of A^{-1}, the hypothesis $|trace(A^{-1})| \geq nu^{-1}$ says that $|\sum_{i=1}^{n} \xi_i| \geq nu^{-1}$. Thus, applying Theorem 13.2 we obtain that

$$|\prod_{i=1}^{n} \xi_i| = |\prod_{i=1}^{n} \lambda_i^{-1}| \geq (1/u)^{\kappa}(1/\ell)^{n-\kappa} \tag{13.8}$$

where

$$\kappa = \frac{n\ell^{-1} - |\sum_{i=1}^{n} \xi_i|}{\ell^{-1} - u^{-1}} = \frac{nu - \ell u|\sum_{i=1}^{n} \xi_i|}{u - \ell} = \frac{nu - \ell u|trace(A^{-1})|}{u - \ell}. \tag{13.9}$$

Taking reciprocals in (13.8) we obtain the claimed upper bound. $\square$

We consider next some applications of Theorem 13.2.

Example 13.1. Let A_n be the $n \times n$ tridiagonal matrix whose diagonal entries are 4, whose superdiagonal entries are 1, and whose subdiagonal entries are 2. For example, if $n = 3$ we have

$$A_3 = \begin{pmatrix} 4 & 1 & 0 \\ 2 & 4 & 1 \\ 0 & 2 & 4 \end{pmatrix}.$$

Corollary 13.4 applies to A_n with $\ell = 1$, $u = 7$, and $\kappa = n/2$. Thus,

$$|\det(A_n)| \geq 7^{n-(\frac{n}{2})} = 7^{\frac{n}{2}}. \tag{13.10}$$

Using the above bound it is possible to show that $\sqrt{2} = 2 - \lim_{n\to\infty} \det(A_n)/\det(A_{n+1})$. More generally, the lower bound can be used as an auxiliary result to give new formulas for the approximation of roots of complex polynomials. By solving a difference equation it can be shown that the sequence $\{\det(A_n)\}_{n=1}^{\infty}$ is asymptotically equivalent to the sequence $\{b_n\}_{n=1}^{\infty}$ where $b_n = \sqrt{2}(2+\sqrt{2})^n$. Thus, $\det(A_n) \approx \sqrt{2}(2+\sqrt{2})^n \geq \sqrt{2}(3.414)^n$ for large n; so, the estimate $7^{\frac{n}{2}} \geq (2.645)^n$ in (13.10) is quite good. If we let B_n be the $n \times n$ symmetric matrix obtained from A_n by replacing each subdiagonal and superdiagonal entry with $\sqrt{2}$, then $\det(A_n) = \det(B_n)$ for each n. By applying Corollary 13.4 with $u = 4+2\sqrt{2}$, $\ell = 4 - 2\sqrt{2}$, and $\kappa = n/2$ we obtain the improved estimate $\det(A_n) \geq 8^{\frac{n}{2}}$.

Example 13.2. Let T_n be the $n \times n$ upper Hessenberg matrix whose subdiagonal entries are $-1/2$, and whose uppertriangular part consists of rows filled with the numbers $1, 0, -1/3!, 0, 1/5!, 0, -1/7!$, etc. starting with the diagonal. For example,

$$T_5 = \begin{pmatrix} 1 & 0 & -\frac{1}{3!} & 0 & +\frac{1}{5!} \\ -\frac{1}{2} & 1 & 0 & -\frac{1}{3!} & 0 \\ 0 & -\frac{1}{2} & 1 & 0 & -\frac{1}{3!} \\ 0 & 0 & -\frac{1}{2} & 1 & 0 \\ 0 & 0 & 0 & -\frac{1}{2} & 1 \end{pmatrix}.$$

Clearly T_n is diagonally dominant. To find a lower bound for $|\det(T_n)|$, we apply Corollary 13.4. In this case $\ell = 1 - \alpha$ and $u = 1 + \alpha$ for some α, while $\kappa = n/2$. If n is even, then

$$\alpha = \frac{1}{2} + \frac{1}{3!} + \frac{1}{5!} + \cdots \leq 0.7.$$

Thus,

$$|\det(T_n)| \geq (1 - \alpha^2)^{\frac{n}{2}} \geq 0.5^{\frac{n}{2}},$$

which is considerably better than the lower bound $(.3)^n$ obtained from Gerschgorin's theorem. Using the above bound it is possible to show that $\pi = 3 \lim_{n\to\infty} \det(T_n)/\det(T_{n+1})$. More generally, in Chapter 14 the lower bound will be used as an auxiliary result to give many new formulas for π.

13.2 An Application in Approximation of Polynomial Root

Consider $f(x)$ a polynomial of degree $n \geq 2$ with complex coefficients. In Chapter 9 we have seen that for any point x_0 which lies in the Voronoi region of a root θ, the corresponding basic sequence $\{B_m(x_0)\}_{m=2}^{\infty}$ converges to θ. Here we provide a general bound on the error, given that x_0 is sufficiently close to θ.

As before we set $D_0(x) \equiv 1$, and for each $m \geq 1$, let

$$D_m(x) = \det \begin{pmatrix} f'(x) & \frac{f''(x)}{2!} & \dots & \frac{f^{(m-1)}(x)}{(m-1)!} & \frac{f^{(m)}(x)}{(m)!} \\ f(x) & f'(x) & \ddots & \ddots & \frac{f^{(m-1)}(x)}{(m-1)!} \\ 0 & f(x) & \ddots & \ddots & \vdots \\ \vdots & \vdots & \ddots & \ddots & \frac{f''(x)}{2!} \\ 0 & 0 & \dots & f(x) & f'(x) \end{pmatrix}$$

and for each $i = m+1, \dots, n+m-1$ define

$$\widehat{D}_{m,i}(x) = \det \begin{pmatrix} \frac{f''(x)}{2!} & \frac{f'''(x)}{3!} & \dots & \frac{f^{(m)}(x)}{(m)!} & \frac{f^{(i)}(x)}{i!} \\ f'(x) & \frac{f''(x)}{2!} & \ddots & \frac{f^{(m-1)}(x)}{(m-1)!} & \frac{f^{(i-1)}(x)}{(i-1)!} \\ f(x) & f'(x) & \ddots & \vdots & \vdots \\ \vdots & \vdots & \ddots & \frac{f''(x)}{2!} & \frac{f^{(i-m+2)}(x)}{(i-m+2)!} \\ 0 & 0 & \dots & f'(x) & \frac{f^{(i-m+1)}(x)}{(i-m+1)!} \end{pmatrix}.$$

Then for each $m \geq 2$, and a root θ of $f(x)$ we have

$$\begin{aligned} B_m(x) &\equiv x - f(x)\ \frac{D_{m-2}(x)}{D_{m-1}(x)} \\ &= \theta + \sum_{i=m}^{m+n-2} (-1)^m \frac{\widehat{D}_{m-1,i}(x)}{D_{m-1}(x)} (x-\theta)^i. \end{aligned} \tag{13.11}$$

To state the main theorem let

$$\delta(x) = \frac{1}{|f'(x)|}\left(|f(x)^{1/2}| + \sum_{i=2}^{n} |f(x)^{i/2}| \frac{|f^{(i)}(x)|}{i!}\right).$$

Define

$$h(x) = \sqrt{\sum_{i=0}^{n} \left(\frac{|f^{(i)}(x)|}{i!}\right)^2}.$$

Theorem 13.5 (Kalantari (1998a)). *Given a root θ of $f(x)$, suppose that x satisfies $\delta(x) < 1$ and*

$$|x-\theta| < \frac{\sqrt{1-\delta^2(x)}\ |f'(x)|}{h(x)}. \tag{13.12}$$

Set

$$r(x) = \frac{|x-\theta| h(x)}{\sqrt{1-\delta^2(x)}|f'(x)|}. \tag{13.13}$$

Then

$$|B_m(x)-\theta| \leq \frac{r^m(x)}{1-|x-\theta|}. \tag{13.14}$$

In particular, given $\epsilon \in (0,1)$, for any

$$m \geq \frac{(\log \frac{1}{\epsilon} + \log \frac{1}{1-|x-\theta|})}{\log \frac{1}{r(x)}} \tag{13.15}$$

we have

$$|B_m(x)-\theta| < \epsilon. \tag{13.16}$$

Proof. From Corollary 4.2 in Chapter 4 $D_m(x)$ can alternatively be written as:

$$D_m(x) = \det \begin{pmatrix} f' & \sqrt{f}\frac{f''}{2!} & \cdots & \sqrt{f}^{m-2}\frac{f^{(m-1)}}{(m-1)!} & \sqrt{f}^{m-1}\frac{f^{(m)}}{(m)!} \\ \sqrt{f} & f' & \ddots & \ddots & \sqrt{f}^{m-2}\frac{f^{(m-1)}}{(m-1)!} \\ 0 & \sqrt{f} & \ddots & \ddots & \vdots \\ \vdots & \vdots & \ddots & \ddots & \sqrt{f}\frac{f''}{2!} \\ 0 & 0 & \cdots & \sqrt{f} & f' \end{pmatrix}.$$

Since $\delta(x) < 1$ and $|f'(x)| \leq h(x)$, $|x-\theta| < 1$. Also it follows that $r(x) < 1$. From $\delta(x) < 1$ it follows that the matrix Δ_m corresponding to the above formulation of $D_m(x)$ is diagonally dominant. From Gerschgorin's theorem it follows that if λ is an eigenvalue of Δ_m, we have

$$l = (1-\delta(x))|f'(x)| \leq \lambda \leq (1+\delta(x))|f'(x)| = u.$$

Since

$$|trace(\Delta_m)| = m|f'(x)| \geq m(1-\delta^2(x))|f'(x)|,$$

Theorem 13.5 applies to Δ_m, where

$$\kappa = \frac{mu - |trace(\Delta_m)|}{u-l} = \frac{m(1+\delta(x))|f'(x)| - m|f'(x)|}{(1+\delta(x))|f'(x)| - (1-\delta(x))|f'(x)|} = \frac{m}{2}.$$

Hence we have:

$$|D_m(x)| \geq \Big((1-\delta(x))|f'(x)|\Big)^{\frac{m}{2}} \Big((1+\delta(x))|f'(x)|\Big)^{\frac{m}{2}}$$

$$= (1-\delta^2(x))^{m/2}|f'(x)|^m.$$

Consider the matrix corresponding to $\widehat{D}_{m,i}(x)$. The norm of each of its columns is bounded above by $h(x)$. From Hadamard's inequality we have

$$|\widehat{D}_{m,i}(x)| \leq h(x)^m.$$

Using the upper bound on $|\widehat{D}_{m,i}(x)|$, the lower bound on $|D_m(x)|$, from (13.11) and the geometric progression formula, we conclude:

$$|B_m(x) - \theta| \leq |\frac{h(x)}{\sqrt{1-\delta^2(x)}f'(x)}(x-\theta)|^m \sum_{i=0}^{\infty} |x-\theta|^i$$

$$\leq |\frac{h(x)}{\sqrt{1-\delta^2(x)}f'(x)}(x-\theta)|^m \frac{1}{1-|x-\theta|}.$$

Hence we have proved (13.16).

The proof of (13.15) is an observation on (13.14). □

13.3 Conclusions

In this chapter we have obtained a nontrivial lower bound for $|\det(\cdot)|$, assuming that lower and upper bounds on the moduli of eigenvalues are available, and a trace inequality is satisfied. In the corollaries we describe several particular application of the bound. Similar results follow from different estimates for ℓ and u. In particular, it is known that if $\|\cdot\|$ is any induced matrix norm, then $\rho(A) \leq \|A\|$ for all $n \times n$ matrices A. Thus, each and every induced matrix norm provides an admissible value for u. For example, the u of Corollary 13.3 is $\max\{\sum_{j=1}^n |a_{ij}| : i \in I_n\}$, which is simply $\|A\|_\infty$, where $\|\cdot\|_\infty$ is the matrix norm induced from the familiar sup-norm on $\mathbb{C}^n$. If $\|\cdot\|_1$ denotes the ℓ_1-norm, then the corresponding u is $\max\{\sum_{i=1}^n |a_{ij}| : j \in I_n\}$. Other admissible estimates can be derived from the inequality $\rho(A) \leq \|A^k\|^{1/k}$ which holds for any induced matrix norm $\|\cdot\|$ and natural number k. If more information is available on the distribution of the moduli of the eigenvalues, then tighter lower bounds may be possible. One expects other applications to be discovered.

Chapter 14

Formulas for Approximation of Pi Based on Root-Finding Algorithms *

In this chapter we derive many new formulas for the approximation of pi. The formulas make use of the sequence of iteration functions of the Basic Family, the determinantal generalization of Taylor's Theorem, as well other ingredients and results presented in the chapter. In one scheme, one evaluates members of the Basic Family, for an appropriately selected function, all at the same input. This scheme generates almost a fixed and pre-selected number of digits in each successive evaluation. The approximations of pi obtained via this scheme are within simple algebraic extensions of the rational field. In a second scheme any specific member of the Basic Family is applied to an appropriately selected trigonometric function. In this scheme for each natural number $m \geq 2$ we prove convergence of order m, starting from the initial point. Some preliminary computational results are described. Analogous formulas can be developed to approximate other transcendental numbers. For instance, we also give a formula for the approximation of e. In fact our results give new formulas and arbitrary high-order methods for the approximation of roots of certain analytic functions.

14.1 Introduction

The present chapter exploits the Basic Family for the approximation of roots of certain analytic functions. In particular, the results give many new formulas for the approximation of π. Other transcendental numbers can also be approximated using the main theorems of this chapter. Our determinantal approximations of π are very different from the existing methods.

---

*Part of this chapter has been reprinted from New Formulas for Approximations of pi and Other Transcendental Numbers, *Numerical Algorithms*, Vol. 24 (2000) 59–81, B. Kalantari. With kind permission of Springer Science and Business Media.

For the amazing history and many formulas for the approximation of this beautiful and magical number of nature, see Pi: A Source Book, by Berggren *et al.* (1997). Some of the existing formulas produce a fixed number of correct decimal digits in each iteration. There are also a variety of high-order methods that have resulted in the approximation of π to billions of digits, see Bailey (1988), Bailey *et al.* (1997), and Takahashi and Kanada (1998). Some of these high-order methods are based on the arithmetic-geometric mean iteration and other results, see Salamin (1976), Brent (1976), and Berggren *et al.* (1997). Other high-order methods have been obtained by converting slowly converging, but amazing Ramanujan-type formulas, derived from modular equations, into iterative methods of high-order, see Borwein *et al.* (1989), Chudnovsky and Chudnovsky (1987). The high-order methods require very few iterations, where some or most operations need to be implemented with the desired accuracy of approximation. For a very interesting and brief account of the history of pi see Bailey *et al.* (1997). For an earlier book on the history of pi see Beckmann (1971). The approximation of π is a fascinating area that has demanded deep theoretical and mathematical results, computing techniques (FFT), as well as supercomputers. The search for new formulas and the computation of more and more digits of pi will most likely remain to be an interesting and challenging problem.

One category of the formulas derived in the present chapter approximate π based on the evaluation of the Basic Family, $B_m(x_0)$, for appropriately selected input x_0, as well as an appropriately selected function $f(x)$. This scheme results in formulas producing about a fixed number of correct decimal digits in each successive evaluation of $B_m(x_0)$. In a second category, we prove that for any $m \geq 2$, the fixed-point iteration $x_{k+1} = B_m(x_k)$, as applied to an appropriately selected function $f(x)$, will converge to a constant multiple of π with order of convergence equal to m, and starting with any input within a specified interval. For example, we prove that this is the case with $f(x) = \sin x - 0.5$, whose root is $\pi/6$, starting with any x_0 in $[0, \pi/4]$.

Our results potentially offer yet a third approach in approximation of π: first use any high-order method to arrive at a satisfactory estimate x_0 of π. Then using this estimate, proceed with the evaluation of $B_m(x_0)$. While high-order methods become prohibitive after a few iterations, the evaluation of $B_m(x_0)$ will result in obtaining further accuracy with moderate computational effort.

The following sections of the chapter are organized as follows: In the 2nd section we state three main theorems. In the 3rd section we describe the necessary results needed to prove the theorems. In the 4th section we prove the three theorems. In the 5th section we specialize the main theorems for the approximation of π. In the 6th section we make use of the results to derive some special formulas for the approximation of π and describe an experimental result with Maple. In the 7th section we apply the Basic Family for the approximation of π. In the 8th section we derive a formula for the approximation of e. In the final section we discuss computational considerations and our concluding remarks.

14.2 Main Results

Theorem 14.1 (Determinantal Expansion, Kalantari (2000b)). *Let $f(x)$ be infinitely differentiable on an interval $I = [a, b]$. Assume that f has a zero $\theta \in I$. Also assume that f' does not vanish in I, and for all $x \in I$,*

$$\alpha(x) = \sum_{i=0}^{\infty} \left|\frac{f^{(i)}(x)}{i!}\right| < \infty. \tag{14.1}$$

Set $D_0(x) = 1$, and for each natural number m define

$$D_m(x) = \begin{vmatrix} f'(x) & \frac{f''(x)}{2!} & \dots & \frac{f^{(m-1)}(x)}{(m-1)!} & \frac{f^{(m)}(x)}{(m)!} \\ f(x) & f'(x) & \ddots & \ddots & \frac{f^{(m-1)}(x)}{(m-1)!} \\ 0 & f(x) & \ddots & \ddots & \vdots \\ \vdots & \vdots & \ddots & \ddots & \frac{f''(x)}{2!} \\ 0 & 0 & \dots & f(x) & f'(x) \end{vmatrix}.$$

Then,

$$D_m(x) = \sum_{i=1}^{m} (-1)^{i-1} f^{i-1}(x) \frac{f^{(i)}(x)}{i!} D_{m-i}(x). \tag{14.2}$$

Let $I_\theta = [a', b']$ be any subinterval of I, $a' < b'$, containing θ, such that for each $x \in I_\theta$, either $\delta_1(x) < 1$, or $\delta_2(x) < 1$, where

$$\delta_1(x) = \frac{1}{|f'(x)|}\left(|f(x)| + \sum_{i=2}^{\infty} \frac{|f^{(i)}(x)|}{i!}\right), \tag{14.3}$$

$$\delta_2(x) = \frac{1}{|f'(x)|}\left(\sqrt{|f(x)|} + \sum_{i=2}^{\infty} \frac{|f^{(i)}(x)|}{i!} |f(x)|^{\frac{i-1}{2}}\right). \tag{14.4}$$

For any $m \geq 2$, set

$$B_m(x) \equiv x - f(x) \ \frac{D_{m-2}(x)}{D_{m-1}(x)}. \tag{14.5}$$

Then, such interval I_θ exists, and for each $x \in I_\theta$, $D_{m-1}(x) \neq 0$. Hence $B_m(x)$ is well-defined. Moreover, for each natural number $n \geq 2$, and $x \in I_\theta$, we have

$$B_m(x) = \theta + \sum_{i=m}^{m+n-3} (-1)^m \frac{\widehat{D}_{m-1,i}(x)}{D_{m-1}(x)} (x-\theta)^i$$

$$+(-1)^m \frac{\widehat{\Delta}_{m-1,m+n-2}(x)}{D_{m-1}(x)} (x-\theta)^{m+n-2}, \tag{14.6}$$

where for each $i \geq m$,

$$\widehat{D}_{m-1,i}(x) = \begin{vmatrix} \frac{f''(x)}{2!} & \frac{f'''(x)}{3!} & \dots & \frac{f^{(m-1)}(x)}{(m-1)!} & \frac{f^{(i)}(x)}{i!} \\ f'(x) & \frac{f''(x)}{2!} & \ddots & \frac{f^{(m-2)}(x)}{(m-2)!} & \frac{f^{(i-1)}(x)}{(i-1)!} \\ f(x) & f'(x) & \ddots & \vdots & \vdots \\ \vdots & \vdots & \ddots & \frac{f''(x)}{2!} & \frac{f^{(i-m+3)}(x)}{(i-m+3)!} \\ 0 & 0 & \dots & f'(x) & \frac{f^{(i-m+2)}(x)}{(i-m+2)!} \end{vmatrix},$$

and $\widehat{\Delta}_{m-1,m+n-2}(x)$ (called the error determinant) is obtained by replacing the last column of the matrix corresponding to $\widehat{D}_{m-1,m+n-2}(x)$ with the vector

$$[\frac{f^{(m+n-2)}(\xi_1)}{(m+n-2)!}, \frac{f^{(m+n-3)}(\xi_2)}{(m+n-3)!}, \dots, \frac{f^{(n)}(\xi_{m-1})}{n!}]^T,$$

where ξ_i's are numbers lying between x and θ.

Theorem 14.2 (A One-Point Formula and Determinantal Series). *Under the assumptions of Theorem 14.1, for each $x_0 \in I_\theta$, we have*

$$|B_m(x_0) - \theta| \leq c(x_0)\Big(\mu(x_0)|x_0-\theta|\Big)^m, \tag{14.7}$$

with

$$c(x) = \gamma\sqrt{1-\delta(x)^2}\frac{|f'(x)|}{\beta(x)},$$

where $\delta(x)$ represents either $\delta_1(x)$ or $\delta_2(x)$, so long as $\delta_i(x) < 1$,

$$\gamma = \max\Big\{\sqrt{\sum_{i=2}^{\infty} |\frac{f^{(i)}(x)}{i!}|}: \ \ x \in I_\theta\Big\}, \tag{14.8}$$

$$\mu(x) = \sqrt{\frac{\beta(x)}{|f'(x)|^2(1-\delta^2(x))}}, \tag{14.9}$$

and

$$\beta(x) = \sum_{i=0}^{\infty} |\frac{f^{(i)}(x)}{i!}|^2. \tag{14.10}$$

In particular, if

$$\mu(x_0)|x_0 - \theta| < 1, \tag{14.11}$$

we have

$$\lim_{m\to\infty} B_m(x_0) = \theta. \tag{14.12}$$

Also, if $\mu(x_0)|x_0 - \theta| < 1$*, the following determinantal series is valid:*

$$B_m(x_0) = \theta + \sum_{i=m}^{\infty} (-1)^m \frac{\widehat{D}_{m-1,i}(x_0)}{D_{m-1}(x_0)}(x-\theta)^i. \tag{14.13}$$

Theorem 14.3 (High-Order Iteration Functions). *Consider the same assumptions on* f *and* I_θ *as in Theorem 14.1. Let*

$$\mu = \max\{\mu(x) : x \in I_\theta\}, \quad c = \max\{c(x) : x \in I_\theta\}, \tag{14.14}$$

with $\mu(x)$ *and* $c(x)$ *as defined in Theorem 14.2. Given* $m \geq 2$*, Let* $I_\theta^{(m)} = [\theta - \epsilon, \theta + \epsilon]$ *be an interval contained in* I_θ *such that for each* $x_0 \in I_\theta^{(m)}$*, we have*

$$c^{\frac{1}{m}} \mu |x_0 - \theta|^{\frac{m-1}{m}} < 1. \tag{14.15}$$

Then, given any $x_0 \in I_\theta^{(m)}$*, the sequence of fixed-point iterates*

$$x_{k+1} = B_m(x_k), \quad k = 0, 1, 2, \ldots, \tag{14.16}$$

lies in $I_\theta^{(m)}$*, and it converges to* θ *having order* m *satisfying,*

$$|x_{k+1} - \theta| \leq c\mu^m |x_k - \theta|^m. \tag{14.17}$$

Also,

$$\lim_{k\to\infty} \frac{(\theta - x_{k+1})}{(\theta - x_k)^m} = (-1)^{m-1} \frac{\widehat{D}_{m-1,m}(\theta)}{D_{m-1}(\theta)} = (-1)^{m-1} \frac{\widehat{D}_{m-1,m}(\theta)}{f'(\theta)^{m-1}}. \tag{14.18}$$

Moreover, if $\widehat{D}_{0,1}(\theta) \equiv 0$*, then*

$$\widehat{D}_{m-1,m}(\theta) = \sum_{i=1}^{m-1} (-1)^{i-1} f'(\theta)^{i-1} \frac{f^{(i+1)}(\theta)}{(i+1)!} \widehat{D}_{m-i-1,m-i}(\theta). \tag{14.19}$$

14.3 Auxiliary Results

The proof of Theorems 14.1, 14.2, and 14.3 require several ingredients. We list these as lemmas, some of which have been derived in previous chapters. The first lemma gives a local expansion formula valid in an existential interval containing a simple root where D_{m-1} does not vanish. It corresponds to Corollary 10.2, Chapter 10.

Lemma 14.1. *Suppose that $f(x)$ is infinitely differentiable on an interval $I = [a, b]$. Assume that θ is a simple zero of f in I. Let m be a natural number greater than one. Let $\bar{I}_\theta = [\bar{a}, \bar{b}]$ be any subinterval of I, containing θ, $\bar{a} < \bar{b}$, such that $D_{m-1}(x)$ does not vanish. Then, $\bar{I}_\theta$ is a nonempty interval, and for each $n \geq 2$, and each $x \in \bar{I}_\theta$, we have*

$$B_m(x) = \theta + \sum_{i=m}^{m+n-3} (-1)^m \frac{\widehat{D}_{m-1,i}(x)}{D_{m-1}(x)} (x-\theta)^i$$

$$+(-1)^m \frac{\widehat{\Delta}_{m-1,m+n-2}}{D_{m-1}(x)} (x-\theta)^{m+n-2}.$$

□

The recursive formula for $D_m(x)$ described in Theorem 14.1 is a consequence of the following more general result.

Lemma 14.2 (4.1, Chapter 4). *Let $a_0, \ldots, a_m$ be arbitrary complex numbers, Define $d_0 = 1$, and given a natural number m, define*

$$d_m = \begin{vmatrix} a_1 & a_2 & \cdots & a_{m-1} & a_m \\ a_0 & a_1 & \ddots & \ddots & a_{m-1} \\ 0 & a_0 & \ddots & \ddots & \vdots \\ \vdots & \vdots & \ddots & \ddots & a_2 \\ 0 & 0 & \cdots & a_0 & a_1 \end{vmatrix}.$$

Then,

$$d_m = \sum_{i=1}^{m} (-1)^{i-1} a_0^{i-1} a_i d_{m-i}.$$

□

The next lemma gives an alternative representation of $D_m(x)$ which enables us to describe a non-trivial interval $\bar{I}_\theta$ containing θ such that $D_{m-1}(x)$ does not vanish.

Lemma 14.3 (4.2, Chapter 4). *The quantity $D_m(x)$ can alternatively be written as*

$$\begin{vmatrix} f' & \sqrt{f}\frac{f''}{2!} & \dots & \sqrt{f}^{m-2}\frac{f^{(m-1)}}{(m-1)!} & \sqrt{f}^{m-1}\frac{f^{(m)}}{(m)!} \\ \sqrt{f} & f' & \ddots & \ddots & \sqrt{f}^{m-2}\frac{f^{(m-1)}}{(m-1)!} \\ 0 & \sqrt{f} & \ddots & \ddots & \vdots \\ \vdots & \vdots & \ddots & \ddots & \sqrt{f}\frac{f''}{2!} \\ 0 & 0 & \dots & \sqrt{f} & f' \end{vmatrix}.$$

The next result is a general result that gives a lower bound to determinants. In particular, it can be used to give a lower bound to $|D_m(x)|$ (Lemma 14.5).

Lemma 14.4 (13.2, Chapter 13). *Let A be an $m \times m$ complex matrix. Assume we are given positive numbers l and u such that if λ is an eigenvalue of A, then $l \leq |\lambda| \leq u$. Let*

$$\kappa = \kappa(l, u) \equiv \frac{m \cdot u - |trace(A)|}{u - l}. \tag{14.20}$$

If $|trace(A)| \geq m \cdot l$, then we have

$$|\det(A)| \geq l^{\kappa} u^{m-\kappa}.$$

□

Lemma 14.5. *Let $\delta_1(x)$ and $\delta_2(x)$ be the functions as defined in Theorem 14.1. Given $x_0 \in I$, suppose that $\delta \equiv \delta_i(x_0) < 1$, $i = 1$ or $i = 2$. Then*

$$|D_m(x_0)| \geq \sqrt{1-\delta^2}^{\,m} |f'(x_0)|^m. \tag{14.21}$$

Proof. Suppose that $\delta = \delta_2(x_0)$. Let $A = (a_{ij})$ be the matrix corresponding to $D_m(x_0)$ as given in Lemma 14.3. For each row i of A we have

$$\sum_{j=1, j\neq i}^{m} |a_{ij}| \leq \sqrt{|f(x)|} + \sum_{i=2}^{\infty} \frac{|f^{(i)}(x)|}{i!} |f(x)|^{\frac{i-1}{2}}$$

$$= \delta_2(x_0)|f'(x_0)| = \delta |f'(x_0)|.$$

From Gerschgorin's Theorem it follows that if λ is an eigenvalue of this matrix, then

$$l = (1-\delta)|f'(x_0)| \leq |\lambda| \leq (1+\delta)|f'(x_0)| = u.$$

Since $|trace(A)| = m|f'(x_0)| \geq m(1-\delta)|f'(x_0)|$, Lemma 14.4 applies to A, where

$$\kappa = \frac{m \cdot u - |trace(A)|}{u - l} = \frac{m(1+\delta)|f'(x_0)| - m|f'(x_0)|}{(1+\delta)|f'(x_0)| - (1-\delta)|f'(x_0)|} = \frac{m}{2}.$$

Thus, from Lemma 14.4 we have,

$$|D_m(x_0)| \geq \Big((1-\delta)|f'(x_0)|\Big)^{\frac{m}{2}} \Big((1+\delta)|f'(x_0)|\Big)^{\frac{m}{2}} = \Big(\sqrt{1-\delta^2}\Big)^m f'(x_0)^m.$$

If $\delta = \delta_1(x_0)$, we let A be the matrix corresponding to $D_m(x_0)$ as defined in Theorem 14.1. □

The next result gives an upper bound on the error determinant.

Lemma 14.6. *Assume that $f(x)$ and $\bar{I}_\theta$ are as in Lemma 14.1. Assume*

$$\alpha(x) = \sum_{i=0}^{\infty} \Big|\frac{f^{(i)}(x)}{i!}\Big| < \infty. \tag{14.22}$$

Let

$$\gamma_{m+n-2} = \max\left\{\sqrt{\sum_{i=m+n-2}^{\infty} \Big|\frac{f^{(i)}(x)}{i!}\Big|^2} : \quad x \in \bar{I}_\theta\right\}, \tag{14.23}$$

$$\beta(x) = \sum_{i=0}^{\infty} \Big|\frac{f^{(i)}(x)}{i!}\Big|^2. \tag{14.24}$$

Set $\gamma = \gamma_2$. For each $x \in \bar{I}_\theta$, we have

$$|\widehat{\Delta}_{m-1,m+n-2}(x)| \leq \gamma_{m+n-2}\beta(x)^{\frac{m-2}{2}} \leq \gamma\beta(x)^{\frac{m-2}{2}}. \tag{14.25}$$

Proof. The fact that $\alpha(x)$ is finite implies that γ_{m+n-2} and $\beta(x)$ are finite. Given a $k \times k$ matrix $A = (a_{ij})$, Hadamard's inequality gives:

$$|\det(A)| \leq \prod_{j=1}^{k} \left(\sum_{i=1}^{k} |a_{ij}|^2\right)^{1/2}.$$

Applying Hadamard's inequality to the $m-1 \times m-1$ matrix corresponding to $\widehat{\Delta}_{m-1,m+n-2}(x)$, and noting that the norm of its last column is bounded by γ_{m+n-2}, and that the norm of its other columns is bounded above by $\sqrt{\beta(x)}$, the proof of the first upper bound $\widehat{\Delta}_{m-1,m+n-2}(x)$ follows. □

14.4 Proof of Main Theorems

To prove Theorem 14.1, first note that since $\alpha(x) < \infty$ on I and $\delta_2(\theta) = 0$, a non-trivial interval I_θ exists. Now Lemma 14.5 implies that $B_m(x)$ is well-defined on I_θ. Then by Lemma 14.1 the finite expansion formula for $B_m(x)$ is valid on I_θ. This completes the proof of Theorem 14.1. To prove

Theorem 14.2, given $x_0 \in I_\theta$, from Theorem 14.1 and while setting $n = 2$, we have

$$B_m(x_0) = \theta + (-1)^m \frac{\widehat{\Delta}_{m-1,m}(x_0)}{D_{m-1}(x_0)}(x_0 - \theta)^m.$$

Using the lower bound on $|D_{m-1}(x_0)|$ from Lemma 14.5, we have

$$\frac{1}{|D_{m-1}(x_0)|} \leq \frac{1}{\left(\sqrt{1-\delta^2(x_0)}|f'(x_0)|\right)^{m-1}}.$$

Also, using the upper bound on $\widehat{\Delta}_{m-1,m}(x_0)$ from Lemma 14.6, we get

$$\begin{aligned} |B_m(x_0) - \theta| &= \left|\frac{\widehat{\Delta}_{m-1,m}(x_0)}{D_{m-1}(x_0)}(x_0 - \theta)^m\right| \\ &\leq \frac{\gamma\sqrt{\beta(x_0)}^{\,m-2}|x_0 - \theta|^m}{\left(\sqrt{1-\delta^2(x_0)}|f'(x_0)|\right)^{m-1}}. \end{aligned}$$

But the above upper bound coincides with

$$\gamma\sqrt{1-\delta^2(x_0)}\frac{|f'(x_0)|}{\beta(x_0)}\left(\mu(x_0)|x_0 - \theta|\right)^m.$$

Finally,

$$c(x_0) = \gamma\sqrt{1-\delta^2(x_0)}\frac{|f'(x_0)|}{\beta(x_0)}.$$

The fact that $B_m(x_0)$ converges to θ if $\mu(x_0)|x_0 - \theta| < 1$ is obvious from this bound. To establish the infinite determinantal series for $B_m(x)$, we can show analogously, using the bound on $|\widehat{\Delta}_{m-1,m+n-2}(x_0)|$ (from Lemma 14.6) that

$$\left|\frac{\widehat{\Delta}_{m-1,m+n-2}(x_0)}{D_{m-1}(x_0)}(x_0 - \theta)^m\right| \leq c(x_0)\left(\mu(x_0)|x_0 - \theta|\right)^m |x_0 - \theta|^{n-2}.$$

Suppose that $\mu(x_0)|x_0 - \theta| < 1$. Since $\mu(x_0) > 1$ implies that $|x_0 - \theta| < 1$, the left-hand side of the above inequality converges to zero as n approaches infinity. Hence the infinite series holds. This completes the proof of Theorem 14.2.

To prove Theorem 14.3, i.e., the m-th order convergence for any point within $I_\theta^{(m)}$, first note that such interval must exist. Now if $x_1 = B_m(x_0)$, then we have

$$|x_1 - \theta| \leq c(\mu|x_0 - \theta|)^m.$$

Since $c^{\frac{1}{m}}\mu|x_0-\theta|^{\frac{m-1}{m}}<1$, we have

$$c(\mu|x_0-\theta|)^m = (c^{\frac{1}{m}}\mu|x_0-\theta|^{\frac{m-1}{m}})^m|x_0-\theta|.$$

It follows that

$$|x_1-\theta|<|x_0-\theta|.$$

Hence x_1 is in $I_\theta^{(m)}$. The fact that $x_{k+1}=B_m(x_k)\in I_\theta^{(m)}$ and it converges to θ follows by induction. The asymptotic constant of convergence is the consequence of the expansion formula for $B_m(x)$. Also, the claimed recursive formula for $\widehat{D}_{m-1,m}(x)$ is an application of Lemma 14.2. This completes the proof. □

14.5 Applications in Approximation of π

In this section we specialize Theorems 14.1 and 14.2 for approximation of π.

Theorem 14.4. *Let θ be any number in $I=[0,b]$, $b<\frac{\pi}{2}$. Let $f(x)=\sin x-\sin\theta$. Define*

$$\alpha_1=\sum_{i=1}^{\infty}\frac{1}{(2i)!}\leq 0.55, \tag{14.26}$$

$$\alpha_2=\sum_{i=1}^{\infty}\frac{1}{(2i+1)!}\leq 0.18. \tag{14.27}$$

Set

$$\delta_1(x)=\frac{|\sin x-\sin\theta|}{\cos x}+\alpha_1\tan x+\alpha_2, \tag{14.28}$$

$$\delta_2(x)=\frac{\sqrt{|\sin x-\sin\theta|}}{\cos x}$$

$$+\tan x\sum_{i=1}^{\infty}\frac{\sqrt{|\sin x-\sin\theta|}^{2i-1}}{(2i)!}+\sum_{i=1}^{\infty}\frac{\sqrt{|\sin x-\sin\theta|}^{2i}}{(2i+1)!}, \tag{14.29}$$

$$\delta_3(x)=\frac{\sqrt{|\sin x-\sin\theta|}}{\cos x}\Big(\alpha_1\tan x+\alpha_2\Big). \tag{14.30}$$

Let $I_\theta = [a', b']$ be any subinterval of I, containing θ, $a' < b'$, such that for each $x \in I_\theta$, $\delta_i(x) < 1$, for some $i \in \{1, 2, 3\}$. Then, for any $a \in I_\theta$ and any $m \geq 1$,

$$D_m(a) = \begin{vmatrix} \cos a & \frac{-\sin a}{2!} & \cdots & \frac{f^{(m-1)}(a)}{(m-1)!} & \frac{f^{(m)}(a)}{(m)!} \\ \sin a - \sin\theta & \cos a & \ddots & \ddots & \frac{f^{(m-1)}(a)}{(m-1)!} \\ 0 & \sin a - \sin\theta & \ddots & \ddots & \vdots \\ \vdots & \vdots & \ddots & \ddots & \frac{-\sin a}{2!} \\ 0 & 0 & \cdots & \sin a - \sin\theta & \cos a \end{vmatrix}$$

is nonzero. Let $u = \sin a$, and $v = \cos a$, $u_0 = \sin\theta$. Set $D_0(a) = 1$, and $D_{-1}(a) = D_{-2}(a) = 0$. Then $D_m(a)$ can alternatively be written as

$$D_m(a) = v\left(\sum_{i=0}^{\lceil m/2 \rceil} (-1)^i \frac{(u-u_0)^{2i}}{(2i+1)!} D_{m-(2i+1)}(a)\right)$$

$$+u\left(\sum_{i=1}^{\lceil m/2 \rceil} (-1)^i \frac{(u-u_0)^{2i-1}}{2i!} D_{m-2i}(a)\right) = vD_{m-1}(a) + \frac{uf(a)}{2!} D_{m-2}(a)$$

$$-\frac{vf^2(a)}{3!} D_{m-3}(a) - \frac{uf^3(a)}{4!} D_{m-4}(a) \cdots . \tag{14.31}$$

Moreover, for any $m \geq 2$,

$$B_m(a) = a - (\sin a - \sin\theta)\ \frac{D_{m-2}(a)}{D_{m-1}(a)}$$

$$= \theta + \sum_{i=m}^{\infty} (-1)^m \frac{\widehat{D}_{m-1,i}(a)}{D_{m-1}(a)} (a-\theta)^i, \tag{14.32}$$

($\widehat{D}_{m-1,i}(a)$ is the determinant corresponding to $f(x) = \sin x - \sin\theta$) and we have

$$|B_m(a) - \theta| \leq c(a)\Big(\mu(a)|a-\theta|\Big)^m, \tag{14.33}$$

where

$$c(a) \leq 0.53 \frac{\cos a}{\beta(a)} \leq 0.613.$$

With $\delta(a)$ representing either $\delta_1(a)$, $\delta_2(a)$, or $\delta_3(a)$, whichever is less than one, and

$$\mu(a) = \sqrt{\frac{\beta(a)}{\cos^2 a(1-\delta^2(a))}}, \tag{14.34}$$

$$\beta(a) = (\sin a - \sin\theta)^2 + \alpha_3 \sin^2 a + \alpha_4 \cos^2 a, \tag{14.35}$$

$$\alpha_3 = \sum_{i=1}^{\infty} \frac{1}{(2i)!^2} \leq 0.26, \tag{14.36}$$

$$\alpha_4 = \sum_{i=1}^{\infty} \frac{1}{(2i-1)!^2} \leq 1.1. \tag{14.37}$$

In particular, if

$$\mu(a)|a - \theta| < 1, \tag{14.38}$$

then

$$\lim_{m\to\infty} B_m(a) = \theta. \tag{14.39}$$

Proof. Proof follows from Theorems 14.1 and 14.2. To get the bound on $c(a)$ note that

$$\gamma = \max\left\{\sqrt{\sum_{i=2}^{\infty} \left|\frac{f^{(i)}(x)}{i!}\right|^2} : \ x \in I_\theta\right\} \leq \sqrt{\sum_{i=2}^{\infty} \left|\frac{1}{i!}\right|^2} \leq 0.53.$$

Also, when $a \in I$ it is easy to check that

$$\frac{\cos a}{\beta(a)} \leq \frac{\cos a}{\frac{1}{4}\sin^2 a + \cos^2 a} \leq 1.1548.$$

Thus,

$$\gamma\frac{\cos a}{\beta(a)} \leq .53 \times 1.1548 \leq .613.$$

The introduction of $\delta_3(x)$ which is an upper bound on $\delta_2(x)$ is only for computational convenience and will be used in the next section. □

14.6 Special Formulas for Approximation of π

In this section we will employ Theorem 14.4 to derive new formulas for approximation of π. These formulas can be viewed as approximations of pi obtained within simple algebraic extensions of the rational field.

In what follows the numbers α_i, $i = 1, 2, 3, 4$ are as defined in Theorem 14.4. Let D_m be the determinant of the $m \times m$ Toeplitz matrix whose subdiagonal entries are $-1/2$, the remaining lower triangular part is filled with zeros, and for each row the entries starting with the diagonal entry are filled with the numbers $1, 0, -1/3!, 0, 1/5!, 0, -1/7!$, etc. For instance

$$D_5 = \begin{vmatrix} 1 & 0 & -\frac{1}{3!} & 0 & +\frac{1}{5!} \\ -\frac{1}{2} & 1 & 0 & -\frac{1}{3!} & 0 \\ 0 & -\frac{1}{2} & 1 & 0 & -\frac{1}{3!} \\ 0 & 0 & -\frac{1}{2} & 1 & 0 \\ 0 & 0 & 0 & -\frac{1}{2} & 1 \end{vmatrix}.$$

Theorem 14.5 (Kalantari (2000b)).

$$\pi = 3 \lim_{m\to\infty} \frac{D_{m-2}}{D_{m-1}}. \qquad (14.40)$$

More precisely,

$$|\pi - 3\frac{D_{m-2}}{D_{m-1}}| \le 6\Big(\frac{\sqrt{3}\pi}{6}\Big)^m. \qquad (14.41)$$

Proof. In Theorem 14.4 set $a = 0$, $\theta = \pi/6$. Then,

$$\delta_1(a) = \frac{1}{2} + \alpha_2 < 0.7,$$

$$\beta(a) = \frac{1}{4} + \alpha_4 \le 1.35.$$

Thus,

$$\mu(a) \le \sqrt{\frac{\beta(a)}{1 - \delta_1(a)^2}} \le \sqrt{3}.$$

Since $c(a) < 1$ we may write

$$|B_m(a) - \theta| = |B_m(a) - \frac{\pi}{6}| = |\frac{\pi}{6} - \frac{1}{2}\frac{D_{m-2}}{D_{m-1}}| \le \Big(\sqrt{3}\frac{\pi}{6}\Big)^m.$$

This completes the proof. □

Next we construct a second sequence convergent to π. Let D_m be $\frac{1}{2}$ times the determinant of $m \times m$ Toeplitz matrix whose subdiagonal entries are $1 - \sqrt{3}$, the remaining lower triangular part is filled with zeros, and for each row the entries starting with the diagonal entry are filled with the numbers $\sqrt{3}, -1/2!, +\sqrt{3}/3!, -1/4!$, etc. For instance, for $m = 5$, we have

$$D_5 = \frac{1}{2}\begin{vmatrix} \sqrt{3} & -\frac{1}{2!} & +\frac{\sqrt{3}}{3!} & -\frac{1}{4!} & +\frac{\sqrt{3}}{5!} \\ 1-\sqrt{3} & \sqrt{3} & -\frac{1}{2!} & +\frac{\sqrt{3}}{3!} & -\frac{1}{4!} \\ 0 & 1-\sqrt{3} & \sqrt{3} & -\frac{1}{2!} & +\frac{\sqrt{3}}{3!} \\ 0 & 0 & 1-\sqrt{3} & \sqrt{3} & -\frac{1}{2!} \\ 0 & 0 & 0 & 1-\sqrt{3} & \sqrt{3} \end{vmatrix}.$$

Theorem 14.6 (Kalantari (2000b)).

$$\pi = 3(\sqrt{3} - 1) \lim_{m \to \infty} \frac{D_{m-2}}{D_{m-1}}. \tag{14.42}$$

More precisely,

$$|\pi - 3(\sqrt{3} - 1)\frac{D_{m-2}}{D_{m-1}}| \leq 6\left(\frac{5\pi}{24}\right)^m. \tag{14.43}$$

Proof. In Theorem 14.4 set $a = \pi/6$, $\theta = \pi/3$. If we compute $\delta_1(a)$ we obtain

$$\delta_1(a) = \frac{\sqrt{3} - 1}{\sqrt{3}} + \frac{\sqrt{3}}{3}\alpha_1 + \alpha_2 \approx 0.92.$$

This will not be sufficiently good to prove convergence. Rather we compute $\delta_3(a)$:

$$\delta_3(a) = \frac{\sqrt{\sqrt{3}/2 - 1/2}}{\sqrt{3}/2}(\alpha_1 \frac{1}{\sqrt{3}} + \alpha_2) \leq 0.35.$$

We have $c(a) < 1.0$.

$$\frac{\beta(a)}{\cos^2 a} = \frac{|\sqrt{3} - 1|^2}{3} + \frac{1}{3}\alpha_3 + \alpha_4 \leq 1.37.$$

Thus,

$$\mu(a) = \sqrt{\frac{\beta(a)}{\cos^2 a(1 - \delta(a)^2)}} \leq 1.25.$$

We have,

$$B_m(a) - \frac{\pi}{3} = \frac{\pi}{6} - (\frac{1}{2} - \frac{\sqrt{3}}{2})\ \frac{D_{m-2}}{D_{m-1}} - \frac{\pi}{3} = -\left(\frac{\pi}{6} - \frac{\sqrt{3} - 1}{2}\frac{D_{m-2}}{D_{m-1}}\right).$$

Thus,

$$|B_m(a) - \frac{\pi}{3}| = |\frac{\pi}{6} - \frac{(\sqrt{3} - 1)}{2}\frac{D_{m-2}}{D_{m-1}}| \leq \left(1.25\frac{\pi}{6}\right)^m.$$

This completes the proof. □

Next we describe an infinite family of formulas indexed by k.

Theorem 14.7 (Kalantari (2000b)). *Consider the function* $f(x) = \sin x - \sin\theta$. *Let* k *be a natural number,* $a = \frac{\pi}{4}\left(\frac{2^k - 1}{2^k}\right)$, $\theta = \frac{\pi}{4}$. *Then,*

$$\left|\pi - 2^{k+2} f(a)\frac{D_{m-2}(a)}{D_{m-1}(a)}\right| \leq 2^{k+2}\left(\frac{\pi}{2^{k+1}}\right)^m. \tag{14.44}$$

Proof. Before proving the theorem we remark that for any k, $\sin a$ and $\cos a$ are algebraic numbers. For instance, for $k = 1$, using half-angle formula we have $\sin a = \sin(\pi/8) = \frac{1}{2}\sqrt{2-\sqrt{2}}$, and $\cos a = \cos(\pi/8) = \frac{1}{2}\sqrt{2+\sqrt{2}}$. Thus, for example for $m = 4$ we have

$$D_4(a) = \frac{1}{2}\begin{vmatrix} \sqrt{2+\sqrt{2}} & -\frac{1}{2!}\sqrt{2-\sqrt{2}} & +\frac{1}{3!}\sqrt{2+\sqrt{2}} & -\frac{1}{4!}\sqrt{2-\sqrt{2}} \\ \sqrt{2-\sqrt{2}}-\sqrt{2} & \sqrt{2+\sqrt{2}} & -\frac{1}{2!}\sqrt{2-\sqrt{2}} & +\frac{1}{3!}\sqrt{2+\sqrt{2}} \\ 0 & \sqrt{2-\sqrt{2}}-\sqrt{2} & \sqrt{2+\sqrt{2}} & -\frac{1}{2!}\sqrt{2-\sqrt{2}} \\ 0 & 0 & \sqrt{2-\sqrt{2}}-\sqrt{2} & \sqrt{2+\sqrt{2}} \end{vmatrix}.$$

We now proceed with the proof. Let $I_\theta = [a, \pi/4]$ in Theorem 14.4. We will work with $\delta_3(a)$. To bound this, from the mean value theorem we have

$$|\sin a - \sin\theta| \le |a-\theta| = \frac{\pi}{2^{k+2}}.$$

Since $\cos a \ge \sqrt{2}/2$, and $\tan a \le 1$, we have

$$\delta_3(a) \le \sqrt{\frac{\pi}{2^{k+2}}\frac{2}{\sqrt{2}}}(\alpha_1+\alpha_2) \le 0.73\sqrt{\frac{\pi}{2^{k+1}}} \le 0.65.$$

Also,

$$\frac{\beta(a)}{\cos^2 a} = \frac{(\sin a - \sin\theta)^2}{\cos^2 a} + \alpha_3\tan^2 a + \alpha_4 \le 2(\frac{\pi}{2^{k+2}})^2 + 1.37 \le 2.0.$$

Thus, we have

$$\mu(a) \le \sqrt{\frac{\beta(a)}{\cos^2 a(1-\delta_3(a)^2)}} \le 2.0.$$

Using $c(a) \le 1$ we may write

$$|B_m(a) - \frac{\pi}{4}| = |\frac{\pi}{4}(\frac{2^k-1}{2^k}) - f(a)\frac{D_{m-2}(a)}{D_{m-1}(a)} - \frac{\pi}{4}|$$

$$= |-\frac{\pi}{2^{k+2}} - f(a)\frac{D_{m-2}(a)}{D_{m-1}(a)}| \le \left(\mu(a)\frac{\pi}{2^{k+2}}\right)^m \le \left(\frac{\pi}{2^{k+1}}\right)^m.$$

This completes the proof. □

The sequence

$$\pi_m^{(k)} = -2^{k+2}f(a)\frac{D_{m-2}(a)}{D_{m-1}(a)}, \quad m = 2, \ldots,$$

in the above theorem can alternatively be described as follows: Pick any natural number k. Let

$$z = \left(\frac{\sqrt{2}}{2} + i\frac{\sqrt{2}}{2}\right)^{\frac{2^k-1}{2^k}}, \quad i = \sqrt{-1},$$

$$u = \mathrm{Im}(z), \quad v = \mathrm{Re}(z), \quad f(a) = u - \frac{\sqrt{2}}{2}.$$

Set $D_0(a) = 1$, and for $m \geq 2$ define

$$D_m(a) = vD_{m-1}(a) + \frac{uf(a)}{2!}D_{m-2}(a)$$

$$-\frac{vf^2(a)}{3!}D_{m-3}(a) - \frac{uf^3(a)}{4!}D_{m-4}(a)\cdots.$$

The fact that u and v coincide with $\sin a$ and $\cos a$ follows from de Moivre's formula:

$$z = \big(\cos\frac{\pi}{4} + i\sin\frac{\pi}{4}\big)^{\frac{2^k-1}{2^k}} = \exp\big(i\frac{\pi}{4}\frac{2^k-1}{2^k}\big)$$

$$= \big(\cos\frac{\pi}{4}\frac{2^k-1}{2^k} + i\sin\frac{\pi}{4}\frac{2^k-1}{2^k}\big).$$

In each successive evaluation of $\pi_m^{(k)}$, depending upon the value of k, we gain about a fixed number of digits in approximating π. For instance, using Maple Table 14.1 establishes the relationship between some values of k and the approximate number of correct digits.

Table 14.1 Number of correct digits in each successive evaluation of $\pi_m^{(k)}$.

k	1	10	100	500	1000	2000	3300
Correct Digits	1	4	30	150	300	600	1000

14.7 Approximation of π Via the Basic Family

Here we establish the following result:

Theorem 14.8 (Kalantari (2000b)). *Consider* $f(x) = \sin x - \frac{1}{2}$. *Let* $I = [0, \frac{\pi}{4}]$. *Given any* $x_0 \in I$, *the sequence of fixed-point iterations*

$$x_{k+1} = B_m(x_k), \quad k = 0, 1, 2, \ldots, \tag{14.45}$$

lies in I, *and it converges to* $\theta = \frac{\pi}{6}$ *having order* m *satisfying,*

$$|x_{k+1} - \theta| \leq c(1.38|x_k - \theta|)^m, \quad c \leq \begin{cases} 0.91, & \textit{if } m-1 \textit{ is even;} \\ 0.613, & \textit{otherwise,} \end{cases}$$

and

$$\lim_{k\to\infty} \frac{(\theta - x_{k+1})}{(\theta - x_k)^m} = (-1)^{m-1}\frac{\widehat{D}_{m-1,m}(\theta)}{D_{m-1}(\theta)} = (-1)^{m-1}\frac{\widehat{D}_{m-1,m}(\theta)}{f'(\theta)^{m-1}}. \quad (14.46)$$

Moreover, if we set $\widehat{D}_{0,1}(\theta) = 0$, *then*

$$\widehat{D}_{m-1,m}(\theta) = \sum_{i=1}^{m-1} (-1)^{i-1} f'(\theta)^{i-1} \frac{f^{(i+1)}(\theta)}{(i+1)!} \widehat{D}_{m-i-1,m-i}(\theta). \quad (14.47)$$

Proof. Consider the function $\delta_3(x)$ defined in Theorem 14.4. It can be shown that the maximum of this quantity over I is attained at $\pi/4$, and

$$\delta_3(x) \leq 0.48, \quad \forall \ x \in I.$$

Also, it can be shown that the maximum of $\beta(x)/\cos^2 x$ over I is attained at $\pi/4$, and

$$\frac{\beta(x)}{\cos^2 x} \leq 1.46, \quad \forall \ x \in I.$$

It follows that

$$\mu(x) = \sqrt{\frac{\beta(x)}{\cos^2 x(1 - \delta^2(x))}} \leq 1.38, \quad \forall x \in I.$$

Also, $c(x) \leq c \equiv 0.613$. Thus, from Theorem 14.4 given $x_0 \in I$ we have

$$|B_m(x_0) - \frac{\pi}{6}| \leq c(x_0)\left(1.38|\frac{\pi}{6} - x_0|\right)^m \leq c\left(1.38\frac{\pi}{6}\right)^m.$$

For each $m \geq 2$, the condition $c^{1/m}\mu|x_0 - \pi/6|^{\frac{m-1}{m}} < 1$ of Theorem 14.3 is satisfied with $\mu \leq 1.38$. However, since the interval I is not symmetric about $\pi/6$, we need to argue that the first iterate, $B_m(x_0)$ lies in $[\pi/12, \pi/4]$. For $m \geq 4$, $.613(1.38\frac{\pi}{6})^m \leq \pi/4 - \pi/6 = \pi/12$. Hence, for $m \geq 4$, $B_m(x_0)$ lies in $[\pi/12, \pi/4]$. For $m = 2$, $0.613(1.38\frac{\pi}{6})^m$ is not less than $\pi/12$ so that the general bound does not imply $B_2(x_0) \in [\pi/12, \pi/4]$. However, in this case by considering the graph of $f(x)$, it can easily be verified that for any $x_0 \in I$, $B_2(x_0)$, i.e. the Newton iterate at x_0, lies in $[\pi/12, \pi/4]$. Likewise for $m = 3$, as in the case of $m = 2$, it can be verified that $B_3(x_0) \in [\pi/12, \pi/4]$. We avoid a formal verification of this. This completes the proof of the theorem. □

14.8 A Formula for Approximation of e

In this section we employ Theorems 14.1 and 14.2 to derive a new formula for approximation of e. Let $D_m(a)$ be the determinant of the $m \times m$ Toeplitz matrix corresponding to $f(x) = \ln x - 1$ at $x = a$. Note that for $i \geq 1$, $f^{(i)}(a)/i! = (-1)^{i-1}/(ia^i)$. For instance,

$$D_5(a) = \begin{vmatrix} a^{-1} & \frac{-1}{2}a^{-2} & \frac{1}{3}a^{-3} & \frac{-1}{4}a^{-4} & \frac{1}{5}a^{-5} \\ \ln a - 1 & a^{-1} & \frac{-1}{2}a^{-2} & \frac{1}{3}a^{-3} & \frac{-1}{4}a^{-4} \\ 0 & \ln a - 1 & a^{-1} & \frac{-1}{2}a^{-2} & \frac{1}{3}a^{-3} \\ 0 & 0 & \ln a - 1 & a^{-1} & \frac{-1}{2}a^{-2} \\ 0 & 0 & 0 & \ln a - 1 & a^{-1} \end{vmatrix}.$$

Theorem 14.9 (Kalantari (2000b)).

$$e = 2\sqrt{2} - (\frac{3}{2}\ln 2 - 1) \lim_{m\to\infty} \frac{D_{m-2}(2\sqrt{2})}{D_{m-1}(2\sqrt{2})}. \tag{14.48}$$

More precisely,

$$|2\sqrt{2} - (\frac{3}{2}\ln 2 - 1)\ \frac{D_{m-2}(2\sqrt{2})}{D_{m-1}(2\sqrt{2})} - e| \leq \left(1.5(2\sqrt{2} - e)\right)^m. \tag{14.49}$$

Proof. In Theorem 14.1, take $f(x) = \ln x - 1$. Thus, $\theta = e$. For any $a \geq 2$ such that $|\ln a - 1| \leq 1$, we have

$$\delta_2(a) \leq a\sqrt{|\ln a - 1|}\left(1 + \sum_{i=2}^{\infty} \frac{1}{i2^i}\right) \leq 1.19a\sqrt{|\ln a - 1|}.$$

Let $a = 2\sqrt{2}$. Then, $\delta_2(a) \leq 0.68$. We can choose $I_\theta = [e, a]$. Then,

$$\gamma \leq \sum_{i=2}^{\infty} \frac{1}{i2^i} \leq 0.19.$$

Also, we have

$$\frac{|f'(a)|}{\beta(a)} \leq \frac{1}{f'(a)} = a.$$

Hence it follows that $c(a) < 1$. Moreover,

$$\frac{\beta(a)}{f'(a)^2} = a^2 \sum_{i=0}^{\infty} \frac{1}{i^2 a^{2i}} \leq 1.1.$$

Thus, $\mu(a) \leq 1.5$. From Theorem 14.2 it follows that

$$|B_m(a) - \theta| = |B_m(a) - e| = |a - (\ln a - 1)\ \frac{D_{m-2}(a)}{D_{m-1}(a)} - e| \leq \left(1.5(a - e)\right)^m.$$

This completes the proof. □

14.9 Concluding Remarks

In this chapter we have derived many new formulas for the approximation of pi. In one scheme we showed that by evaluating members of the Basic Family, $B_m(x)$, for an appropriately selected function, as well as input a, it is possible to generate almost a fixed and pre-selected number of digits in each successive evaluation. The computation of $B_m(a)$ amounts to the evaluation of a recursive formula and is equivalent to the computation of special Toeplitz matrix determinants. The approximations of pi obtained via this scheme are within finite algebraic extensions (hence simple extensions) of $\mathbb{Q}$, the field of rational numbers. Theorem 14.5 gives an approximation of π within $\mathbb{Q}$ itself. Theorem 14.6 gives an approximation of π within $\mathbb{Q}(\sqrt{3})$. Theorem 14.7 gives approximation of π within algebraic extensions $\mathbb{Q}(\theta)$, where $deg(\theta)$, the degree of the minimal polynomial of θ, increases with the parameter k. Thus, using this scheme it is possible to obtain approximations of π within a given algebraic extension of $\mathbb{Q}$ to a desired accuracy. One can then convert this algebraic approximation to a decimal approximation.

In a second scheme, one applies the fixed-point iteration to any fixed member of the Basic Family, while selecting an appropriate function. In this scheme, analyzed in Theorem 14.8, we proved high-order of convergence from the initial point. In this scheme one needs to approximate the elementary function $\sin x$ at each iteration. And this may not be desirable. Our results potentially offer yet a third approach in the approximation of π: first use any high-order method to arrive at a satisfactory estimate a of π. Then using this estimate proceed with the evaluation of $B_m(a)$. While any high-order method becomes prohibitive after a few iterations, the evaluation of $B_m(a)$ will result in obtaining further accuracy with moderate computational effort. Needless to say that this approach also requires high accuracy estimate of $\sin a$. It should be mentioned that when a is appropriately selected, e.g. as in Theorem 14.7, $\sin a$ is a special algebraic number so that its approximation may be easier than the case of arbitrary a.

Once for a given a the quantities $\sin a$ and $\cos a$ are computed, $D_m(a)$ can be evaluated in $O(m \log^2 m)$ arithmetic operations. One way to argue the faster time complexity is as follows: The determinant of an $m \times m$ Toeplitz matrix that is strongly nonsingular, i.e. all its leading principal submatrices are nonsingular, can be computed in $O(m \log^2 m)$ arithmetic operations, see Bini and Pan (1994). While this can be justified, using that $D_m(a)$ satisfies a recurrence relation, it can be computed in $O(m \log^2 m)$ arithmetic operations, see Fiduccia (1985).

It should be clear that Theorem 14.4 is simply one application of the first two theorems of the chapter. By considering different trigonometric functions, it is possible to arrive at many other determinantal formulas of the type described in Theorems 14.5, 14.6, and 14.7. This is also the case with respect to high-order methods for the approximation of π. Finally, as demonstrated in Theorem 14.9, our determinantal formulas can also approximate other transcendental numbers. More generally, as demonstrated in Theorems 14.1, 14.2, 14.3, it can be used to approximate roots of many analytic functions. These in particular justify extensions of polynomiography techniques to more general analytic functions. In a subsequent chapter we shall give visualizations corresponding to approximation of π.

Chapter 15

Bounds on Roots of Polynomials and Analytic Functions *

In this chapter we make use of the Basic Family to derive an infinite family of lower bounds on the gap between two distinct zeros of a given analytic function $f(z)$ (Kalantari (2005b)). We then use the bounds to compute lower bounds on the distance from an arbitrary complex point to the nearest root of $f(z)$. In particular, when $f(z)$ is a polynomial, for each $m \geq 2$ we give explicit upper and lower bounds, U_m and L_m on the modulus of zeros. These bounds are efficiently computable and have many theoretical and practical applications, for instance in Weyl's classical quad-tree algorithm for computing all roots of a complex polynomial. McNamee and Olhovsky (2005) computational comparison shows even U_4 is superior to more than 45 existing bounds in the literature (see also McNamee (2007)). Even for $m = 2$ our estimate of lower bound is more than twice as good as Smale's bound, Smale (1986), or its improved version given in Blum *et al.* (1998). A significant property of these bounds, as proved by Jin (2006), is their asymptotic convergence to the radii of tightest annulus containing the zeros. Jin has also given an efficient, $O(mn)$-time algorithm, for the computation of the first m bounds for a polynomial of degree n.

15.1 Introduction

Computing a priori bound on the zeros of polynomials is an interesting and important problem with many theoretical and practical applications. There is a vast literature on this topic, for example see the recent book McNamee (2007). As a consequence of the main results in this chapter we will show that if $f(z) = a_n z^n + \cdots + a_1 z + a_0$ is a polynomial with $a_n a_0 \neq 0$, for each

*Part of this chapter has been reprinted from An Infinite Family of Bounds on Zeros of Analytic Functions and Relationship to Smale's Bound, *Mathematics of Computation*, Vol. 74 (2005) 841–852, B. Kalantari; with permission from the American Mathematical Society.

$m \geq 2$ we can state upper (and lower) bounds on its zeros the first few of which are given below.

Let $r_m \in [1/2, 1)$ be the unique positive root of the polynomial $q(t) = t^{m-1} + t - 1$. Assume ξ is any root of $f(z)$. Then for $m = 2$, $r_2 = 0.5$ and we have

$$|\xi| \leq \frac{1}{r_2} \max\left\{ \left| \frac{a_{n-k+1}}{a_n} \right|^{\frac{1}{k-1}} : k = 2, \ldots, n+1 \right\}.$$

For $m = 3$, $r_3 = 0.618034$ and we have

$$|\xi| \leq \frac{1}{r_3} \max\left\{ \left| \frac{1}{a_n{}^2} \det \begin{pmatrix} a_{n-1} & a_{n-k+1} \\ a_n & a_{n-k+2} \end{pmatrix} \right|^{\frac{1}{k-1}} : k = 3, \ldots, n+2 \right\},$$

where $a_{-1} = 0$.

For $m = 4$, $r_4 = 0.682328$ and we have

$$|\xi| \leq \frac{1}{r_4} \max\left\{ \left| \frac{1}{a_n{}^3} \det \begin{pmatrix} a_{n-1} & a_{n-2} & a_{n-k+1} \\ a_n & a_{n-1} & a_{n-k+2} \\ 0 & a_n & a_{n-k+3} \end{pmatrix} \right|^{\frac{1}{k-1}} : k = 4, \ldots, n+3 \right\},$$

where $a_{-1} = a_{-2} = 0$.

For $m = 5$, $r_5 = 0.724492$ and we have

$$|\xi| \leq \frac{1}{r_5} \max\left\{ \left| \frac{1}{a_n{}^4} \det \begin{pmatrix} a_{n-1} & a_{n-2} & a_{n-3} & a_{n-k+1} \\ a_n & a_{n-1} & a_{n-2} & a_{n-k+2} \\ 0 & a_n & a_{n-1} & a_{n-k+3} \\ 0 & 0 & a_n & a_{n-k+4} \end{pmatrix} \right|^{\frac{1}{k-1}} : k = 5, \ldots, n+4 \right\},$$

where $a_{-1} = a_{-2} = a_{-3} = 0$.

15.2 Estimate to Zeros of Analytic Functions

Let $f(z)$ be a complex-valued function analytic everywhere on the complex plane. Consider Newton's iteration function

$$N(z) = z - \frac{f(z)}{f'(z)}. \tag{15.1}$$

Define

$$\gamma(z) = \sup\left\{ \left| \frac{f^{(k)}(z)}{f'(z)k!} \right|^{1/(k-1)}, k \geq 2 \right\}. \tag{15.2}$$

From Smale's analysis of the one-point theory for Newton's method the following theorem is deducible:

Theorem 15.1 (Smale (1986)). *If ξ, ξ' are distinct zeros of f, ξ a simple zero, then they are separated by a distance according to*

$$|\xi - \xi'| \geq \frac{3 - \sqrt{7}}{2\gamma(\xi)} \approx \frac{.177}{\gamma(\xi)}. \tag{15.3}$$

The following stronger lower bound is given in Blum *et al.* (1998) (Corollary 1, page 158):

$$|\xi - \xi'| \geq \frac{5 - \sqrt{17}}{4\gamma(\xi)} \approx \frac{.219}{\gamma(\xi)}. \tag{15.4}$$

Such theorems are referred as *separation theorems.* Dediue (1997) gives separation theorems for system of complex polynomials and in particular polynomials in one complex variable.

In this chapter we will derive a family of lower bounds indexed by an integer $m \geq 2$ on the gap of Theorem 15.3 which in particular when $m = 2$ improves (15.1) as well as (15.4) by replacing their lower bounds with $1/(2\gamma(\xi))$ which is more than twice as good. Our results make use of the Basic Family, $\{B_m(z), m = 2, \dots\}$.

The chapter is organized as follows: In the 2nd section, we describe the Basic Family and its significant relevant properties for complex polynomials. We then extend these to the case of analytic functions. In the 3rd section, we make use of the Basic Family to derive lower bounds on the distance from a simple zero of f to its nearest distinct zero. In the 4th section, we make use of the preceding lower bounds to derive lower bounds on the distance between an arbitrary point and the nearest root of f. In particular using the latter result we show that given a complex polynomial f, for each $m \geq 2$ we can compute an annulus containing the roots. In the 5th section, we consider the application of the bounds on the modulus of roots within Weyl's algorithm. We conclude the chapter in the 6th section.

15.3 The Basic Family for General Analytic Functions

Assume that $f(z)$ is a complex polynomial of degree n. Consider the Basic Family:

$$B_m(z) \equiv z - f(z) \ \frac{D_{m-2}(z)}{D_{m-1}(z)}, \tag{15.5}$$

where for each $m \geq 2$, $D_0(z) = 1$, and for each $m \geq 1$

$$D_m(z) = \det \begin{pmatrix} f'(z) & \frac{f''(z)}{2!} & \dots & \frac{f^{(m-1)}(x)}{(m-1)!} & \frac{f^{(m)}(z)}{(m)!} \\ f(z) & f'(z) & \ddots & \ddots & \frac{f^{(m-1)}(z)}{(m-1)!} \\ 0 & f(z) & \ddots & \ddots & \vdots \\ \vdots & \vdots & \ddots & \ddots & \frac{f''(z)}{2!} \\ 0 & 0 & \dots & f(z) & f'(z) \end{pmatrix}. \tag{15.6}$$

If ξ is a root of f, $D_m(\xi) = f'(\xi)^m$. Thus, whether or not ξ is a simple root of f it is a fixed-point of B_m since we have

$$B_m(\xi) = \xi - f(\xi)\frac{f'(\xi)^{m-2}}{f'(\xi)^{m-1}} = \xi - \frac{f(\xi)}{f'(\xi)} = \xi.$$

With

$$\widehat{D}_{m,k}(z) = \det \begin{pmatrix} \frac{f''(z)}{2!} & \frac{f'''(z)}{3!} & \dots & \frac{f^{(m)}(z)}{(m)!} & \frac{f^{(k)}(z)}{k!} \\ f'(z) & \frac{f''(z)}{2!} & \ddots & \frac{f^{(m-1)}(z)}{(m-1)!} & \frac{f^{(k-1)}(z)}{(k-1)!} \\ f(z) & f'(z) & \ddots & \vdots & \vdots \\ \vdots & \vdots & \ddots & \frac{f''(z)}{2!} & \frac{f^{(k-m+2)}(z)}{(k-m+2)!} \\ 0 & 0 & \dots & f'(z) & \frac{f^{(k-m+1)}(z)}{(k-m+1)!} \end{pmatrix}, \tag{15.7}$$

where $m \geq 1$, and $k \geq (m+1)$, the following theorem is already proved in Chapter 10 (Corollary 10.2) and is a consequence of the main determinantal theorem.

Theorem 15.2. *Assume that $f(z)$ is a complex polynomial of degree n. Let ξ be a root of $f(z)$. Then, except for finitely many values of $z \in C$, $B_m(z) \in C$, and*

$$B_m(z) = \xi + \sum_{k=m}^{m+n-2} (-1)^m \frac{\widehat{D}_{m-1,k}(z)}{D_{m-1}(z)}(\xi - z)^k. \tag{15.8}$$

We will now proceed by proving a more general version of Theorem 15.2.

Theorem 15.3. *Let $f(z)$ be a complex-valued function analytic over the entire complex plane. For each $m \geq 2$, define $B_m(z)$ as in (15.5). Then,*

$$B_m(z) = \xi + \sum_{k=m}^{\infty} (-1)^m \frac{\widehat{D}_{m-1,k}(z)}{D_{m-1}(z)}(\xi - z)^k. \tag{15.9}$$

Proof. (Sketch of Proof) Since $f(z)$ is analytic and ξ is a root, from Taylor's theorem we have

$$0 = f(\xi) = \sum_{k=0}^{\infty} \frac{f^{(k)}(z)}{k!}(\xi - z)^k. \tag{15.10}$$

Adding and subtracting the quantity $z - f(z) = \xi - (\xi - z) - f(z)$ to both sides of (15.10) we get

$$B_1(z) \equiv z - f(z) = \xi + (f'(z) - 1)(\xi - z) + \sum_{k=2}^{\infty} \frac{f^{(k)}(z)}{k!}(\xi - z)^k. \tag{15.11}$$

From (15.10) one can also easily obtain the following expansion for Newton's iteration which coincides with (15.9) for $m = 2$.

$$B_2(z) \equiv z - \frac{f(z)}{f'(z)} = \xi + \sum_{k=2}^{\infty} \frac{f^{(k)}(z)}{k! f'(z)}(\xi - z)^k. \tag{15.12}$$

Subtracting (15.12) from (15.11) we get

$$B_1(z) - B_2(z) = -f(z) + \frac{f(z)}{f'(z)} \tag{15.13}$$

$$= (f'(z) - 1)(\xi - z) + \sum_{k=2}^{\infty} \frac{(f'(z) - 1) f^{(k)}(z)}{k! f'(z)}(\xi - z)^k.$$

Multiplying (15.10) by $-(f'(z) - 1)(\xi - z)$, and (15.13) by $f(z)$ and then adding the results we get

$$f(z)(B_1(z) - B_2(z)) = f(z)^2 \frac{(1 - f'(z))}{f'(z)} = \sum_{k=2}^{\infty} u_k(z)(\xi - z)^k, \tag{15.14}$$

where for $k \geq 2$

$$u_k(z) = (f'(z) - 1)\left(\frac{f(z) f^{(k)}(z)}{k! f'(z)} - \frac{f^{(k-1)}(z)}{(k-1)!}\right). \tag{15.15}$$

Multiplying (15.14) by

$$-\frac{f''(z)}{2f'(z)} \frac{1}{u_2(z)} = -\frac{f''(z)}{(1 - f'(z))(2f'(z)^2 - f(z) f''(z))} \tag{15.16}$$

and adding the results to (15.12) and simplifying we get a new iteration function

$$B_3(z) \equiv z - \frac{f(z)}{f'(z)} - \frac{f(z)^2 f''(z)}{f'(z)(2f'(z)^2 - f(z) f''(z))} \tag{15.17}$$

$$= z - f(z)\frac{f'(z)}{f'(z)^2 - f(z)f''(z)/2} = z - f(z)\frac{D_1(z)}{D_2(z)}.$$

The corresponding expansion for $B_3(z)$ becomes

$$B_3(z) = \xi - \sum_{k=3}^{\infty}\left(\frac{f^{(k)}(z)}{k!f'(z)} - \frac{f''(z)}{2f'(z)}\frac{u_k(z)}{u_2(z)}\right)(\xi - z)^k \tag{15.18}$$

$$= \xi - \sum_{k=3}^{\infty}\frac{\left(\frac{f^{(k)}(z)f'(z)}{k!} - \frac{f''(z)f^{(k-1)}(z)}{2(k-1)!}\right)}{\left(f'(z)^2 - \frac{1}{2}f(z)f''(z)\right)}(\xi - z)^k$$

$$= \xi - \sum_{k=3}^{\infty}\frac{\widehat{D}_{2,k}(z)}{D_2(z)}(\xi - z)^k.$$

This proved the theorem for $m = 3$ since it coincides with (15.9).

In general we can recursively derive $B_{m+1}(z)$ and its corresponding expansion, given $B_m(z)$, $B_{m-1}(z)$ and their respective expansions in a similar fashion as obtaining $B_3(z)$ from $B_2(z)$ and $B_1(z)$. The fact that this process is well-defined, together with the claimed expansion has already been proved for polynomials in Chapter 7 and Chapter 10. We thus avoid reproducing a complete proof that would replace polynomials with analytic functions. □

15.4 Application of Basic Family in Separation Theorems

In this section we will make use of the Basic Family and Theorem 15.3 to derive a family of lower bounds on the distance between a given zero of $f(z)$ to its nearest distinct zero. Define

$$\gamma_m(z) = \sup\left\{\left|\frac{\widehat{D}_{m-1,k}(z)}{D_{m-1}(z)}\right|^{1/(k-1)} : k \geq m\right\}. \tag{15.19}$$

The following is a key result.

Proposition 15.1. *Assume ξ is a root of $f(z)$. For a given z set*

$$u(z) = |z - \xi|\gamma_m(z).$$

If $u(z) < 1$, then

$$|B_m(z) - \xi| \leq \frac{u^{m-1}(z)}{1 - u(z)}|\xi - z| = \frac{\gamma_m^{m-1}(z)|\xi - z|^m}{1 - u(z)}.$$

Proof. From the definition of $\gamma_m(z)$ and from the expansion formula (15.8) for $B_m(z)$ in Theorem 15.3 we have

$$|B_m(z) - \xi| \leq \sum_{k=m}^{\infty} \gamma_m(z)^{k-1}|\xi - z|^k = |\xi - z| \sum_{k=m}^{\infty} u^{k-1}(z).$$

Since $u(z) < 1$,

$$\sum_{k=m}^{\infty} u^{k-1}(z) = u^{m-1}(z) \sum_{k=0}^{\infty} u^k(z) = \frac{u^{m-1}(z)}{(1 - u(z))}.$$

Thus we get

$$|B_m(z) - \xi| \leq \frac{u^{m-1}(z)}{(1 - u(z))}|\xi - z| = \frac{\gamma_m^{m-1}(z)|\xi - z|^m}{(1 - u(z))}.$$

Hence the proof. □

We now state the lower bound.

Theorem 15.4 (Separation Theorem, Kalantari (2005b)). *Assume ξ, ξ' are zeros of f, and ξ a simple zero. Then*

$$|\xi - \xi'| \geq \frac{r_m}{\gamma_m(\xi)},$$

where $r_m \in [1/2, 1)$ is the unique positive root of the polynomial

$$q(t) = t^{m-1} + t - 1.$$

In particular, if $m = 2$, $r_m = 1/2$, $\gamma_m(\xi)$ coincides with $\gamma(\xi)$ (see (15.2)), *and we get*

$$|\xi - \xi'| \geq \frac{1}{2\gamma(\xi)}.$$

Proof. The polynomial $q(t) = t^{m-1} + t - 1$ has a unique positive root $r_m < 1$. The existence, uniqueness, and bound follow from the inequalities $q(0) < 0$, $q(1) > 1$, and the fact that since $q'(t) > 0$, $q(t)$ is monotonically increasing for positive t.

Since $r_m < 1$, it follows that if $u(\xi) = \gamma_m(\xi)|\xi' - \xi| \geq 1$, then the lower bound in the statement is already satisfied.

Thus we assume that $u(\xi) = \gamma_m(\xi)|\xi' - \xi| < 1$. Then in Proposition 15.1 substituting ξ' for ξ, and ξ for z, also using that $B_m(\xi') = \xi'$ we get

$$|B_m(\xi) - \xi'| = |\xi - \xi'| \leq \frac{u^{m-1}(\xi)}{(1 - u(\xi))}|\xi - \xi'|.$$

Dividing the above by $|\xi - \xi'|$ yields

$$u^{m-1}(\xi) \geq 1 - u(\xi).$$

Equivalently

$$q(u(\xi)) = u^{m-1}(\xi) + u(\xi) - 1 \geq 0.$$

Thus by monotonicity of $q(t)$ we conclude

$$u(\xi) = \gamma_m(\xi)|\xi - \xi'| \geq r_m.$$

Equivalently,

$$|\xi - \xi'| \geq \frac{r_m}{\gamma_m(\xi)}.$$

In particular if $m = 2$, then $r_2 = 1/2$ and we get

$$|\xi - \xi'| \geq \frac{1}{2\gamma_2(\xi)} = \frac{1}{2\gamma(\xi)}.$$

Hence the proof. □

Remark 15.1. For polynomials of degree n and small values of m, given ξ, the quantity $\gamma_m(\xi)$ is computable within the same complexity as that of the normalized derivatives of $f(z)$ which can be done in $O(n \log n)$ (see Pan (1997)).

Remark 15.2. For $m = 2$ our lower bound is more than twice as good as Smale (1986) lower bound (15.1) as well as Blum *et al.* (1998) bound (15.4). It should be mentioned that these bounds were obtained as auxiliary results within the context of the analysis of Smale's one-point theory on Newton's method.

Example 15.1. We now examine the quality of our lower bound on a simple example. We will consider $f(z) = z^2 - 1$. More generally one could consider $z^2 - \xi^2$ (or in fact an arbitrary quadratic polynomial), but the results will be analogous to the case of $z^2 - 1$. Then $\xi = 1$, $\xi' = -1$. Then $f'(\xi) = 2$, $f''(\xi)/2! = 1$, and $f'''(\xi) = 0$. For any $m \geq 2$, $\widehat{D}_{m-1,k}(\xi) = 0$, whenever $k > m$. Thus (15.19) implies:

$$\gamma_m(\xi) = \left(\frac{\widehat{D}_{m-1,m}(\xi)}{D_m(\xi)}\right)^{1/(m-1)}.$$

But it is easy to see that $\widehat{D}_{m-1}(\xi) = f'(\xi)^{m-1} = 2^{m-1}$, and $D_{m-1,m}(\xi) = 1$. Thus, $\gamma_m(\xi) = 1/2$, and according to Theorem 15.4

$$|\xi - \xi'| \geq 2r_m.$$

When $m = 2$, $r_m = 1/2$ and thus the lower bound is only 1. When $m = 3$, r_m is the positive root of $u^2 + u - 1$, i.e. the lower bound is $\sqrt{5} - 1$. As m goes to infinity r_m converges to 1 and hence the lower bounds converge to the actual gap, namely 2.

Remark 15.3. Estimation of the quantity $\gamma_m(\xi)$ (see (15.19)) for m large and for a general polynomial or analytic function can be cumbersome. For this reason we may consider an estimate of this quantity which can be computed efficiently. If for each k we have an upper bound M_k on the modulus of column k of the matrix corresponding to $\widehat{D}_{m-1,k}(\xi)$, then from Hadamard's bound on determinants we have

$$|\widehat{D}_{m-1,k}(\xi)| \leq \prod_{k=1}^{m-1} M_k.$$

Also $D_{m-1}(\xi) = f'(\xi)^{m-1}$. Since M_k can easily be calculated we see that an efficiently computable upper bound on $\gamma_m(\xi)$ may be available.

15.5 Estimate to Nearest Zero and Bounds on Zeros

The previous section gives lower bounds on the distance between a given root of $f(z)$ to its nearest root. In this section we show that the above results can be used to estimate the distance to the nearest zero of $f(z)$ for an arbitrary point in the complex plane. Before doing so let us recall the quantities $\gamma_m(z)$, $D_{m-1}(z)$, $\widehat{D}_{m-1,k}(z)$ defined in (15.19), (15.6), (15.7), respectively. We will first represent these as two variable functions

$$\gamma_m(z, f), \quad D_{m-1}(z, f), \quad \widehat{D}_{m-1,k}(z, f),$$

to indicate that they are defined with respect to a given z and a given function f.

We next prove:

Theorem 15.5 (Estimate to Zeros, Kalantari (2005b)). *Let z_0 be a given complex number different from a root of $f(z)$. Let $F(z) = f(z)(z - z_0)$. For a given $m \geq 2$ define*

$$\gamma_m(z, F) = \sup\left\{\left|\frac{\widehat{D}_{m-1,k}(z, F)}{D_{m-1}(z, F)}\right|^{1/(k-1)} : k \geq m\right\}. \tag{15.20}$$

Then if ξ is any root of $f(z)$, we have

$$|z_0 - \xi| \geq \frac{r_m}{\gamma_m(z_0, F)},$$

where $r_m \in [1/2, 1)$ *is the unique positive root of the polynomial* $t^{m-1}+t-1$. *In particular, for* $m = 2$, $r_m = 1/2$,

$$\gamma_2(z_0, F) = \sup\left\{\left|\frac{f^{(k)}(z_0)}{f(z_0)k!}\right|^{1/k}, k \geq 1\right\},$$

and

$$|z_0 - \xi| \geq \frac{1}{2\gamma_2(z_0, F)};$$

and for $m = 3$, $r_m = (\sqrt{5} - 1)/2$,

$$\gamma_3(z_0, F) = \max\left\{\left|\frac{1}{f(z_0)^2} \det\begin{pmatrix} f'(z_0) & \frac{f^{(k)}(z_0)}{k!} \\ f(z_0) & \frac{f^{(k-1)}(z_0)}{(k-1)!}\end{pmatrix}\right|^{1/k} : k \geq 2\right\},$$

and

$$|z_0 - \xi| \geq \frac{\sqrt{5} - 1}{2\gamma_3(z_0, F)}.$$

Proof. The first assertion is merely the application of Theorem 15.4 to $F(z)$ at its root z_0. To prove the formula for $\gamma_2(z_0, F)$ requires to verify that

$$F^{(k)}(z_0) = kf^{(k-1)}(z_0). \tag{15.21}$$

But this can be shown easily. This gives

$$\gamma_2(z_0, F) = \sup\left\{\left|\frac{F^{(k)}(z_0)}{F'(z_0)k!}\right|^{1/(k-1)}, k \geq 2\right\}$$

$$= \sup\left\{\left|\frac{f^{(k-1)}(z_0)}{f(z_0)(k-1)!}\right|^{1/(k-1)}, k \geq 2\right\}.$$

Now changing $k - 1$ to k we obtain the claimed result for $\gamma_2(z_0, F)$ and hence the lower bound on the gap $|z_0 - \xi|$.

For $m = 3$, r_m is the positive root $t^2 + t + 1$. This is as claimed. Next we need to compute $\gamma_3(z_0, F)$. From (15.21) we have

$$D_2(z_0, F) = \det\begin{pmatrix} F'(z_0) & \frac{F''(z_0)}{2!} \\ F(z_0) & F'(z_0)\end{pmatrix} = \det\begin{pmatrix} f(z_0) & f'(z_0) \\ 0 & f(z_0)\end{pmatrix} = f(z_0)^2.$$

For each $k \geq 3$ from (15.21) we have

$$\widehat{D}_{m-1,k}(z_0, F) = \det\begin{pmatrix} \frac{F''(z_0)}{2!} & \frac{F^{(k)}(z_0)}{k!} \\ F'(z_0) & \frac{F^{(k-1)}(z_0)}{(k-1)!}\end{pmatrix} = \det\begin{pmatrix} f'(z_0) & \frac{f^{(k-1)}(z_0)}{(k-1)!} \\ f(z_0) & \frac{f^{(k-2)}(z_0)}{(k-2)!}\end{pmatrix}.$$

Next changing $k - 1$ to k we get the claimed quantity for $\gamma_3(z_0, F)$. □

We will now state two corollaries of the above giving upper and lower bounds on the modulus of polynomial roots. We will refer to the first as a *second-order bound* since it is based on $m = 2$ and the next as *third-order bound* since it is based on $m = 3$. More generally, for any natural number $m \geq 2$ we can state an *mth-order bound.*

Corollary 15.1. (2nd-order lower bound) *Assume that* $f(z) = a_n z^n + a_{n-1}z^{n-1} + \cdots + a_0$, $a_n a_0 \neq 0$. *Then the modulus of each root of* f *is bounded below by the quantity*

$$L_2 \equiv \frac{1}{2}\left(\max\left\{|\frac{a_k}{a_0}|^{1/k} : k = 1, \ldots, n\right\}\right)^{-1}.$$

Proof. Let $m = 2$, set $z_0 = 0$ and apply Theorem 15.5 using that $f(0) = a_0$, $f^{(k)}(0)/k! = a_k$. □

Corollary 15.2. (3rd-order lower bound) *Assume that* $f(z) = a_n z^n + a_{n-1}z^{n-1} + \cdots + a_0$, $a_n a_0 \neq 0$. *Then the modulus of each root of* f *is bounded below by the quantity*

$$L_3 \equiv \frac{\sqrt{5}-1}{2}\left(\max\left\{\left(\frac{|a_1 a_{k-1} - a_0 a_k|}{|a_0^2|}\right)^{1/k} : k = 2, \ldots, n+1\right\}\right)^{-1},$$

where $a_{n+1} \equiv 0$.

Proof. Let $m = 3$, set $z_0 = 0$ and apply Theorem 15.5 to compute $\gamma_3(z_0, F)$. First observe that

$$\det\begin{pmatrix} f'(0) & \frac{f^{(k)}(0)}{k!} \\ f(0) & \frac{f^{(k-1)}(0)}{(k-1)!}\end{pmatrix} = \det\begin{pmatrix} a_1 & a_k \\ a_0 & a_{k-1}\end{pmatrix} = a_1 a_{k-1} - a_0 a_k.$$

The index k ranges from 2 to $n+1$ and since $f^{(n+1)}(0) = 0$, we can set $a_{n+1} = 0$. Also $f(0)^2 = a_0^2$. Hence we have

$$\gamma_3(z_0, F) = \max\left\{\left(\frac{|a_1 a_{k-1} - a_0 a_k|}{|a_0^2|}\right)^{1/k} : k = 2, \ldots, n+1\right\}.$$

From this and Theorem 15.5 we get the desired lower bound on $|z_0 - \xi|$. □

Applying the above corollaries to the polynomial

$$g(z) = z^n f(\frac{1}{z}) = a_0 z^n + a_1 z^{n-1} + \cdots + a_{n-1}z + a_n$$

we derive second and third order lower bounds to the roots of $g(z)$. But since the roots of $g(z)$ are the reciprocal of the roots of $f(z)$ we get the following second and third order upper bounds to the modulus of the roots of $f(z)$.

Corollary 15.3. (2nd-order upper bound) *Assume that* $f(z) = a_n z^n + a_{n-1}z^{n-1} + \cdots + a_0$, $a_n a_0 \neq 0$. *Then the modulus of each root of* f *is bounded above by the quantity*

$$U_2 \equiv 2\max\left\{ |\frac{a_{n-k}}{a_n}|^{1/k} : k = 1, \ldots, n \right\}.$$

Remark 15.4. The upper bound given in Corollary 15.3 was derived through different means in Henrici (1974) (Corollary 6.4k, page 457). Also a more relaxed version of it is given in Blum *et al.* (1998) (Lemma 2, page 170).

Corollary 15.4. (3rd-order upper bound) *Assume that* $f(z) = a_n z^n + a_{n-1}z^{n-1} + \cdots + a_0$, $a_n a_0 \neq 0$. *Then the modulus of each root of* f *is bounded above by the quantity*

$$U_3 \equiv \frac{\sqrt{5}+1}{2} \max\left\{ \left(\frac{|a_{n-1}a_{n-k+1} - a_n a_{n-k}|}{|a_n^2|} \right)^{1/k} : k = 2, \ldots, n+1 \right\},$$

where $a_{-1} \equiv 0$.

More generally we may state

Theorem 15.6 (Bound on Zeros, Kalantari (2005b)). *For each natural number* $m \geq 2$, *let* $r_m \in [1/2, 1)$ *be the unique positive root of the polynomial* $t^{m-1} + t - 1$. *For any root* θ *of the polynomial* $p(z)$ *we have*

$$L_m \leq \theta \leq U_m,$$

$$U_m = \frac{1}{r_m} \max_{m \leq k \leq n+m-1} \left\{ \left| \frac{\delta_k}{a_n{}^{m-1}} \right|^{1/(k-1)} \right\},$$

$$\delta_k = \det \begin{pmatrix} a_{n-1} & a_{n-2} & \cdots & a_{n-m+2} & a_{n-k+1} \\ a_n & a_{n-1} & \cdots & a_{n-m+3} & a_{n-k+2} \\ 0 & a_n & \ddots & \vdots & \vdots \\ \vdots & \vdots & \ddots & a_{n-1} & a_{n-k+m-2} \\ 0 & 0 & \cdots & a_n & a_{n-k+m-1} \end{pmatrix},$$

where $a_{-1} = a_{-2} = \cdots = a_{-m+2} = 0$. *The quantity* L_m *is obtained by computing* U_m *corresponding to the reciprocal polynomial, namely*

$$a_0 z^n + a_1 z^{n-1} + \cdots + a_n. \qquad \square$$

Example 15.2. We consider a simple example to indicate the utility of the last four corollaries. Let $f(z) = z^n - 1$ where the roots are roots of unity. From the above four corollaries we deduce the following 2nd and 3rd order upper and lower bounds:

$$\{z : L_2 \leq |z| \leq U_2\} = \{z : \frac{1}{2} \leq |z| \leq 2\},$$

$$\{z : L_3 \leq |z| \leq U_3\} = \{z : \frac{\sqrt{5}-1}{2} \leq |z| \leq \frac{\sqrt{5}+1}{2}\}.$$

We see that the annulus reduces in size as we go from $m = 2$ to $m = 3$. It is tempting to conclude that as m increases, the annulus converges to the circle of radius one, i.e. the tightest possible annulus containing the roots. This can be established by proving that for each $m \geq 2$ the roots lie in the annulus

$$\{z : L_m \leq |z| \leq U_m\} = \{z : r_m \leq |z| \leq r_m^{-1}\}.$$

15.6 Applications, Asymptotic Analysis, Computational Efficiency and Comparisons

Here we discuss some applications of the above bounds. Clearly it is always desirable to have lower and upper bound estimates on the roots of a given analytic function, especially polynomials. One direct application is in Weyl's algorithm for computing all roots of a given complex polynomial (see Henrici (1974), Pan (1997), Weyl (1924)). Weyl's algorithm can be viewed as a two-dimensional version of the bisection algorithm. It begins with an initial *suspect square* containing all the roots. Given a suspect square we partition it into four congruent subsquares. At the center of each of the four subsquares we perform a *proximity test*, i.e. we estimate the distance from the center to the nearest zero. If the proximity test guarantees that the distance exceeds half of the length of the diagonal of the square, then the square cannot contain any zeros and it is discarded. The remaining squares are called *suspect* and each of them will recursively be partitioned into four congruent subsquares and the process repeated. We see that our lower bounds can be used as alternative proximity tests to the existing ones, possibly much more effective since we have exhibited infinitely many lower bounds. Moreover our upper bounds can be used to obtain a tight initial suspect square.

Next we state without proof two important results summarized as a single theorem and for proofs refer the reader to his article.

Theorem 15.7 (Jin (2006)). *If θ_i, $i = 1, \ldots, n$ are the roots of $f(z)$, then*

$$\lim_{m \to \infty} U_m = \max\{|\theta_i|, i = 1, \ldots, n\},$$

$$\lim_{m \to \infty} L_m = \min\{|\theta_i|, i = 1, \ldots, n\}.$$

Moreover, the computation of L_j, U_j, $j = 1, \ldots, m$ can be established in $O(mn)$ arithmetic operations. □

It should be clear that for small m the computation of L_m and U_m can be done in $O(n)$ arithmetic operations. In particular, McNamee and Olhovsky (2005) (see also McNamee (2007)) have carried out extensive computational comparison of 45 different existing formulas in the literature for upper bounds on modulus of zeros of polynomials. These included the bounds U_2, U_3, and U_4. They found U_4 to give the most accurate result in comparison to the 45 bounds tested. These results are testimonial to the significance and uniqueness of the general bounds.

15.7 Concluding Remarks

In this chapter we have made use of the Basic Family to arrive at infinitely many new lower bounds on the distance between a zero of a complex polynomial or an analytic function, to its nearest distinct zero. We then showed how to derive from these lower bounds an estimate of the distance from an arbitrary point in the complex plane to the nearest zero. Moreover, we showed that for each natural number $m \geq 2$ we can compute mth-order upper and lower bounds, U_m and L_m, on the modulus of the roots of complex polynomials. These bounds in particular suggest new proximity test within Weyl's algorithm for the computation of all roots of complex polynomials, a task well worth implementation and testing. From the practical point of view, even for small values of m, our bounds are capable of producing very good estimate, more effective and accurate than all the perviously known bounds. Indeed in the vast literature on bounds on zeros of polynomials there appears to be no previously known class of bounds that can even theoretically close the gap between the upper bounds and the maximum modulus of the roots. Jin's asymptotic analysis proves the optimality of the bounds with respect to the convergence of the gap to zero. Moreover, his algorithm also points to the efficiency of their computation, growing linearly with m. While in some practical situations it would be sufficient to compute U_m for some small m, such as $m = 4$, it is conceivable that in some theoretical situations one may need to compute U_m, or L_m, for very large m. One would expect that many practical and theoretical applications will emerge.

Problem 1. Consider all polynomials $f(z)$ of degree n with coefficients that are 0, or ± 1. Let $u^*_{\max}(n)$ denote the maximum of the moduli of the root of all such polynomials. Give an exact or approximate value of this in terms of n.

Problem 2. Exclude from the class of polynomials in the previous problem all the ones that have a root with modulus equal to one. Then denote by $u^*_{\min}(n)$ the minimum of all the roots that have modulus larger than one. Give an exact or approximate value of this in terms of n.

Problem 3. The Mahler measure of a polynomial (see e.g. Mossinghoff (1998))

$$f(x) = a_n x^n + \cdots + a_1 x + a_0 = a_n \prod_{i=1}^{n} (x - \alpha_i)$$

is defined as

$$M(f) = |a_n| \prod_{i=1}^{n} \max\{1, |\alpha_i|\}.$$

Assume $a_n a_0 \neq 0$. Restricting the coefficient to integers $M(f) \geq 1$. It is known that $M(f) = 1$ if and only if $f(x)$ is the product of cyclotomic polynomials. There is much literature on Mahler measure of integer polynomials. In particular, Lehmer's conjecture is that if the Maher measure of an integer polynomial is not one, then it cannot be arbitrary close to one. More specifically, Lehmer's conjecture is that the following polynomial has the least Mahler measure greater than unity among all integer polynomials:

$$l(x) = x^{10} + x^9 - x^7 - x^6 - x^5 - x^4 - x^3 + x + 1$$

having Mahler measure $M(l) \geq 1.1762808$.

Let f be a polynomial with coefficients that are 0, $+1$, or -1, having no root with modulus equal one. Let f^* be its reciprocal polynomial. Prove

$$\max\{M(f), M(f^*)\} \geq (u^*_{\min}(n))^{\lceil \frac{n}{2} \rceil}.$$

Problem 4. Consider the integer polynomial above and bounds $L_m \leq |\alpha_i| \leq U_m$. Note $\prod_{i=1}^{n} |\alpha_i| = |a_0|$. For each $m \geq 2$ consider the optimization problem

$$G_m = \min\{\prod_{i=1}^{n} \max\{1, \beta_i\} : \prod_{i=1}^{n} \beta_i = |a_0|, L_m \leq \beta_i \leq U_m, i = 1, \ldots, n\}.$$

Clearly, for each $m \geq 2$ we have

$$(U_m)^n \geq M(f) \geq G_m.$$

But the optimization of G_m is a linear programming:

$$G_m = \min\{\prod_{i=1}^{n} w_i : \prod_{i=1}^{n} \beta_i = |a_0|, w_i \geq 1, w_i \geq \beta_i, L_m \leq \beta_i \leq U_m, i = 1, \ldots, n\}.$$

Now taking logarithm of the objective function and defining

$$y_i = \ln w_i, \quad z_i = \ln \beta_i, \quad l_m = \ln L_m, \quad u_m = \ln U_m, \quad g_m = G_m, \quad c = \ln |a_0|,$$

we get the following linear programming problem

$$g_m = \min\{\sum_{i=1}^{n} y_i : \sum_{i=1}^{n} z_i = c, y_i \geq 0, y_i \geq z_i, l_m \leq z_i \leq u_m, i = 1, \ldots, n\}.$$

Investigate this linear programming problem and apply it to specific polynomials, computing G_m for different values of m. More generally investigate this strategy in estimating $M(f)$ for an arbitrary polynomial.

Problem 5. For a given polynomial $f(z) = a_n z^n + \cdots + a_0$ of degree n with maximum modulus root $\alpha_{\max}$, given $\epsilon > 0$, estimate a number m_f such that for all $m \geq m_f$ the following inequality would hold $U_m - \alpha_{\max} \leq \epsilon$.

Chapter 16

A Geometric Optimization and its Algebraic Offsprings: Gauss-Lucas, Maximum Modulus, and Novelties

In this chapter we will prove several algebraic-geometric properties of a complex polynomial by posing a single optimization problem. Corollaries are stronger versions of the classical Gauss-Lucas Theorem as well as the Maximum Modulus Principle for polynomials. The proofs and geometric insights also offer algorithmic points of view into the Maximum Modulus problem as well as generalizations that include novel and interesting problems in computational geometry, and in dynamical systems, in two or more-dimensional Euclidean spaces. In particular, we define a problem in computational geometry, the *Algebraic Art Gallery Problem*, and an iteration function we call *Gauss-Lucas iteration function.* The Gauss-Lucas iteration function in dimension two gives rise to a quadratically convergent algorithm for computing the square root of a positive number, and more general polynomials. We also give some polynomiography of Gauss-Lucas iteration function.

16.1 Introduction

The classical Maximum Modulus Principle asserts that the modulus of a non-constant complex analytic function over an open domain is exclusively attained at a boundary point. Equivalently, no interior point can be a local optimal solution of the modulus function. This significant theorem of complex analysis can in particular be utilized to prove the Fundamental Theorem of Algebra. The Gauss-Lucas Theorem, which is based on the Fundamental Theorem of Algebra, asserts: the convex hull of the zeros of a nonconstant complex polynomial must contain the critical points. Although the Gauss-Lucas Theorem and the Maximum Modulus Principle, when stated for polynomials, are closely related statements with respect to the same Euclidean modulus function, apparently the two problems have not been studied jointly. As we shall argue, both theorems come to view when posing a single natural question about this function. What connects the Euclidean modulus function to a complex polynomial is of course the Fundamental Theorem of Algebra for which there are innumerable proofs without the need to resort to the Maximum Modulus Principle. In fact, a consequence of our analysis is a proof of the Maximum Modulus Principle for polynomials, using the Fundamental Theorem of Algebra.

Proof of the Gauss-Lucas Theorem or its generalization based on the higher dimensional Euclidean modulus function can be found in Goodman (1975), Marden (1966), Sheil-Small (2002). Dimitrov (1998) proves a refinement. Indeed the Gauss-Lucas Theorem can be viewed as a restatement of the general result on the Euclidean modulus function, but in terms of a polynomial, as made possible by the Fundamental Theorem of Algebra. In contrast, proving the Maximum Modulus Principle for a polynomial, equivalently the Euclidean modulus function, or generalization to higher dimensions is not straightforward.

In this chapter we limit our attention to complex polynomials and prove some algebraic-geometric properties whose corollaries include stronger versions of both the Gauss-Lucas Theorem and the Maximum Modulus Principle for polynomials. Our proof makes use of elementary analysis, but are based on the interplay between polynomial modulus and its equivalent Euclidean representation. The proof technique gives rise to several interesting generalizations in terms of complex analytic functions, or the optimization of a higher dimensional Euclidean modulus function. The results also offer an algorithmic point of view into these problems.

Consider a collection of points in the Euclidean plane, (a_j, b_j), $j = 1, \dots, n$, not necessarily distinct, and define the Euclidean *modulus function*:

$$F(x, y) = \prod_{j=1}^{n} d_j(x, y), \quad d_j(x, y) = [(x - a_j)^2 + (y - b_j)^2] \qquad (16.1)$$

i.e. the product of the square of the distances from the point (x, y) to n given points. Clearly, the minimum and the maximum of $F(x, y)$ are 0 and ∞. However, it is natural to ask: Does $F(x, y)$ have any nontrivial extremum?

This simple and natural question not only connects the Maximum Modulus Principle for polynomials to the Gauss-Lucas Theorem, but motivates stronger versions of both theorems.

It is easy to argue that no point outside of the convex hull of the n points can be a local optimal solution:

Let L be any line which separates (x, y) from the convex hull. Such line must exist. Now using the property of obtuse triangles, it is straightforward to see that moving away from the convex hull along the line perpendicular to L increases the distance to any (a_j, b_j), hence increases $F(x, y)$ itself.

Thus, any local optimal solution must lie in the convex hull. From this simple argument it also follows that when the points are collinear there is no local maximum in the complex plane, although along the containing line $F(x, y)$ could have many local maxima. In the collinear case the analysis and optimization of $F(x, y)$ along the line is trivial and is reducible to the case of a polynomial with real roots. Moreover, in this case the existence of critical points is merely a consequence of the Rolle's theorem. From the classical optimality condition a local optimum solution (x, y) is necessarily a stationary point:

$$\frac{\partial F}{\partial x} = 2F(x,y)\sum_{j=1}^{n}\frac{x-a_j}{d_j(x,y)} = 0, \quad \frac{\partial F}{\partial y} = 2F(x,y)\sum_{j=1}^{n}\frac{y-b_j}{d_j(x,y)} = 0. \quad (16.2)$$

From the above arguments a local optimal solution is a stationary point that also lies within the convex hull. However, from (16.2) it follow trivially that any stationary point must necessarily lie in the convex hull. This proves the Gauss-Lucas Theorem stated in terms of the Euclidean modulus function. The Fundamental Theorem of Algebra then allows its restatement in terms of a polynomial.

16.2 Elementary Proof of the Gauss-Lucas Theorem and the Maximum Modulus Principle

In what follows we consider a nontrivial stationary point of $F(x, y)$ and assume without loss of generality that it is the origin. It follows trivially that the origin must lie in the relative interior of the convex hull of the points. The main results are to prove that the origin is not a local optimal solution and the algebraic-geometric characterizations of descent and ascent directions. In Lemma 16.1, we consider the case when the origin is a non-degenerate stationary point. In this case we exhibit directions of ascent and descent as well as a simple characterization of all such directions. In Theorem 16.1 we restate this key result in terms of a polynomial having the origin as a simple critical point and give algebraic-geometric representation of directions of ascent and descent for polynomial modulus in terms of polynomial coefficients, rather than its roots. Equivalently, these characterization can be stated with regard to the zeros of the reciprocal polynomial of $p(z)$. Given any critical point that may or may not be degenerate, in Theorem 16.2 we give an algorithm for the explicit computation within each neighborhood of the critical point another point that proves the non-optimality of the critical point. Theorems 16.1 and 16.2 when combined with the proof that a non-critical point cannot have optimal modulus, Proposition 16.1 and Lemma 16.2, result in elementary proof of the Maximum Modulus Principle for polynomials, giving stronger and more revealing results than do general proofs for complex analytic functions.

Lemma 16.1. *Consider a collection of points in the Euclidean plane, (a_j, b_j), $j = 1, \ldots, n$. Assume $(0,0)$ is a nontrivial stationary point of the Euclidean modulus function $F(x,y)$ as defined in (16.1). For $j = 1, \ldots, n$, define the "reciprocal points" (α_j, β_j), where*

$$\alpha_j = \frac{a_j}{(a_j^2 + b_j^2)}, \quad \beta_j = \frac{b_j}{(a_j^2 + b_j^2)}.$$

Then

$$\sum_{j=1}^{n} \alpha_j = 0, \quad \sum_{j=1}^{n} \beta_j = 0. \tag{16.3}$$

In particular, $(0,0)$ lies in the convex hull of the points. Moreover, $(0,0)$ is in the relative interior of the convex hull.

Suppose the following conditions are not simultaneously satisfied

$$\sum_{j}^{n} \alpha_j^2 = \sum_{j=1}^{n} \beta_j^2, \qquad \sum_{j=1}^{n} \alpha_j \beta_j = 0. \tag{16.4}$$

Then, $(0,0)$ is not a local optimal of F. Specifically, let

$$A = \sum_{j=1}^{n} \alpha_j^2 - \sum_{j=1}^{n} \beta_j^2, \quad B = \sum_{j=1}^{n} \alpha_j \beta_j,$$

and $q(\theta) = A\theta^2 - 4B\theta - A$. Then $q(\theta)$ has real roots, and for each θ satisfying $q(\theta) > 0$ ($q(\theta) < 0$), $(0,0)$ is a strict local maximum (minimum) of F along the line $y = \theta x$.

Before proving the lemma we describe geometric interpretations. Note that if $\alpha = (\alpha_1, \cdots, \alpha_n)$, $\beta = (\beta_1, \cdots, \beta_n)$, and $e = (1, \cdots, 1)$, then (16.3) is equivalent to $e^T\alpha = e^T\beta = 0$, also (16.4) is equivalent to $\|\alpha\| = \|\beta\|$, $\alpha^T\beta = 0$.

Figure 16.1 shows a set of points that satisfies condition (16.3). The figure shows the points (a_j, b_j) and (α_j, β_j). The figure also shows the corresponding convex hulls. Only the original points are labeled. Note that if $A \neq 0$, the directions of ascent and descent are determined by the roots of $q(\theta)$,

$$\frac{2B \pm \sqrt{4B^2 + A^2}}{A}.$$

Since the products of the roots are -1 the directions of ascent or descent are specified by the four orthants determined by the corresponding lines. If $B = 0$ the roots of $q(\theta)$ are ± 1. Thus the regions of ascent and descent can be characterized as depicted in the left image in Figure 16.2, either by the shaded area or the plain area, respectively, or the opposite case. If $A = 0$ the directions of ascent-descent will be specified as in the right image in Figure 16.2.

Proof. The proof of (16.3) is immediate from (16.2). To prove containment of $(0,0)$ in the convex hull, we simply normalize the equation in (16.3) through division by $\sum_{j=1}^{n} 1/(a_j^2 + b_j^2)$. To prove $(0,0)$ lies in the relative interior of the convex hull, assume otherwise. Then $(0,0)$ must lie on an edge of the convex hull and since the convex hull can be rotated without affecting distances, we may assume the edge lies on the real axis with all the n points lying on or above it. Thus $b_j \geq 0$ for all j. If the points are collinear then $b_j = 0$ for all j and since $(0,0)$ is a non-trivial critical point it lies in the relative interior of the convex hull. On the other hand, if the points are not collinear their convex hull has nonempty interior, implying the existence of j with $b_j > 0$. This contradicts that $\sum_{j=1}^{n} \beta_j = 0$, proving $(0,0)$ lies within the relative interior.

To prove $(0,0)$ is not a local optimal solution, consider $F(x,y)$ along a typical line through the origin, $y = \theta x$, θ real. If the points are not collinear, for any θ the line remains feasible to the convex hull. Define $f_\theta(x) = F(x, \theta x)$ and set $\delta_j(x) = d_j(x, \theta x)$, $j = 1, \ldots, n$. We have

$$f_\theta(x) = \prod_{j=1}^{n} \delta_j(x), \quad f'_\theta(x) = f_\theta(x) \sum_{j=1}^{n} \frac{\delta'_j(x)}{\delta_j(x)}, \tag{16.5}$$

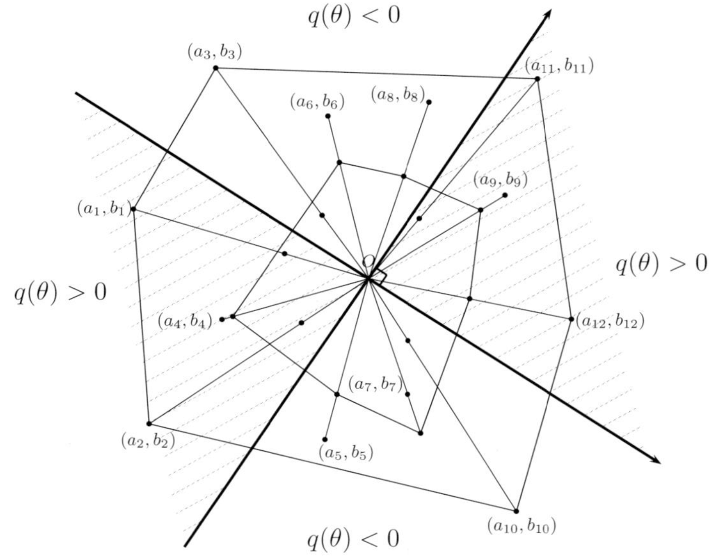

Fig. 16.1 A collection of points (a_j, b_j) with their reciprocals (unlabeled) having the origin as a critical point of $F(x, y)$.

$$f''_\theta(x) = f'_\theta(x) \sum_{j=1}^{n} \frac{\delta'_j(x)}{\delta_j(x)} + f_\theta(x) \sum_{j=1}^{n} \frac{\delta''_j(x)\delta_j(x) - \delta'_j(x)^2}{\delta_j(x)^2}. \tag{16.6}$$

We note

$$\delta_j(x) = x^2(1+\theta^2) - 2x(a_j + \theta b_j) + (a_j^2 + b_j^2),$$

$$\delta'_j(x) = 2(1+\theta^2)x - 2(a_j + \theta b_j), \quad \delta''_j(x) = 2(1+\theta^2).$$

Evaluating these at $x = 0$, substituting into (16.5) and (16.6), and using (16.3) we get

$$f'_\theta(0) = 2f_\theta(0) \sum_{j=1}^{n} \frac{-(a_j + \theta b_j)}{a_j^2 + b_j^2} = -2f_\theta(0)(\sum_{j=1}^{n} \alpha_j + \theta \sum_{j=1}^{n} \beta_j) = 0, \tag{16.7}$$

where the last equality is as expected since $(0, 0)$ is a stationary point of F. Moreover,

$$f''_\theta(0) = f_\theta(0) \sum_{j=1}^{n} \frac{\delta''_j(0)\delta_j(0) - \delta'_j(0)^2}{\delta_j(0)^2}$$

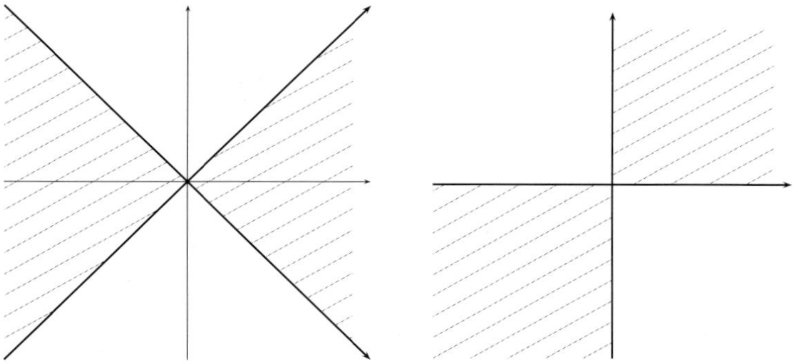

Fig. 16.2 Directions of ascent and descent for the cases $B = 0$ (left) and $A = 0$ (right).

$$= 2f_\theta(0) \sum_{j=1}^{n} \frac{(1+\theta^2)(a_j^2 + b_j^2) - 2(a_j + \theta b_j)^2}{(a_j^2 + b_j^2)^2}. \tag{16.8}$$

Setting $q(\theta) = f_\theta''(0)/2f_\theta(0)$, and evaluating in terms of α_j, β_i, A and B, from (16.8) we get

$$q(\theta) = \theta^2 \sum_{j=1}^{n} (\alpha_j^2 - \beta_j^2) - 4\theta \sum_{j=1}^{n} \alpha_j \beta_j + \sum_{j=1}^{n} (\beta_j^2 - \alpha_j^2) = A\theta^2 - 4B\theta - A. \tag{16.9}$$

The discriminant of q is $\Delta = 16B^2 + 4A^2$. Thus if A and B are not both zero, q can have only real roots. This implies the existence of θ such that $f_\theta''(0) > 0$, as well as θ for which $f_\theta''(0) < 0$, hence establishing the existence of directions of ascent and descent for $F(x, y)$ at the origin. □

Remark 16.1. Lemma 16.1 not only gives a constructive proof that $(0, 0)$ is not a local optimal solution of F, but characterizes directions of ascent and descent in terms of roots of a quadratic equation. It also reveals the surprising characterization of the regions of ascent and descent. In a sense if we choose a direction at random, there is fifty percent chance of being a direction of ascent. The computation of an explicit direction of ascent or descent also extends to the case when the n given points are implicit roots of a polynomial, known only through its coefficients. The following is a restatement of the lemma in terms of polynomials.

Theorem 16.1. *Suppose* $p(z) = p_n z^n + \cdots + p_2 z^2 + p_0$ *with* $p_n p_2 p_0 \neq 0$. *Then* $(0, 0)$ *is not a local optimal solution with respect to* $|p(z)|$. *More* *specifically, if*

$$A = -\mathrm{Re}(\frac{p_2}{p_0}), \qquad B = \frac{1}{2}\mathrm{Im}(\frac{p_2}{p_0}), \qquad q(\theta) = A\theta^2 - 4B\theta - A, \quad (16.10)$$

then q has real roots and for each θ satisfying $q(\theta) > 0$ ($q(\theta) < 0$), the origin is a strict local maximum (minimum) of $|p(z)|$ along the corresponding line $y = \theta x$.

Proof. From the Fundamental Theorem of Algebra $p(z)/p_n = \prod_{j=1}^{n}(z - z_j)$, where $z = x+iy$, $i = \sqrt{-1}$, and $z_j = a_j+ib_j$, $j = 1, \ldots, n$ are the roots. Considering $F(x, y)$ corresponding to the points (a_j, b_j), $j = 1, \ldots, n$, and using properties of complex conjugation, we have:

$$|p(z)|^2 = p(z)\overline{p}(z) = \prod_{j=1}^{n}(z-z_j)(\overline{z}-\overline{z}_j) = \prod_{j=1}^{n}[(x-a_j)^2+(y-b_j)^2] = F(x, y). \quad (16.11)$$

We also have

$$p'(z) = \sum_{j=1}^{n} \frac{p(z)}{(z - z_j)}, \quad p''(z) = \sum_{j=1}^{n} \frac{p'(z)(z - z_j) - p(z)}{(z - z_j)^2}. \quad (16.12)$$

Setting α_j, β_j, A, and B as defined in Lemma 16.1, and since $p(0) = p_0 \neq 0$, $p'(0) = 0$, and $p''(0) = p_2 \neq 0$, from (16.12) we get

$$-\frac{p''(0)}{p(0)} = -\frac{p_2}{p_0} = \sum_{j=1}^{n} \frac{1}{z_j^2} = \sum_{j=1}^{n} \frac{\overline{z}_j{}^2}{(z_j\overline{z}_j)^2} = \sum_{j=1}^{n} \frac{a_j^2 - b_j^2 - 2ia_jb_j}{(a_j^2 + b_j^2)^2}$$

$$= \sum_{j=1}^{n} \alpha_j^2 - \sum_{j=1}^{n} \beta_j^2 - 2i\sum_{j=1}^{n} \alpha_j\beta_j = A - 2iB.$$

Thus $p_2/p_0 \neq 0$ implies A and B cannot both be zero. Hence Lemma 16.1 applies to $|p(z)|^2$. □

Remark 16.2. Aside from the fact that Theorem 16.2 implies that directions of ascent and descent of $|p(z)|$ can be computed trivially, it also reveals some intrinsic geometric properties of the roots and in connection with critical points. If the origin is a simple root of $p'(z)$ the vectors $\alpha = (\alpha_1, \ldots, \alpha_n)$ and $\beta = (\beta_1, \ldots, \beta_n)$ cannot simultaneously be orthogonal and have identical Euclidean norm. In this case the theorem gives complete characterization of ascent and descent in terms of a pair of orthogonal lines. It is also interesting to note that the points (α_j, β_j) correspond to the roots of the reciprocal polynomial:

$$p^*(z) = z^n\overline{p(1/\overline{z})} = \overline{p}_0 z^n + \overline{p}_2 z^{n-2} + \cdots + \overline{p}_{n-1}z + \overline{p}_n,$$

hence justifying the name, reciprocal points. The roots are,

$$\frac{1}{\overline{z}_j} = \frac{z_j}{z_j \overline{z}_j} = \frac{a_j + ib_j}{a_j^2 + b_j^2} = \alpha_j + i\beta_j, \quad j = 1, \ldots, n.$$

Proposition 16.1. *For any $z = (x + iy)$ we have*

$$\frac{\partial F(x,y)}{\partial x} + i\frac{\partial F(x,y)}{\partial y} = 2p(z)\overline{p'(z)}.$$

In particular, z is a critical point or a zero of $p(z)$ if and only if (x, y) is a stationary point of $F(x, y)$.

Proof. Using the well-known identity

$$\frac{p'(z)}{p(z)} = \sum_{j=1}^{n} \frac{1}{(z - z_j)} \tag{16.13}$$

and the equation for partial derivatives of F, see (16.2), we have

$$\frac{\partial F(x,y)}{\partial x} + i\frac{\partial F(x,y)}{\partial y} = 2F(x,y)\sum_{j=1}^{n} \frac{(x+iy) - (a_j + ib_j)}{d_j(x,y)}$$

$$= 2|p(z)|^2 \sum_{j=1}^{n} \frac{(z - z_j)}{(z - z_j)(\overline{z - z_j})}$$

$$= 2|p(z)|^2 \overline{\left(\frac{p'(z)}{p(z)}\right)} = 2|p(z)|^2 \frac{\overline{p'(z)}}{\overline{p(z)}} = 2p(z)\overline{p'(z)}.$$

□

Lemma 16.2. *Let $p(z) = p_n z^n + \cdots + p_1 z + p_0$. Suppose $z_0 = x_0 + iy_0$ is not a zero or a critical point of $p(z)$. Let $z_0' = p(z_0)\overline{p'(z_0)}$. Then there exists $\alpha_0 > 0$ such that for all $0 < \alpha \leq \alpha_0$, $|p(z_0 + \alpha z_0')| > |p(z_0)|$ and $|p(z_0 - \alpha z_0')| < |p(z_0)|$.*

Proof. From Proposition 16.1 it follows that (x_0, y_0) is not a stationary point of F. Moreover, the steepest ascent direction at (x_0, y_0) is (x_0', y_0'), where $z_0' = x_0' + iy_0'$. Thus there exists $\alpha_0 > 0$ such that for all $0 < \alpha \leq \alpha_0$ we have

$$F(x_0 + \alpha x_0', y_0 + \alpha y_0') > F(x_0, y_0), \qquad F(x_0 - \alpha x_0', y_0 - \alpha y_0') < F(x_0, y_0). \tag{16.14}$$

Using (16.11) we may rewrite (16.14) to get the desired inequalities in terms of $p(z)$.

□

Corollary 16.1 (Weak Maximum Modulus Principle). *Assume all critical points of a complex polynomial $p(z)$ in an open domain U are simple roots of $p'(z)$. Then the maximum of $|p(z)|$ over U is attained at its boundary.*

Proof. If $z_0 \in U$ is neither a zero nor a critical point of $p(z)$, then by Lemma 16.2 it is not a local maximum. Suppose $p'(z_0) = 0$. Through the change of variable $u = (z - z_0)$, we may assume $z_0 = 0$ and since by assumption it is a simple critical point, Theorem 16.1 implies that it cannot be a local maximum. □

Remark 16.3. To prove the Maximum Modulus Principle for a general analytic function, the traditional approach (see remark in Bak and Newman (1997) p. 77) is to demonstrate a direction of ascent for the modulus, whether or not a given point is a critical point or even a degenerate critical point. While this general approach has its advantages it is surprising that the steepest direction does not seem to have been considered before. Indeed using the Cauchy-Riemann equations, Proposition 16.1 can be extended to a general analytic function $f(z)$ and if $F(x,y) = |f(z)|^2$, the second derivative test discriminant at a critical point - which coincides with the determinant of the Hessian of F - can be shown to give (Hajja (2006))

$$\nabla^2 F(x,y) = -|f(z)f''(z)|. \tag{16.15}$$

This extended formula not only gives an algorithmic proof of the non-optimality of the modulus function when $f(z)f''(z) \neq 0$, but comes handy from the computational point of view. Classical proofs of the Maximum Modulus Principle never seem to have taken the actual computational point of view into consideration. We will address some natural and interesting problems later. The formula (16.15) together with the extension of Proposition 16.1 for analytic functions, already does give a very simple proof of the Maximum Modulus Principle when the underlying domain does not contain a critical point with $f''(z) = 0$. Thus the Weak Maximum Modulus Principle for polynomials, Corollary 16.1, also extends to analytic functions. In summary, a simple critical point of an analytic function f that is not its zero, is necessarily a saddle point of F. This fact was independently shown in Bak *et al.* (2007) and in more generality. Their proof however is based on the Maximum Modulus Principle for analytic function.

We will next describe an algorithm to compute within each neighborhood of a given point z_0 that is not a root, a new point where the polynomial modulus is greater (or smaller) than the modulus at z_0. Next we state and prove a theorem that also implies the general case of Corollary 16.1.

Theorem 16.2. *Let* $p(z) = p_n z^n + \cdots + p_k z^k + p_0$, $k \geq 2$ *with* $p_n p_k p_0 \neq 0$. *Let* r *be the minimum of the moduli of roots of* $p(z)$ *and those of* $p'(z)/z^{k-1}$. *Let* $\bar{r}$ *be the minimum of the moduli of roots of the polynomial*

$$h(z) = n^2 p_n z^{n-k} + \cdots + k^2 p_k.$$

For any $\rho < \min\{r, \bar{r}\}$, *let* $D_\rho = \{z : |z| \leq \rho\}$, *and let* $z^*(\rho)$ *and* $z_*(\rho)$ *satisfy*

$$|p(z^*(\rho))| = \max\left\{|p(z)| : z \in D_\rho\}, \quad |p(z_*(\rho))| = \min\{|p(z)| : z \in D_\rho\right\}.$$

Then

$$|p(z_*(\rho))| < |p(0)| < |p(z^*(\rho))|.$$

In particular, the origin is not a local optimal solution of $|p(z)|$.

Proof. Given $\epsilon > 0$, define

$$p_\epsilon(z) = p(z) + \epsilon z^2.$$

Thus, for all $\epsilon > 0$ the origin is a critical point of $p_\epsilon(z)$. We claim the disk D_ρ contains no degenerate critical point of $p_\epsilon(z)$. To prove this note that other than the origin, any other critical point of $p_\epsilon(z)$ that is non-degenerate must be a root of $p'_\epsilon(z)/z$, as well as a root of $p''_\epsilon(z)$. Hence it must satisfy

$$\frac{p'_\epsilon(z)}{z} - p''_\epsilon(z) = -(n^2 p_n z^{n-2} + \cdots + k^2 p_k z^{k-2}) = 0. \tag{16.16}$$

But (16.16) is equivalent to the equation $z^{k-2}h(z) = 0$, hence the proof of claim follows from the definition of $\bar{r}$.

Now given any $\epsilon > 0$, from Corollary 16.1 it follows that the maximum of $|p_\epsilon(z)|$ over D_ρ is attained at a boundary point of D_ρ, say $z_\epsilon(\rho)$. Thus

$$|p_\epsilon(z_\epsilon(\rho))| > |p_\epsilon(0)| = |p(0)|. \tag{16.17}$$

Thus we get

$$\lim_{\epsilon \to 0} |p_\epsilon(z_\epsilon(\rho))| \geq |p(0)|. \tag{16.18}$$

Moreover, we must have

$$\lim_{\epsilon \to 0} |p_\epsilon(z_\epsilon(\rho))| = \max\{|p(z)| : z \in D_\rho\} = |p(z^*(\rho))|. \tag{16.19}$$

We claim that the inequality in (16.18) is strict. If not, pick $\rho' < \rho$, see Figure 16.3. From (16.17) and since from Corollary 16.1, $|p_\epsilon(z^*(\rho)| \geq |p_\epsilon(z^*(\rho')|$, it follows that

$$|p(z^*(\rho'))| = |p(z^*(\rho))| = |p(0)|.$$

But this implies $z^*(\rho')$ is a critical point of $p(z)$. Otherwise, since $z^*(\rho')$ is not a zero of $p(z)$ either, by Lemma 16.2 and the fact that $\rho' < \rho$, there exists $\alpha > 0$ such that if $z_0 = z^*(\rho')$, and $z_0' = p(z_0)\overline{p'(z_0)}$, then $z_0 + \alpha z_0'$ lies inside D_ρ, satisfying $|p(z_0 + \alpha z_0')| > |p(z^*(\rho))|$, see Figure 16.3. But in view of (16.19) this is a contradiction. Hence $z^*(\rho')$ must be a critical point of $p(z)$. But this too is a contradiction by the choice of r and $\bar{r}$. Hence the inequality in (16.18) is strict for any $\rho < \min\{r, \bar{r}\}$. Thus the origin is not a local maximum of $|p(z)|$. Similar arguments apply with respect to local minimum. □

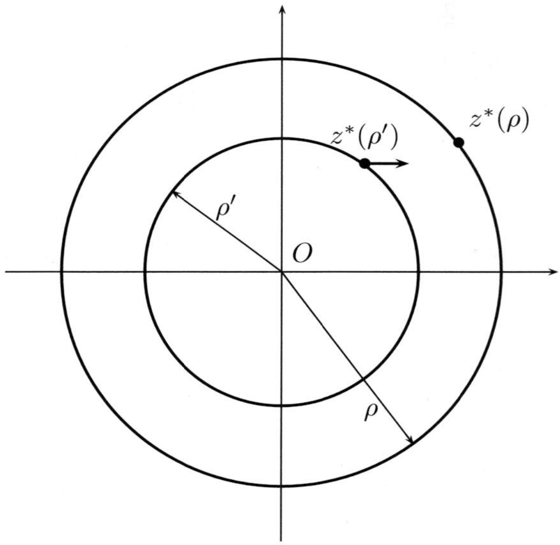

Fig. 16.3 Maximization of $|p(z)|$ over circles of radius ρ and ρ'.

Remark 16.4. The quantities r and $\bar{r}$ used in Theorem 16.2 can be replaced with any lower bound on these numbers. Moreover, we can effectively compute such lower bounds (see Chapter 15).

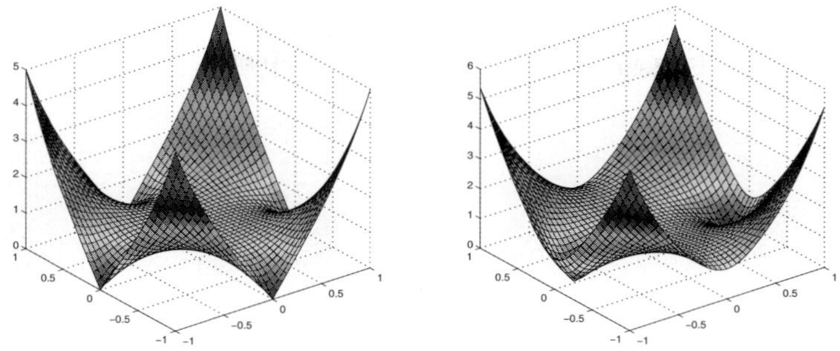

Fig. 16.4 The graphs of modulus for $p(z) = z^4 - 1$ (left) and $p_\epsilon(z) = z^4 + \epsilon z^2 - 1$ (right).

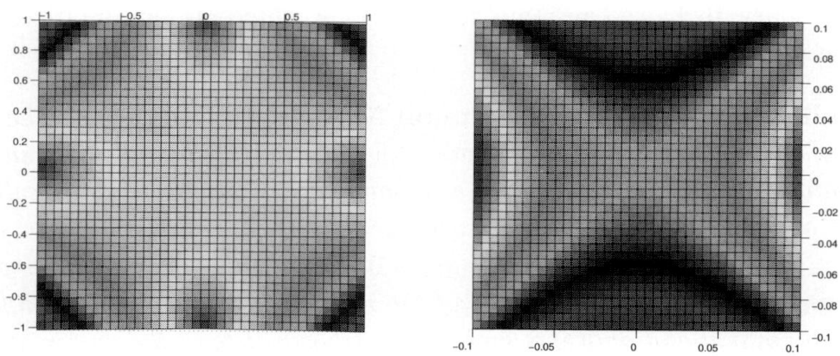

Fig. 16.5 The top view of the graphs of modulus with appropriate scaling.

Example 16.1. Consider the case where $p(z) = z^4 - 1$. The modulus over the convex hull of the four roots is maximized at the four points with real and imaginary parts equal to $\pm\sqrt{2}/2$. Figure 16.4 shows the graph of the modulus of $p(z)$ and $p_\epsilon(z) = z^4 + \epsilon z^2 - 1$. Figure 16.5 shows the top view. Figure 16.6 shows the flow graph of the directions of ascent and descent with respect to the two corresponding graphs. The graph for p_ϵ is zoomed in appropriately to demonstrate the anticipated behavior. Note that while for $p(z)$ there are 8 regions of ascent and descent, for $p_\epsilon(z)$ there are 4 regions as predicted by Theorem 16.1. It is interesting to note that in $p_\epsilon(z)$ it turns out that for all $\epsilon > 0$ the corresponding $q(\theta) = -\epsilon\theta^2 + \epsilon$ so that $\theta = \pm 1$. Hence the regions of ascent (or descent) remain invariant for all ϵ.

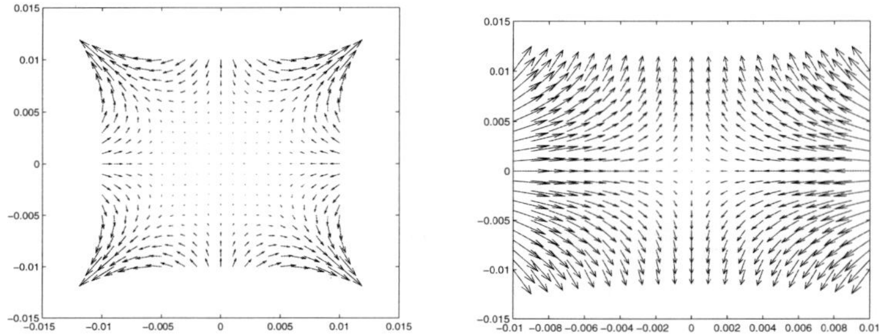

Fig. 16.6 Corresponding flow graphs of ascent and descent for $p(z)$ (8 regions) and for $p_\epsilon(z)$ (4 regions).

Using Theorems 16.1 and 16.2 and analogous arguments as in the proof of Corollary 16.1, we can state the following which also summarizes some of the main results:

Corollary 16.2 (Strong Maximum Modulus Principle). *The maximum of the modulus of a nonconstant complex polynomial over an open domain U is exclusively attained at a boundary point. More specifically, let z_0 be an interior point. Assume without loss of generality that $z_0 = 0$ and $p(z) = p_0 + p_1 z + \cdots p_n z^n$, with $p_0 p_n \neq 0$. Let α and β be the vector of real and imaginary parts of the zeros of the reciprocal polynomial. Then 0 is a critical point if and only if α and β are orthogonal to e, the vector of ones. If $p_1 \neq 0$, the direction $p_0 \overline{p}_1$ is the steepest descent with respect to $|p(z)|^2$, hence 0 is non-optimal. If $p_1 = 0$, then $p_2 \neq 0$ if and only if the vectors α, β are not simultaneously orthogonal having and have identical Euclidean norm. When $p_2 \neq 0$, the directions of ascent of $|p(z)|$ are completely characterizable in terms of opposite orthants determined by all θ where $q(\theta) = A\theta^2 - 4B\theta - A$ is positive, $A = -\mathrm{Re}(p_2/p_0)$, $B = 0.5\mathrm{Im}(p_2/p_0)$. When $p_1 = 0$, whether or not $p_2 = 0$, for any $\rho < \min\{r, \bar{r}\}$ (see Theorem 16.2 for definition of r, $\bar{r}$) with $D_\rho \subset U$, the maximum of $|p(z)|$ over the circle of radius ρ will result in a point $z^*(\rho)$ with larger modulus than $|p(0)| = |p_0|$.* □

Corollary 16.3 (Stronger Gauss-Lucas Theorem). *A critical point of a nonconstant complex polynomial that is not a root of the polynomial must lie in the relative interior of the convex hull of its roots. Moreover, it is not a local optimal solution of the polynomial modulus function in any* *disk around it.* □

16.3 The Gauss Lucas Iteration Function and Extensions of the Maximum Modulus Principle

In this section we will consider several extensions of the Gauss Lucas Theorem and the Maximum Modulus Principle for polynomials. In particular, we give extension of the Maximum Modulus Theorem in terms of the Euclidean modulus function in higher dimensions. These give rise to interesting and challenging problems. In particular, the maximization of the Euclidean modulus function over the convex hull of its defining points in two, three, or higher dimensional Euclidean space is a problem that falls within the realm of computational geometry and find their own applications. Also, motivated by the Gauss-Lucas theorem we define a novel iteration function we call the Gauss-Lucas iteration function. We will study some of the properties of this iteration function and also give some corresponding polynomiography.

Definition 16.1. Given a set of points $a_j \in \mathbb{R}^m$, $j = 1, \ldots, n$, not necessarily distinct, the *modulus function* is defined as

$$F(x) = \prod_{j=1}^{n} \|x - a_j\|^2.$$

The *Gauss-Lucas iteration function* is defined as

$$G(x) = \sum_{j=1}^{n} \alpha_j(x) a_j,$$

where

$$\alpha_j(x) = \frac{\|x - a_j\|^{-2}}{\sum_{j=1}^{n} \|x - a_j\|^{-2}}.$$

The functions $F(x)$ and $G(x)$ give rise to many interesting questions and problems. Indeed our goal in this chapter is to give a flavor of some of these and postpone their detailed analysis to a future work.

Clearly the given points are the minimizers of $F(x)$. Moreover, while $\alpha_j(a_j)$ is not defined, it can be taken to be one which coincides with the limit of $\alpha_j(x)$ as x approaches a_j. Thus each a_j is a fixed point of $G(x)$. We call these the trivial fixed points.

Next we state a proposition whose proof is straightforward and will be omitted.

Proposition 16.2. *A point $x^* \in \mathbb{R}^m$ is a stationary point of $F(x)$ if and only if it is a fixed point of $G(x)$.* □

Note that $G(x)$ maps every $x \in \mathbb{R}^m$ into the convex hull of the points $a_j, j = 1, \ldots, n$. Using this Proposition 16.2 implies:

Theorem 16.3 (Gauss-Lucas Theorem). *Every stationary point of F lies in the convex hull of n given points.*

Next we state a property of $F(x)$ and a problem in computational geometry. The proof of the theorem and other characterizations will be given in a future article.

Theorem 16.4. *Any local maximum of $F(x)$ over the convex hull of its defining points, $a_1, \ldots, a_n$, is attained at a boundary point of the convex hull and never in its interior.* □

Problem 16.1 (The Algebraic Art Gallery Problem). *Given a set of n points $a_j \in \mathbb{R}^m$, $j = 1, \ldots, n$, find a point x^* in their convex hull such that*

$$F(x^*) = \max\{F(x) : x = \sum_{j=1}^{n} \alpha_j a_j, \quad \alpha_j \geq 0, \quad \sum_{j=1}^{n} \alpha_j = 1\}.$$

Moreover, give the complexity in computation of such a point when $m = 2, 3$.

This problem is interesting even in dimensions two and three. In particular, consider the problem in dimension three. The justification behind the name is as follows: if we wish to place a camera in an art gallery to have in view several priceless artworks, it would be more appealing to place it as far away from the artworks as possible so as not to annoy the typical gallery visitor. Then maximizing the product of the distances is the most sensible objective function. If instead we would maximize the sum of the distances, by concavity of the objective function, it would occur at a vertex, namely one of the n given points. Thus the above problem is a meaningful optimization problem. It falls within the realm of computational geometry, though it does not appear to be a typical problem in this field.

Other optimization problems can be stated, e.g. the computation of the maxima of the two or higher dimensional modulus function over a given ball centered at a critical point, or at an arbitrary point.

We will next analyze the Gauss-Lucas iteration function for the two dimensional case of the problem.

Theorem 16.5. *If* $m = 2$, $a_j = (x_j, y_j)$, $j = 1, \ldots, n$, $z_j = x_j + iy_j$, $z = (x, y)$, *and* $p(z) = \prod_{j=1}^{n}(z - z_j)$, *then*

$$\beta_j(z) = \frac{1}{|z - z_i|^2}$$

and

$$G(z) = z - \frac{\overline{p}'(z)}{\overline{p}(z)} \times \frac{1}{\sum_{j=1}^{n} \beta_j(z)}.$$

Proof. Since we may write

$$\alpha_j(z) = \frac{\beta_j(z)}{\sum_{j=1}^{n} \beta_j(z)},$$

we have

$$G(z) = \sum_{j=1}^{n} \alpha_j(z) z_j.$$

Thus

$$z - G(z) = \sum_{j=1}^{n} \alpha_j(z) z - \sum_{j=1}^{n} \alpha_j(z) z_j = \sum_{j=1}^{n} \alpha_j(z)(z - z_j)$$

$$= \sum_{j=1}^{n} \frac{\beta_j(z)}{\sum_{j=1}^{n} \beta_j(z)}(z - z_j) = \frac{1}{\sum_{j=1}^{n} \beta_j(z)} \sum_{j=1}^{n} \beta_j(z)(z - z_j)$$

$$= \frac{1}{\sum_{j=1}^{n} \beta_j(z)} \sum_{j=1}^{n} \frac{\overline{(z - z_j}}{|z - z_j|^2}.$$

From (16.13) we have

$$\sum_{j=1}^{n} \frac{\overline{(z - z_j}}{|z - z_j|^2} = \sum_{j=1}^{n} \frac{1}{\overline{z - z_j}} = \frac{\overline{p}'(z)}{\overline{p}(z)}.$$

This completes the proof. □

Corollary 16.4. *The only fixed points of* $G(z)$ *are the roots of* $p(z)$ *and* $p'(z)$.

We next consider two examples and their polynomiography.

Example 16.2. Consider the Gauss-Lucas function for the case when $p(z) = z^2 - \alpha$, where $\alpha > 0$. It is easy to show that

$$G(z) = \frac{\alpha(z + \bar{z})}{(z\bar{z} + \alpha)} = \frac{2\alpha x}{x^2 + y^2 + \alpha}.$$

From the above it is evident that given an initial input z_0, $z_1 = G(z_0)$ is real and will remain real. The real function

$$G(x) = \frac{2\alpha x}{x^2 + \alpha}$$

has three fixed points 0 and $\pm\sqrt{\alpha}$.

$$G'(x) = \frac{2\alpha(\alpha - x^2)}{(x^2 + \alpha)^2}.$$

Thus $\pm\sqrt{\alpha}$ are superattractive and 0 is repelling. Figure 16.7, left image, gives the polynomiography of $G(z)$.

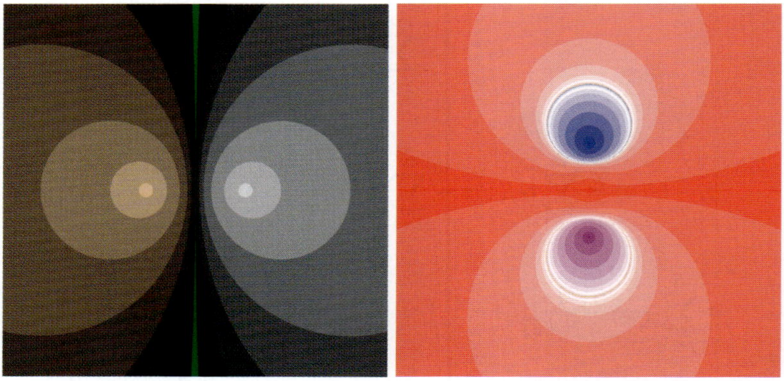

Fig. 16.7 Polynomiography of $G(z)$ for $p(z) = (z^2 - 1)$ (left image) and for $p(z) = z(z^2 + 1)$.

Example 16.3. Consider $G(z)$ for $p(z) = z(z^2 + 1)$. In this example it can be shown

$$G(z) = \frac{2z\bar{z}(z - \bar{z})}{3z^2\bar{z}^2 + (z + \bar{z})^2 + 1} = \frac{4y(x^2 + y^2)}{3(x^2 + y^2)^2 + 4x^2 + 1}.$$

For any z_0, $z_1 = G(z_0)$ will lie on the imaginary axis so that we may write

$$G(y) = \frac{4y^3}{3y^4 + 1}i.$$

The fixed points of $G(z)$ as expected are:

$$z = 0, \quad z = \pm i, \quad z = \pm i\frac{\sqrt{3}}{3}.$$

Figure 16.7, right image, gives the polynomiography corresponding to $G(z)$. The interior of the two black circles, drawn in heavy lines, are regions of convergence of $G(z)$ corresponding to the fixed points $\pm i$. The black circles are inverse images of the critical points of $p(z)$. All other points converge to $z = 0$.

We end this chapter with an observation about the Gauss-Lucas Theorem. While given a polynomial $p(z)$ it does not seem to be straightforward to give a point w whose corresponding Basic Sequence would be convergent to a root of $p(z)$, it is an immediate consequence of Gauss-Lucas to define a point in the convex hull of the roots.

Proposition 16.3. *Let $p(z) = a_n z^n + \cdots + a_1 z + a_0$, $a_n \neq 0$. If $a_i = 0$, for some $i = 0, \ldots, n-1$, then $w = 0$ lies in the convex hull of the roots. Otherwise,*

$$w = -\frac{a_{n-1}}{n a_n}$$

lies in the convex hull.

Proof. If no coefficient is zero, then

$$p^{(n-1)}(z) = n! a_n z + (n-1)! a_{n-1},$$

and the solution of this by repeated application of Gauss-Lucas theorem must lie in the convex hull. Otherwise, one of the higher derivatives of p must have 0 as a root. □

If such point is not equidistant to two distinct roots (which is not the case when the coefficients are real), then the corresponding Basic Sequence is guaranteed to converge to a root of p.

16.4 Conclusions

In this chapter we have defined a single geometric optimization problem that has given rise to several important theorems and interesting problems, including two classical problems, the Gauss-Lucas Theorem and the Maximum Modulus Principle for polynomials and their extensions. Restriction to complex polynomials has not only resulted in algorithmic point of view into the polynomial case of these theorems, but also has motivated generalizations. When the critical point of a polynomial is simple, picking a point at random has fifty percent chance of being an ascent direction. The classification of the ascent and descent direction for the general case is an interesting problem, as is that of general analytic function at a simple critical point. The Gauss-Lucas iteration function is an interesting function in its own right. In fact polynomiography of a polynomial (Kalantari (2002b), Kalantari (2004c)) and its connection to that of its derivative were the motivation behind the results presented in the chapter. On the other hand, the polynomial case gives rise to interesting questions from the point of view of fixed point theory, dynamical systems and iteration functions.

Problem 1. The Gauss-Lucas function $G(x)$ maps the convex hull of the defining points into itself. Goodman (1975) argues that the function $F(x)$ must have at least one non-trivial stationary point. How many non-trivial fixed points can $G(x)$ have?

Problem 2. When is G an onto map of the convex hull into itself?

Problem 3. Are the nontrivial fixed points of $G(x)$ topologically repelling? What about the special case of $G(z)$?

Problem 4. Do the repelling fixed points of $G(z)$ coincide with saddle points of the corresponding $F(x)$?

Problem 5. Although $G(z)$ is not analytic in the sense that it requires conjugation, one may ask is it generally convergent for its corresponding polynomial?

Problem 6. Give a classification of the direction of ascent and descent for a polynomial at a non-simple critical point.

Chapter 17

Polynomiography: Algorithms for Visualization of Polynomial Equations

Polynomiography is the art and science of visualizing approximation of the zeros of complex polynomials using iteration functions. Informally speaking, polynomiography allows one to create colorful images of polynomials. These images can subsequently be re-colored in many ways, using one's own creativity and artistry, need, and particular situation at hand. The term has been defined such as in Kalantari (2002b, 2004c,a). The above definition of polynomiography may appear to be vague because one may interpret this, for instance, as visualization via any rational iteration function whose fixed points are the root of a given polynomial $p(z)$. For instance if

$$F(z) = p(z) + z,$$

its fixed points are precisely the roots of $p(z)$. But the fixed points are not necessarily attractive. Thus although using such function is interesting even in the case of quadratics, as it is the basis for the much studied Mandelbrot set, by polynomiography we shall mean the use of rational iteration functions such that the roots of the underlying polynomial are attractive or superattractive fixed points. Indeed in the context of root-finding such is the definition of iteration functions. This definition is good but it may exclude interesting cases where the iteration function may not be a rational function. We have in this book seen example of such cases. Thus we may broaden the definition of polynomiography to visualizations that make use of iteration functions where the roots of the underlying polynomial form topologically attractive fixed points of the underlying function whose fixed points are being sought. Even with this broader definition readers are inclined to think that one may only use an individual iteration function in order to generate a polynomiograph. This stems from the fact that such techniques, whether generated by amateurs for art and fun or by experts in the goal of understanding the dynamics of iteration functions, have been the source of producing images. The resulting images have commonly been termed as "fractal" which convey some properties such as self-similarity, but is otherwise vague and carries no other meanings. Typical fractal images while perhaps interesting to a viewer, typically offer no information on what the original goal or purpose may have inspired them. Indeed a significance of the term polynomiography is that it immediately offers an explanation that the image corresponds to a polynomial equation.

17.1 A Basic Coloring Algorithm

While making use of an individual iteration function $F(z)$ for a particular polynomial $p(z)$, a typical algorithm consists of exploring or visualizing regions of convergence described in the following steps.

(1) Let θ_i, $i = 1, \ldots, n$, be the roots of $p(z)$. Let $\mathcal{C}$ be a bijection from $\{1, \ldots, n\}$ to a set of n distinct colors. Fix accuracy ϵ, maximum number of iterations K, and the initial point z_0. Create an empty table called *color table* (later to be used as associative map *point* $\rightarrow i \in \{1, \ldots, n\}$).

(2) While $k \leq K$, do $z_k = F(z_{k-1})$. If $|p(z_k)| < \epsilon$, see if z_k is close to a point in the color table. If yes, find the index of that point, i. If no, add the point z_k to the table, and use as index i the number of elements in the table. Assign to z_0 the color $\mathcal{C}(i)$ and terminate. In this sense the point z_0 will be considered as a point belonging to the basin of attraction of $z_{(i)}$.

(3) If the maximum number of iterations is reached ($k = K$) and $|p(z_k)| > \epsilon$, assign to z_0 the color white (indicating the algorithm failed to converge).

In practice, we may assign to a point a *hue* of a color $\mathcal{C}(i)$ depending on the number of iterations needed to get in an ϵ neighborhood of θ_i. An image produced through the algorithm above is called a *polynomiograph.*

There are several different ways to change this basic coloring and by changing ϵ we may or may not affect the coloring. It may be possible that the size of ϵ and accuracies may give rise to a larger number of roots than the actual number. While in polynomiography accuracy is of course significant, even the presence of error could produce magnificent images. The Basic Family and its variants offer many choices and other schemes of coloring. We will discuss these next.

17.2 Basic Family and Variants: The Basis of Polynomiography

As we have seen before the Basic Family and its variants, see Figure 17.1, gives rise to numerous iteration functions of interest.

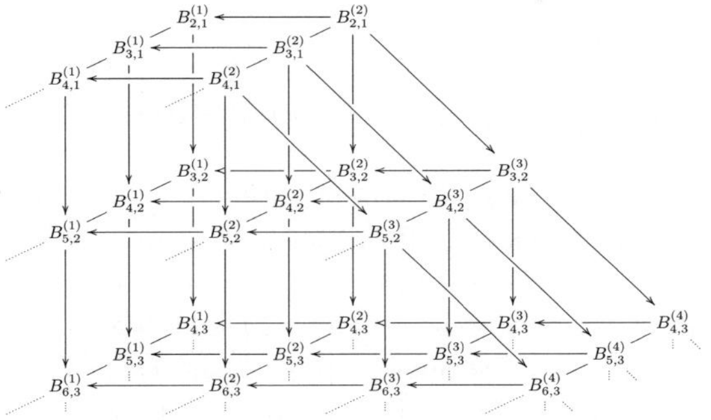

Fig. 17.1 Many variation of Basic Family: Single, multipoint and truncated.

Each of these in turn gives rise to a *parametric* iteration function. The simplest parametric Basic Family member is defined as

$$B_{m,\alpha}(z) = z - \alpha p(z)\frac{D_{m-2}(z)}{D_{m-1}(z)},$$

where α is a constant complex number satisfying the property that each root of $p(z)$ is an attractive or superattractive fixed point of $B_{m,\alpha}(z)$. Clearly, since we have proved that the roots of $p(z)$ are attractive or superattractive fixed of $B_m(z)$, $\alpha = 1$ is one such constant. But in fact there are infinitely many such values. For any $m \geq 2$ and root θ of $p(z)$ it can be shown that

$$B'_{m,\alpha}(\theta) = 1 - \alpha.$$

Thus it suffices to choose any α satisfying

$$|1 - \alpha| < 1.$$

Equivalently, any α in the open unit disk of radius one at $z_0 = 1$:

$$\{(x, y) : (x - 1)^2 + y^2 < 1\}.$$

In particular, for $m = 2$, the parametric Newton's method gives rise to infinitely many images of a polynomial, even of $p(z) = z^3 - 1$.

Another basic coloring scheme can be based on the use of Basic Sequence. That is, the assignment of a color to a point z_0 may be decided using the smallest m such that $B_m(z_0)$ falls within an epsilon neighborhood of a point. This approach gives rise to many choices and interesting images.

In the next section we demonstrate the richness of the Basic Family and its variants, even in the visualization of $z^3 - 1$.

17.3 Many Polynomiographs of Cubic Roots of Unity

In this section we give many polynomiographs of $z^3 - 1$ based on the Basic Family and variants. In fact in all but the last figure shown in the chapter we will restrict the polynomiographs to the case of this cubic polynomial. Furthermore, we will restrict the window through which we view the polynomial to generate the polynomiographs to the case where the real and imaginary parts are restricted to lie within the interval $[-3, 3]$.

In Figure 17.2, top row, we show three polynomiographs the first of which is the classical polynomiograph of $z^3 - 1$ under Newton's method. The second image in that row is a 3D depiction of topological Fatou-Julia graph. Although this is not a usual polynomiograph, it does fall within the category of algorithms that visualize the process of approximation of the roots and thus it justifies being called a polynomiograph. The right-most image in the first row is a polynomiograph under B_m for some large value of m. The image confirms the theoretical performance that the basins of attraction are in fact approximation to the Voronoi regions of the roots and in this case they are very good approximation to these geometric polygons.

In Figure 17.2, second row, we show polynomiographs based on the use of Basic Sequence. Thus these polynomiographs do not fall within the category of fractal polynomiographs. The images do not exhibit fractal property. The last row in the figure shows an assortment of images.

Figure 17.3 shows polynomiographs under parametric Newton method. We see that it is possible to come up with numerous images of this simple polynomial under different values of α. Clearly only the value of $\alpha = 1$, the actual Newton method, has quadratic rate of convergence. The other cases of α give linear rate of convergence, hence slower. From the point of view of dynamics and polynomiography even these give rich class of images and give rise to many questions of theoretical nature.

The polynomiographs in Figure 17.4 show richness and diversity that choice of color would afford the polynomiographer even under the same

Fig. 17.2 Polynomiographs of $z^3 - 1$ under Newton's, a 3D depiction, and B_m with large m, first row; Polynomiographs based on the use of Basic Sequence, second row; an assortment, third row.

mode of Basic Family usage and restriction to the same polynomial.

To describe exactly the possible modes is more like describing a manual for a medium, such as photography, which we avoid giving here as it serves no purpose. Furthermore, in analogy with a camera it is possible to introduce and continue introducing numerous features into a polynomiography software and allow increasing the capabilities of polynomiography software.

Other than the Basic Family members spoken about here one can introduce an even more extended family consisting of affine or convex combination of two or more members. These would result in yet more diversity into the set of potential polynomiographs that can be assigned to even one such simple polynomial as $z^3 - 1$. In essence even the polynomiography of this simple polynomial is endless.

In the next chapter we will extend the Basic Family members to the

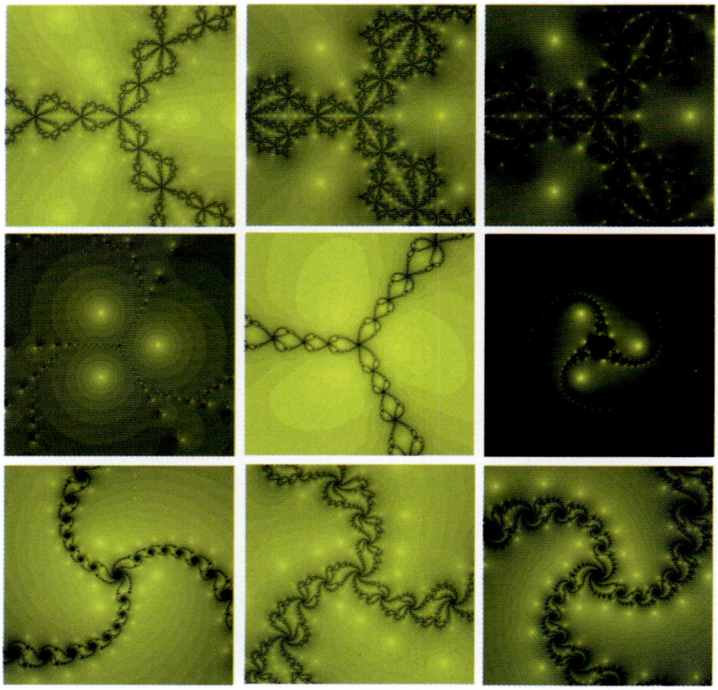

Fig. 17.3 Polynomiographs of $z^3 - 1$ emphasizing Julia sets under parametric Newton.

visualization of homogeneous linear recurrence relations. In a sense polynomiography under the Basic Sequence is a scheme inherent to homogeneous linear recurrence relations than iteration functions. We show in the next chapter how homogeneous linear recurrence relations give rise to iteration functions. These also broaden the horizon of polynomiography, polynomial root-finding, and dynamical systems.

Before moving to the next chapter we will show some polynomiography with respect to the polynomial $p(z) = z^3 - 2z + 2$, studies in detail in Chapter 5 where it was shown that Newton's method would fail to be generally convergent. In Figure 17.5 we show the polynomiographies with $\alpha = .9$ (left image) and closeup near the origin for two values of α close to one. While for $\alpha = 1$, as seen in Chapter 5, Figure 5.3 Newton's method would give white areas of non-convergence near the origin, for $\alpha = .9$ the white area disappear, seemingly resulting in a generally convergent version of Newton's for this polynomial. The next two images show another inter-

Fig. 17.4 Polynomiographs of $z^3 - 1$ under Basic Family variants.

Fig. 17.5 Polynomiographs of $z^3 - 2z + 2$ corresponding to different α values.

esting property: by changing α in a neighborhoods of 1 it seems that near the origin the resulting filled Julia sets are analogous to those corresponding to small perturbation of the constant and linear terms of this polynomial, to be contrasted with Figure 5.4.

Chapter 18

Visualization of Homogeneous Linear Recurrence Relations

In this chapter we show how one may associate visualizations to a homogeneous linear recurrence relation with an arbitrary set of initial conditions. In fact in doing so we shall discover many new families of iteration functions which are induced by the Basic Family, we shall refer to these as *Induced Basic Family*. The first such visualizations were formally described by the author such as in Kalantari (2002b, 2004c). When the initial conditions are Basic Initial Conditions, the visualization of a homogenous linear recurrence relation is justified by association of any polynomiography with respect to the negative reciprocal polynomial of characteristic polynomial. Such a polynomiograph can be obtained via the Basic Family or the Basic Sequence. When the initial conditions are arbitrary we can still associate visualizations with respect to Induced Basic Families to be defined in the chapter. These lead to a new infinite families of iterations functions with high order of convergence and many more tools for polynomiography. In particular, we define two infinite special cases of the Induced Basic Family, the *Fibonacci Family* and the *Lucas Family*. We also give some polynomiographies associated with some special recurrence relations: the generalized Fibonacci sequence and a more general version we call the *Hyper Fibonacci* sequence.

18.1 Introduction

We will first review some results on a homogeneous linear recurrence relation (HLRR), proved earlier in Chapter 9. Consider

$$a_m = c_1 a_{m-1} + c_2 a_{m-2} + \cdots + c_n a_{m-n}, \tag{18.1}$$

where $c_1, \ldots, c_n$ are given complex numbers, $c_n \neq 0$. Its Basic Initial Conditions are

$$a_0 = 1, \qquad a_{-1} = a_{-2} = \cdots = a_{-n+1} = 0.$$

The Fundamental Solution of HLRR is the solution corresponding to Basic Initial Conditions. The characteristic polynomial of HLRR is

$$q(z) = z^n - c_1 z^{n-1} - c_2 z^{n-2} + \cdots - c_{n-1} z - c_n.$$

The characteristic roots are distinct roots of $q(z)$:

$$\{\eta_1, \ldots, \eta_t\}.$$

The negative reciprocal polynomial is

$$p(z) = -z^n q(\frac{1}{z}) = c_n z^n + c_{n-1} z^{n-1} + \cdots + c_1 z - 1.$$

Its roots are

$$\{\theta_1 = \frac{1}{\eta_1}, \ldots, \theta_t = \frac{1}{\eta_t}\}.$$

Given a complex number w, we associate a homogeneous linear recurrence relation HLRR(w), where for each $m \geq 1$ we set

$$a_m(w) = \sum_{i=1}^{n} c_i(w) a_{m-i}(w), \tag{18.2}$$

with

$$c_i(w) = (-1)^{i-1} p^{i-1}(w) \frac{p^{(i)}(w)}{i!}.$$

We consider HLRR(w) with its Basic Initial Conditions

$$a_0(w) = 1, \quad a_j(w) = 0, \quad \forall\ j = -1, \ldots, -n+1. \tag{18.3}$$

The characteristic polynomial associated with the sequence $\{a_m(w)\}$ is the polynomial

$$q_w(z) = z^n - (c_1(w) z^{n-1} + c_2(w) z^{n-2} + \cdots + c_{n-1}(w) z + c_n(w)). \tag{18.4}$$

The corresponding negative reciprocal is

$$p_w(z) = z^n q_w(\frac{1}{z}) = c_n(w) z^n + \cdots + c_1(z) z - 1.$$

At $w = 0$, $p(0) = 1$ and $p^{(i)}(0)/i! = c_i$, thus $c_i(0) = c_i$ so that

$$a_m(0) = a_m, \quad q_0(z) = q(z), \quad p_0(z) = p(z).$$

Thus HLRR(0) coincides with HLRR. The Universal Class of HLRR associated with $p(z)$ is defined to be

$$U(p(z)) = \{HLRR(w) : w \in \mathbb{C}\}.$$

Given $w \in \mathbb{C}$, the Basic Sequence is

$$\left\{ b_m(w) = w - p(w)\frac{a_{m-2}(w)}{a_{m-1}(w)} \right\}_{m=2}^{\infty},$$

and if w lies in Voronoi region of a root θ of $p(z)$, the corresponding Basic Sequence satisfies

$$\lim_{m \to \infty} b_m(w) = \theta.$$

The quantity $a_m(w)$ coincides with $D_m(w)$ where as before

$$D_m(z) = \det \begin{pmatrix} p'(z) & \frac{p''(z)}{2!} & \dots & \frac{p(m-1)(z)}{(m-1)!} & \frac{p^{(m)}(z)}{(m)!} \\ p(z) & p'(z) & \ddots & \ddots & \frac{p(m-1)(z)}{(m-1)!} \\ 0 & p(z) & \ddots & \ddots & \vdots \\ \vdots & \vdots & \ddots & \ddots & \frac{p''(z)}{2!} \\ 0 & 0 & \dots & p(z) & p'(z) \end{pmatrix}.$$

Two HLRR corresponding to w_1 and w_2 are said to be *conjugate* if w_1, w_2 belong to the Voronoi region of the same root θ of $p(z)$. Thus in the sense of approximation of θ, w_1 and w_2 belong to the same equivalence class. This justifies association of polynomiographies with respect to a given HLRR and its Basic Initial Conditions. In the next section we give polynomiographies associated with Fibonacci sequence and its generalizations. Later we shall return to the visualization of an HLRR with respect to an arbitrary set of initial conditions.

18.2 The Generalized Fibonacci, the Hyper Fibonacci, and their Polynomiography

Generalized Fibonacci numbers of order n are defined to be

$$F_m^{(n)} = F_{m-1}^{(n)} + F_{m-2}^{(n)} + \cdots + F_{m-n}^{(n)}.$$

Its Basic Initial Conditions are

$$F_0^{(n)} = 1, \quad F_j^{(n)} = 0, \quad j = -1, \ldots, -n+1.$$

For $n = 2$ this corresponds to the ordinary Fibonacci sequence.

We define the *Hyper Fibonacci* numbers of order n to be

$$F_m^{(n)} = c_1 F_{m-1}^{(n)} + c_2 F_{m-2}^{(n)} + \cdots + c_n F_{m-n}^{(n)},$$

where $c_i \in \{0, 1\}$ and $c_n = 1$ and the Basic Initial Conditions are

$$F_0^{(n)} = 1, \quad F_j^{(n)} = 0, \quad j = -1, \ldots, -n+1.$$

There are 2^{n-1} Hyper Fibonacci of order n one of which is the Generalized Fibonacci sequence of order n. We will give polynomiography for $n = 5$. The 16 images in Figures 18.1 give polynomiographies based on the Basic Sequence.

18.3 The Induced Basic Family and Induced Basic Sequence

Definition 18.1. Given a pair of constants α_0, α_1, not both equal to zero, we define the *Induced Basic Family of order* 2 to be

$$\left\{ \mathfrak{B}_{\mathrm{m}}(z) = z - p(z) \frac{\alpha_1 D_{m-2}(z) + \alpha_0 D_{m-3}(z)}{\alpha_1 D_{m-1}(z) + \alpha_0 D_{m-2}(z)} \right\}_{m=3}^{\infty}.$$

Theorem 18.1. *The induced Basic Family of order* 2 *is a family of iteration functions for* $p(z)$. *Moreover, given a simple root* θ *of* $p(z)$ *the order of convergence of* $\mathfrak{B}_{\mathrm{m}}(z)$ *is at least* $m - 1$.

Proof. We will prove the order of converge of $\mathfrak{B}_{\mathrm{m}}(z)$ is at least k, for $k = 2, 3, 4$ and when m is at least $3, 4, 5$, respectively. The proof of these special cases give a flavor of the approach to prove the general case.

Let θ be a simple root of $p(z)$. For $i \geq 2$, set

$$A_i(z) = \alpha_1 D_{i-1}(z) + \alpha_0 D_{i-2}(z). \tag{18.5}$$

For $i \geq 3$ set

$$U_i(z) = \frac{A_{i-1}(z)}{A_i(z)}. \tag{18.6}$$

Thus

$$\mathfrak{B}_{\mathrm{m}}(z) = z - p(z) U_m(z).$$

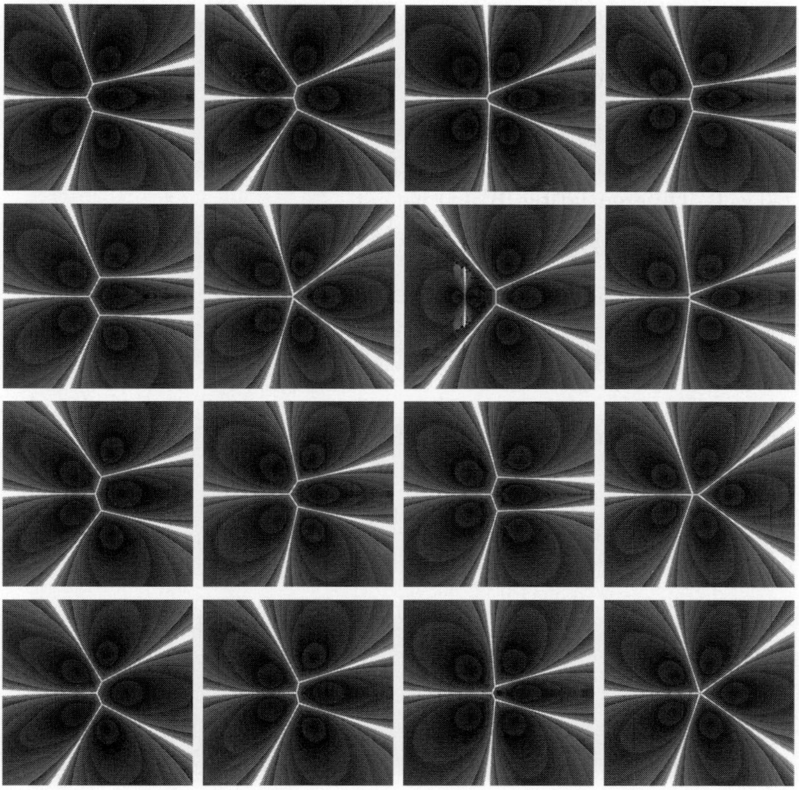

Fig. 18.1 Polynomiography of Hyper Fibonacci of order 5.

Differentiating we have

$$\mathfrak{B}'_{\mathfrak{m}}(z) = 1 - [p'(z)U_m(z) + p(z)U'_m(z)]. \tag{18.7}$$

Recalling the determinantal formula for $D_k(z)$, for $k \geq 1$ we have

$$D_k(\theta) = p'(\theta)^k. \tag{18.8}$$

Thus

$$A_i(\theta) = p'(\theta)^{i-2}(\alpha_1 p'(\theta) + \alpha_0). \tag{18.9}$$

Thus

$$U_i(\theta) = \frac{1}{p'(\theta)}. \tag{18.10}$$

This implies that for $m \geq 3$

$$\mathfrak{B}'_{\mathfrak{m}}(\theta) = 1 - p'(\theta)\frac{1}{p'(\theta)} = 0.$$

Thus $\mathfrak{B}_{\mathfrak{m}}(z)$ has order of convergence at least 2.

To prove its order of convergence is at least 3 when $m \geq 4$, we prove $\mathfrak{B}''_{\mathfrak{m}}(\theta) = 0$. Differentiating $\mathfrak{B}'_{\mathfrak{m}}(z)$ we get

$$\mathfrak{B}''_{\mathfrak{m}}(z) = -[p''(z)U_m(z) + 2p'(z)U'_m(z) + p(z)U''_m(z)]. \tag{18.11}$$

Substituting $z = \theta$ and since $U'_m(\theta) = 1/p'(\theta)$ we get

$$\mathfrak{B}''_{\mathfrak{m}}(\theta) = -[p''(\theta)U_m(\theta) + 2p'(\theta)U'_m(\theta)] = -\left[\frac{p''(\theta)}{p'(\theta)} + 2p'(\theta)U'_m(\theta)\right]. \tag{18.12}$$

We thus need to compute $U'_m(\theta)$. Differentiating $U_m(z) = A_{m-1}(z)/A_m(z)$ we get

$$U'_m(z) = \frac{A'_{m-1}(z)A_m(z)}{A_m(z)^2} - \frac{A_{m-1}(z)A'_m(z)}{A_m(z)^2}$$

$$= \frac{A'_{m-1}(z)A_{m-1}(z)}{A_{m-1}(z)A_m(z)} - \frac{A_{m-1}(z)A'_m(z)}{A_m(z)^2}.$$

For $i \geq 2$ set

$$V_i(z) = \frac{A'_i(z)}{A_i(z)}. \tag{18.13}$$

It follows that

$$U'_m(z) = U_m(z)(V_{m-1}(z) - V_m(z)). \tag{18.14}$$

To evaluate $U'_m(\theta)$ we need to compute $V_i(z)$ and this in turn requires the evaluation of $D'_k(z)$. As proved in Chapter 4 (Theorem 4.5) we have

$$D'_k(z) = (k+1)\sum_{i=2}^{n}(-1)^i\frac{p(z)^{i-2}p^{(i)}(z)}{i!}D_{k+1-i}(z). \tag{18.15}$$

Substituting θ for z all the terms in the summation vanish except for the first term. Substituting for $D_{k-1}(\theta)$ we get

$$D'_k(\theta) = \frac{k+1}{2}p''(\theta)p'(\theta)^{k-1}. \tag{18.16}$$

This implies

$$A'_i(\theta) = \frac{p''(\theta)}{2}p'(\theta)^{i-3}[\alpha_1 i p'(\theta) + \alpha_0(i-1)].$$

Using this and $A_i(\theta)$ we get

$$V_i(\theta) = \frac{p''(\theta)}{2p'(\theta)}\left[\frac{\alpha_1 ip'(\theta) + \alpha_0(i-1)}{\alpha_1 p'(\theta) + \alpha_0}\right]. \tag{18.17}$$

From this we get

$$V_{m-1}(\theta) - V_m(\theta) = \frac{p''(\theta)}{2p'(\theta)}\left[\frac{-\alpha_1 p'(\theta) - \alpha_0}{\alpha_1 p'(\theta) + \alpha_0}\right] = -\frac{p''(\theta)}{2p'(\theta)}. \tag{18.18}$$

Substituting into the formula for $U'_m(z)$ we have

$$U'_m(\theta) = -\frac{p''(\theta)}{2p'(\theta)^2}. \tag{18.19}$$

Finally, we have

$$\mathfrak{B}''_{\mathfrak{m}}(\theta) = -\left[\frac{p''(\theta)}{p'(\theta)} + 2p'(\theta)(-\frac{p''(\theta)}{2p'(\theta)^2})\right] = 0.$$

Hence for $m \geq 4$, $\mathfrak{B}_{\mathfrak{m}}(z)$ has order of convergence at least equal to 3.

To prove the next step we would need to show the third derivative of $\mathfrak{B}_{\mathfrak{m}}$ at θ is zero. Differentiating $\mathfrak{B}''_{\mathfrak{m}}(z)$, it is easy to see that

$$\mathfrak{B}'''_{\mathfrak{m}}(\theta) = -[p'''(\theta)U_m(\theta) + 3p''(\theta)U'_m(\theta) + 3p'(\theta)U''_m(\theta) + p(\theta)U'''_m(\theta)]. \tag{18.20}$$

To compute this it suffices to compute $U''_m(\theta)$. This in turn requires $D''_k(\theta)$. To differentiate $D'_k(z)$ and substitute θ, it suffices to differentiate only the first two terms of the summation formula (18.15), namely:

$$(k+1)\left[\frac{p''(z)}{2}D_{k-1}(z) - \frac{p(z)p'''(z)}{6}D_{k-2}(z)\right].$$

Differentiating and substituting $z = \theta$ we get

$$D''_k(\theta) = (k+1)\left[\frac{p'''(\theta)}{2}D_{k-1}(\theta) + \frac{p''(\theta)}{2}D'_{k-1}(\theta) - \frac{p'(\theta)p'''(\theta)}{6}D_{k-2}(\theta)\right].$$

Substituting for $D_{k-2}(\theta)$, $D_{k-3}(\theta)$, $D'_{k-2}(\theta)$ and simplifying we get

$$D''_k(\theta) = (k+1)p'(\theta)^{k-2}\left[\frac{p'''(\theta)}{3}p'(\theta) + \frac{k}{4}p''(\theta)^2\right]. \tag{18.21}$$

We may now compute $U''_m(\theta)$. First, differentiating $U'_m(z)$ we get

$$U''_m(z) = U'_m(z)(V_{m-1}(z) - V_m(z)) + U_m(z)(V'_{m-1}(z) - V'_m(z)).$$

Substituting for $U'_m(z)$ we get

$$U''_m(\theta) = U_m(\theta)[(V_{m-1}(\theta) - V_m(\theta))^2 + (V'_{m-1}(\theta) - V'_m(\theta))]. \tag{18.22}$$

For $i \geq 2$ set

$$W_i(z) = \frac{A_i''(z)}{A_i'(z)}.$$

Then it follows that

$$V_i'(z) = \frac{A_i''(z)}{A_i'(z)} \frac{A_i'(z)}{A_i(z)} - (\frac{A_i'(z)}{A_i(z)})^2 = V_i(z)W_i(z) - V_i^2(z).$$

From this we may write

$$V_{m-1}'(z) - V_m'(z) = V_m^2(z) - V_{m-1}^2(z) + (V_{m-1}(z)W_{m-1}(z) - V_m(z)W_m(z)). \tag{18.23}$$

To complete the evaluation of $U_m''(\theta)$ we note that

$$V_m^2(\theta) - V_{m-1}^2(\theta) = (V_m(\theta) - V_{m-1}(\theta))(V_m(\theta) + V_{m-1}(\theta)).$$

Substituting for $V_i(\theta)$ and simplifying we get

$$V_m^2(\theta) - V_{m-1}^2(\theta) = \frac{p''(\theta)^2}{4p'(\theta)^2}\left[\frac{\alpha_1(2m-1)p'(\theta) + \alpha_0(2m-3)}{\alpha_1 p'(\theta) + \alpha_0}\right]. \tag{18.24}$$

To compute $W_k(\theta)$, we first note that

$$A_k''(\theta) = \alpha_1 D_{k-1}''(\theta) + \alpha_0 D_{k-2}''(\theta) = \frac{p'''(\theta)}{3} p'(\theta)^{k-3}[\alpha_1 k p'(\theta) + \alpha_0(k-1)]$$

$$+\frac{p''(\theta)}{4} p''(\theta)^{k-4}[\alpha_1(k-1)kp'(\theta) + \alpha_0(k-2)(k-1)].$$

Thus

$$W_k(\theta) = \frac{2p'''(\theta)}{3p''(\theta)} + \frac{p''(\theta)}{2p'(\theta)}\left[\frac{\alpha_1(k-1)kp'(\theta) + \alpha_0(k-2)(k-1)}{\alpha_1 k p'(\theta) + \alpha_0(k-1)}\right].$$

It can now be shown that

$$V_{m-1}(\theta)W_{m-1}(\theta) - V_m(\theta)W_m(\theta)$$

$$= -\frac{p'''(\theta)}{3p'(\theta)} + \frac{p''(\theta)^2}{4p'(\theta)^2}\left[\frac{\alpha_1(-2m+3)p'(\theta) + \alpha_0(-2m+4)}{\alpha_1 p'(\theta) + \alpha_0}\right]. \tag{18.25}$$

Substituting (18.24) and (18.25) as well as $U_m(\theta) = 1/p'(\theta)$ into (18.22) we get

$$U_m''(\theta) = -\frac{p'''(\theta)}{3p'(\theta)^2} + \frac{p''(\theta)^2}{2p'(\theta)^3}.$$

It is now straightforward to show that in (18.20) $\mathfrak{B}_m'''(\theta) = 0$. □

Associated with Induced Basic Family we also define a corresponding *Induced Basic Sequence.*

Definition 18.2. Given a complex number w the *Induced Basic Sequence* is the sequence

$$\{\mathfrak{B}_{\mathrm{m}}(w)\}_{m=3}^{\infty}.$$

The most significant property of the Induced Basic Sequence is the following.

Theorem 18.2. *Let $p(z)$ be a polynomial of degree n and θ a root. Given a complex number w in the Voronoi region of θ the corresponding Induced Basic Sequence $\{\mathfrak{B}_{\mathrm{m}}(w)\}_{m=3}^{\infty}$ converges to θ.*

Proof. From Chapter 9 we know that

$$D_m(w) = \sum_{i=1}^{t} \beta_{i,w}(m)\eta_i(w)^m$$

where $\eta_i(w) = p(w)/(w - \theta_i)$ and $\beta_{i,z}(z)$ is either identically zero or a polynomial of degree at most n_i, the multiplicity of θ_i as a root of $p(z)$. If w lies in the Voronoi region of θ we know that the corresponding coefficient polynomial is not identically zero. From this it is easy to show that

$$\lim_{m\to\infty} \frac{\alpha_1 D_{m-2}(w) + \alpha_0 D_{m-3}(w)}{\alpha_1 D_{m-1}(w) + \alpha_0 D_{m-2}(w)} = \frac{1}{\eta(w)} = \frac{w-\theta}{p(w)}.$$

Hence

$$\lim_{m\to\infty} \mathfrak{B}_{\mathrm{m}}(w) = \theta.$$

□

18.4 The Fibonacci and Lucas Families of Iteration Functions

Two special cases of the Induced Basic Family we will consider here are the following.

Definition 18.3. Given a polynomial $p(z)$, the Fibonacci Family is defined as

$$\left\{\mathfrak{F}_{\mathrm{m}}(z) = z - p(z)\frac{D_{m-2}(z) + D_{m-3}(z)}{D_{m-1}(z) + D_{m-2}(z)}\right\}_{m=3}^{\infty}.$$

The Lucas Family is defined as

$$\left\{\mathfrak{L}_{\mathrm{m}}(z) = z - p(z)\frac{2D_{m-2}(z) - D_{m-3}(z)}{2D_{m-1}(z) - D_{m-2}(z)}\right\}_{m=3}^{\infty}.$$

Clearly these are just two special cases of Induced Basic Family. The justification in their naming lies in the fact that the Fibonacci numbers satisfy, $F_m = F_{m-1} + F_{m-2}$ and that the Lucas numbers satisfy $L_m = 2F_m - F_{m-1}$.

As an example for $m = 3$ we get

$$\mathfrak{F}_3(z) = z - p(z)\frac{D_1(z) + D_0(z)}{D_2(z) + D_(z)} = z - p(z)\frac{p'(z) + 1}{p'(z)^2 - 0.5p''(z)p(z) + p'(z)},$$

and

$$\mathfrak{L}_3(z) = z - p(z)\frac{2D_1(z) - D_0(z)}{2D_2(z) - D_(z)} = z - p(z)\frac{2p'(z) - 1}{2p'(z)^2 - p''(z)p(z) - p'(z)}.$$

These are two iteration functions of order 2, alternatives to Newton's method. In particular we can apply these to any of the Hyper Fibonacci recurrences, equivalently to their corresponding negative reciprocal polynomials.

As an example applying $\mathfrak{F}_3(z)$ and $\mathfrak{L}_3(z)$ to $p(z) = z^3 - 1$ we get polynomiographs of Figure 18.2.

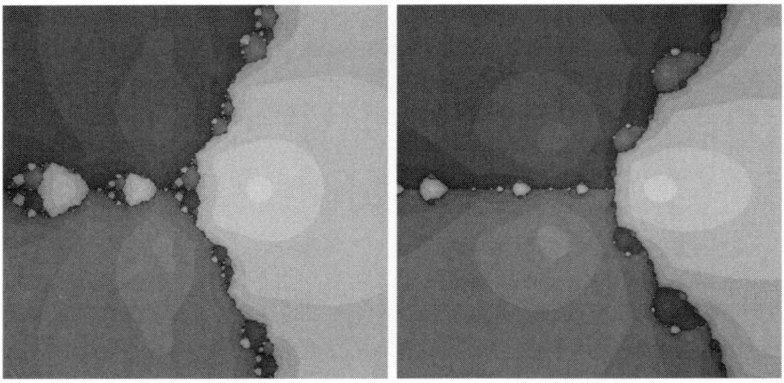

Fig. 18.2 Polynomiography of $z^3 - 1$ under $\mathfrak{F}_3(z)$ and $\mathfrak{L}_3(z)$.

18.5 Visualization of HLRR with Arbitrary Initial Conditions

We have seen that an HLRR with its Basic Initial Conditions inherits the Basic Family associated with its negative reciprocal polynomial $p(z)$. Thus many polynomiographies can be associated to such HLRR. But what if the initial conditions are arbitrary?

Suppose that $\{b_m\}$ satisfies the recurrence relation (18.1), but with a given set of initial conditions $b_0, b_1, \ldots, b_{n-1}$. From the Representation Theorem for HLRR (Theorem 9.54, Chapter 9) we know that

$$b_m = \alpha_{n-1}a_m + \alpha_{n-2}a_{m-1} + \cdots + \alpha_0 a_{m-n+1},$$

where $\alpha_0, \ldots, \alpha_{n-1}$ can be solved from linear equation.

Definition 18.4. Given a set of constants $\alpha_0, \ldots, \alpha_{n-1}$, not all equal to zero, for each $m \geq n-1$

$$\mathfrak{D}_{\mathfrak{m},\mathfrak{n}}(z) = \alpha_{n-1}D_m(z) + \alpha_{n-2}D_{m-1}(z) + \cdots + \alpha_0 D_{m-n+1}(z).$$

The *Induced Basic Family of order n* is defined to be

$$\left\{ \mathfrak{B}_{\mathfrak{m},\mathfrak{n}}(z) = z - p(z)\frac{\mathfrak{D}_{\mathfrak{m}-2,\mathfrak{n}}(z)}{\mathfrak{D}_{\mathfrak{m}-1,\mathfrak{n}}(z)} \right\}_{m=n+1}^{\infty}.$$

We can thus associate polynomiographies with an HLRR with an arbitrary set of initial conditions.

Furthermore, just as we can associate Fibonacci and Lucas Families of iteration functions to an arbitrary polynomial, we can associate Induced Basic Families with respect to other recurrence relations, e.g. Generalized Fibonacci where we would take all $\alpha_i = 1$, or Generalized Lucas and their Hyper versions.

Chapter 19

Applications of Polynomiography in Art, Education, Science and Mathematics

The goal of this chapter is to explore polynomiography as a serious and powerful field and medium for the general artist, for educators and students, for scientists and mathematicians. We argue that polynomiography is a unique and novel medium for creativity and playful learning with numerous applications in education, math, sciences, art and design. Polynomiography is based on sophisticated algorithmic visualization in solving polynomial equations. Using inventive programming it creates a medium where an individual, independent of his/her mathematical background, age, and artistic background, can be playful, experimental, directed, expressive, or research-minded with satisfying outcome. Very significantly in the cases of younger/non-technical individuals, polynomiography helps them learn about concepts in mathematics that they would otherwise be less motivated to learn or would find too dry. Polynomiography as software can be the basis of a technology that would lend itself to the encouragement of creativity in multi-disciplinary teaching and learning experiences, and to the development of curricula for a wide range of educational courses. Prototype polynomiography software in limited settings has already been tested, and it has proven to be an enthusiastically popular medium for the middle school and high school teachers interested in introducing it in their curricula and students. Survey of students, some as young as 11-13, introduced to polynomiography and because of it are demonstrably interested in learning about polynomials which are central to mathematics and science. Thus young students early on get closer to these critical building blocks of sciences and mathematics and related complex notions that are otherwise too distant to them.

In the first section we introduce polynomiography as a medium for artists and in doing so we attempt to give a non-technical description of the subject. In the subsequent section we consider the educational application of polynomiography and why it would serve as a medium for educators. We also provide some evidence as the utility of the subject at various levels, from K-12 and beyond. In the last section we will give some scientific and mathematical applications. The applications considered here are only examples and are by no means intended to be exhaustive, nor necessarily exclusive.

Fig. 19.1 Cover Design, a visualization of $p(z) = 10z^{48} - 11z^{24} + 1$.

19.1 Polynomiography in Art

In this section we introduce polynomiography as a medium for art. Polynomiography provides a tool for artists to create a 2D image - a *polynomiograph* - based on the computer visualization of a *polynomial equation*. The image is dependent upon the solutions of a polynomial equation, various interactive coloring schemes driven by *iteration functions* (algorithms for solving polynomial equations), and several other parameters under the control of the *polynomiographer's* choice and creativity. Polynomiography software can mask all of the underlying mathematics, offering a tool that, though easy to use, affords the polynomiographer infinite artistic capabilities. Polynomiography can lead to intricate designs and images that are reminiscent of abstract art and the most sophisticated human designs. It does so by combining human creativity and computer power to create artwork of great variety and diversity.

Formally, a polynomial, written as $p(z)$, is defined as a linear combination of integral powers of the variable z. As an example Figure 19.1 is a visualization based on a polynomial equation.

A *root* or *zero* of the polynomial is a value of z for which $p(z)$ equals zero, i.e. a *solution* to the polynomial equation

$$p(z) = 0.$$

The *degree* of $p(z)$ - the highest exponent of z - and the *coefficients* of the powers of z describe the polynomial. The Fundamental Theorem of Algebra (FTA) is a magical property that always guarantees at least one solution to any polynomial equation of degree at least one. In fact, there are as many solutions to a polynomial equation as its degree. The solutions need not be distinct. The name of the theorem is due to Gauss, one of the greatest mathematician of all time, but its validity was conjectured long before him. The solutions are *complex numbers*. A complex number is merely a point in the *Euclidean plane*, also called the *complex plane*. It is written as the ordered pair (a, b) which gives its horizontal and vertical coordinates, respectively. Algebraically, this is written as $a + ib$ where $i = \sqrt{-1}$ (i.e. $i^2+1 = 0$). The complex numbers inherit the four elementary operations on the real numbers. From the FTA, it follows that a polynomial equation is an algebraic description of the set of points in the Euclidean plane, namely its roots. Conversely, any set of points in the plane can be written as a polynomial equation having those points and only those points as its solutions.

Polynomiography can be considered as *painting via points*, an art form capable of creating an interesting variety of images by the manipulation of a finite set of points, whether given explicitly, generated by a polynomial equation, or selected with the click of a mouse. This view of polynomiography should not be confused with *pointillism*, a term first used with respect to the work of the artist Georges Seurat. In a sense, polynomiography is a minimalist art form, yet of enormous power. Below we shall be clear why polynomiography should be of interest to the general artist.

Initially, it may not appear artistically interesting to work with a finite set of points. This skepticism would be valid only if the set of points did not offer anything but the shape it defined. However, the magic of polynomiography is that this finite - possibly small - set of points, when combined with one or many iteration functions imposes a coloring scheme on all other points in the plane or a particular region of the plane. Together with the polynomiographer's personal creativity and choice, the great variety of iteration functions - which act as a window and means through which to view a polynomial - and polynomial equations, all amount to a powerful tool for artistic creation. Even with polynomials of small degree, artists can learn to produce interesting images on a laptop computer in reasonable time.

Theoretically, polynomiography can be considered as visual verification of the FTA. However, in polynomiography we are not merely interested in the roots of a polynomial equation, but in the way in which they relate to or influence all the other points within a particular region in the plane. For instance, a rectangular region that includes all or some of the roots. The polynomiography software makes use of these relationships to create artwork. In particular, in the context of visualization and art we can reverse the role of the ancient root-finding problem and select the roots of the polynomial as we wish so as to create desirable designs. Thus polynomiography turns the root-finding problem into a tool of art and design.

Polynomiography software could allow the artist a degree of creativity and control comparable to what may be experienced in painting, or in photography. There is no need to have knowledge of the underlying mathematics. Anyone can learn to do it. Thus, the class of polynomials which are so frequently in use in literally every branch of science, through polynomiography, becomes a valuable tool of art and design. So many significant polynomials have arisen in numerous branches of science. What will their polynomiography look like? And there are so many interesting classes of polynomials yet to be discovered by artists.

Polynomiography's appeal to the artist could stem from several properties: a relatively simple foundation; the ease with which one can generate polynomiographs; the ability to create images and designs of enormous complexity the likes of which have never been seen - or even imagined - before, or images reminiscent of familiar abstract art, at times impossible to detect as computer-generated; the fact that there is meaning and human control behind the images, as opposed to unpredictable or random computer-generated images; the fact that polynomiography techniques can be acquired methodically.

A reader familiar with *fractal* images may be inclined to view polynomiography as an extension of the well-known method for finding and plotting basins of attraction of roots of polynomials. But this view falls quite short of a good or fair description of polynomiography. While some polynomiographs may turn out to be fractal images obtained via plotting of the basins of attraction of roots, many other interesting polynomiographs are neither fractal nor based on such colorings. Even those polynomiographs which are based on the familiar coloring of basins of attractions make intelligent use of special iteration functions capable of producing anticipated shapes, as opposed to random or limited iterations often used to generate fractals.

The infinite class of iteration functions that have given rise to the images in this chapter are *Basic Family* and its variants. In a sense all iteration functions originate from the FTA. For a polynomiographer iteration functions within the Basic Family need not be known explicitly. They are analogous to the lenses of a camera. A photographer needs only learn to work with various lenses, not the physics behind their construction. Likewise, polynomiography software could mask all of the underlying mathematics underlying it.

Although a polynomiograph may turn out to be a fractal image, polynomiography is not a subset of fractals, neither as theory nor as art. Indeed, polynomiography is complementary to what is known as *fractal art* (see Mandelbrot (1983) Mandelbrot (1993)). The polynomiographer, when producing fractal art, is working with a restricted but well-defined class: fractal images coming from special iteration functions designed for root-finding whose properties give rise to tools of design. This method provides a basis for producing fractal art with an underlying foundation, as opposed to random fractal art. One can speak of *fractal polynomiographs.* Such a term would refer to a well-defined subset of fractal art. This method can be used to create many new sets of fractal images, and hence to broaden the horizon of fractal art. On the other hand, an important feature of polynomiography is that it gives rise to a wide range of interesting non-fractal images.

In a sense visualization of polynomial root-finding methods, the best-known of which is Newton's method, was attempted before the advent of computers. Cayley (1897) questioned the behavior of Newton's method for quadratic and cubic polynomials in the complex plane (see Peitgen *et al.*

(1992)). He was only able to find the answer for quadratics. For cubic equations the regions of attraction to a root happens to form shapes the boundaries of which exhibit fractal behavior, known as *Julia* sets. The first computer visualization of this phenomenon was apparently obtained by Hubbard (e.g. see Glick (1988)). Mandelbrot's fractals (see Mandelbrot (1983)) popularized the Julia sets and generated new interest in the computer visualizations of fractal images. Computer technology has not only been a significant tool for fractal art, but other forms of art inspired by mathematics and science, see e.g. Emmer (1993), Peterson (2001).

While polynomiography's theoretical aspects do intersect with both the theory of fractals and dynamical systems, it has its own independent characteristics. For instance, polynomiography not only results in a unified perspective into the theory of root-finding, but will also enable the discovery of new properties of this ancient problem. As an art form, polynomiography is perhaps the most systematic method for the visualization of root-finding algorithms – bringing it to the realm of art and design.

19.1.1 *Polynomiography as a Tool of Art and Design*

There are many similarities between photography and polynomiography. To justify this analogy. A question often posed by those who have viewed the author's personal artworks is:

"Why don't you write down the underlying polynomial equation next to each of your images?"

While this suggestion may be valid with respect to some polynomiographs and a defining equation may be satisfying to the viewer, in many other instances that question would be meaningless. In many artistic paintings or photograph of a person the name of the subject is often completely immaterial. It does not give a complete or fair description of the work. However, the suggestion is a positive reflection on polynomiography, because even a non-technical viewer does possess some understanding and appreciation of the origins of the images, and therefore feels comfortable enough asking questions about the underlying equation. Contrasting this perspective with typical fractal images, the viewer often has no clue as to the source of the image. While typical fractal art does make use of iterative schemes and coloring based on them, it is seldom apparent what the iterative scheme is trying to accomplish, if anything.

In photography there are three main components: the photographer, the camera, and the subject. These three components combine with other parameters to create an interesting photograph. In polynomiography, there are also three main components: the polynomiographer, the computer software, that generates the polynomiographs, and the underlying polynomial equation. As in photography, the final polynomiograph is produced by a combination of these three components as well as many other parameters such as the particular iteration function or collection of iteration functions being used, the region or area through which the polynomial equation is being viewed, the interactive coloring schemes. As in photography or in painting, polynomiography allows a great deal of creativity and choice.

Here are four basic polynomiography techniques:

- Just as a photographer can shoot different pictures of a model using a variety of lenses and angles, a polynomiographer can produce different images of the same polynomial equation and make use of a variety of iteration functions, zooming approaches and interactive coloring schemes until a desirable image is discovered.
- More creatively, an initial polynomiograph, even a very ordinary one, can be turned into a desirable image, based on the user's choice of coloration, individual creativity, and imagination. This is analogous to carving a statue out of stone.
- The polynomiographer may employ the mathematical properties of the iteration functions, or the underlying polynomial, or both. This is truly a marriage of art and mathematics.
- Images can be produced as a collage of two or more polynomiographs created through one of the previous three methods. Many other techniques are possible, either through artistic compositional means, or through computer-assisted design programs.

In what follows we will describe in more detail how to create two different categories of images using polynomiography. As mentioned earlier, polynomiography is based on the manipulation of a finite set of points, whether they are given explicitly or implicitly through their polynomial equation. The first category of images to be described is based on the approximation of the *Voronoi regions* of the solutions of a polynomial equation. To describe Voronoi regions let us assume that we have a rectangular canvas and four random points marked on it as *Blue*, *Green*, *Red*, and *Yellow*. Suppose that we wish to color blue the set of all points on the canvas that are closer to *Blue* than any of the other three colored points. The shape of this region happens to be a polygon and is called the Voronoi region of the point *Blue*. Likewise, Voronoi regions of the other three points can be colored.

We can think of these four points as the roots of a polynomial of degree four. Voronoi regions and a similar coloring scheme can be defined for any arbitrary set of a finite number of points, placed in any arbitrary shape. The Voronoi region of any one of the points, which can be thought of as the *region of attraction* of that point, depends on the position of all the other points. We can create polynomiographs based on this coloring rule. We could do the coloring very precisely, or by approximation of the Voronoi regions. Let us refer to this coloring scheme as *Voronoi coloring*, whether it is done by actual paint on a canvas or on a computer screen.

Here are a few reasons why polynomiography can be a tool of art and design:

- Voronoi coloring produces a diverse set of images with anticipated symmetry or asymmetry.
- Voronoi coloring can be achieved via special iteration functions encoded within polynomiography software.
- With polynomiography software, Voronoi coloring can be established to any precision desired: each member of an infinite class of special iteration functions, readily selected via an index number, generates a different approximation for the Voronoi regions of the same set. As this index increases the approximation improves.
- The boundaries of the approximate Voronoi regions obtained via polynomiography are fractal sets. Thus, polynomiography can also produce fractal images, with a great deal of variety.
- An artist can input any particular set of points, either through a polynomial equation, or explicitly. In particular, it is possible to input certain arrangement of points via a compact and simple formulation. Thus, complex patterns of points can be manipulated conveniently with polynomiography software.
- Polynomiography can help create desirable and anticipated shapes without the knowledge of the underlying mathematical theory. The artist can produce such shapes by learning to use the geometric effects of arithmetic operations on polynomials or complex numbers.

The computer implementation of Voronoi coloring via polynomiography generates one category of polynomiographs based on the manipulation of approximate Voronoi regions, using a single iteration function. Another class of polynomiographs makes use of sets of iteration functions and iterates by moving from one to the next in a *pointwise* fashion. This method does not use repeated iterations as is normally done via iterative methods. For mathematical detail see Kalantari (2004c).

The corresponding images in this second class not fractal and are not based on the coloring of the basins of attraction. Rather, the colorings are defined with respect to a certain proximity to roots as measured by attributes of particular members of the special class of iteration functions in polynomiography, the Basic Family. Again, an artist does not need to know how this process works, only what it produces and how this may be used to produce polynomiographs. For the sake of reference, we will call this second category of images as polynomiographs based on *Levels of Convergence.*

Many other control parameters can be defined with respect to each of these two general categories, but they will not be discussed here. In the next two sections we will show images based on the two categories of polynomiographs described above.

19.1.2 *Polynomiography Based on Voronoi Coloring*

In the first example of polynomiographs based on Voronoi coloring, by placing fewer than a dozen points in the shape of the letter A, and with subsequent coloring, the author produced *Summer variations*, Figure 19.2.

Fig. 19.2 *Summer Variations.*

The artist need not know the formula for the underlying iteration function, nor is it necessary to know an implicit equation of the underlying polynomial. I selected these points with the click of the computer mouse.

Mathematics of a Heart, Figure 19.3 (top-left), is another polynomiograph created with Voronoi region approximation. The initial set of points were placed in the shape of a romantic heart, the coloring achieved using interactive features of polynomiography software and personal choice. The other images in the figure using essentially similar techniques.

Fig. 19.3 Clockwise from top-left: *Mathematics of a Heart, Valentine, Golden Heart, Hearts.*

In Figure 19.17, *Acrobats* uses an orderly arrangement of points. In this example it was necessary, not to mention convenient, to input the explicit formula of the underlying polynomial equation. A polynomial equation can often give a very compact description of a set of points. For instance, $x^{100} - 1 = 0$ describes 100 points equally spaced on the circumference of a circle of radius one unit.

Fig. 19.4 *Circus.*

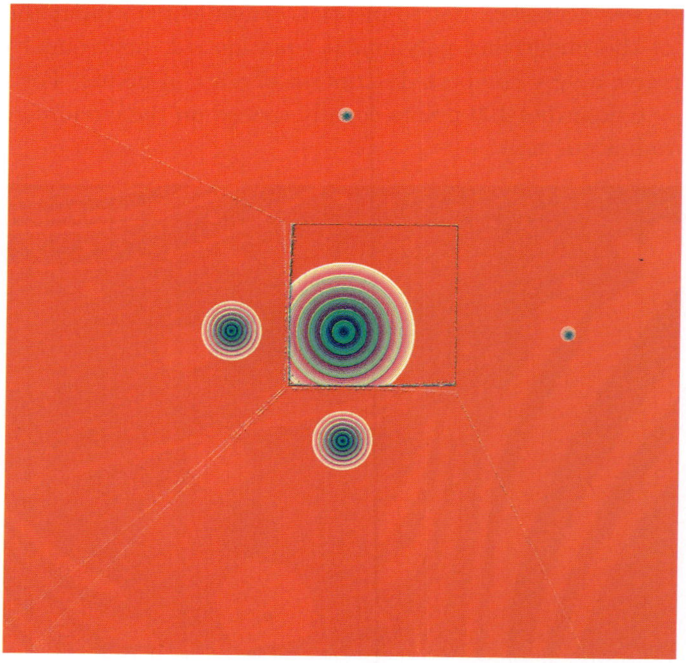

Fig. 19.5 *Squaring the Circle.*

Fig. 19.6 *Waltz* and untitled.

Fig. 19.7 *Mona Lisa in 2001.*

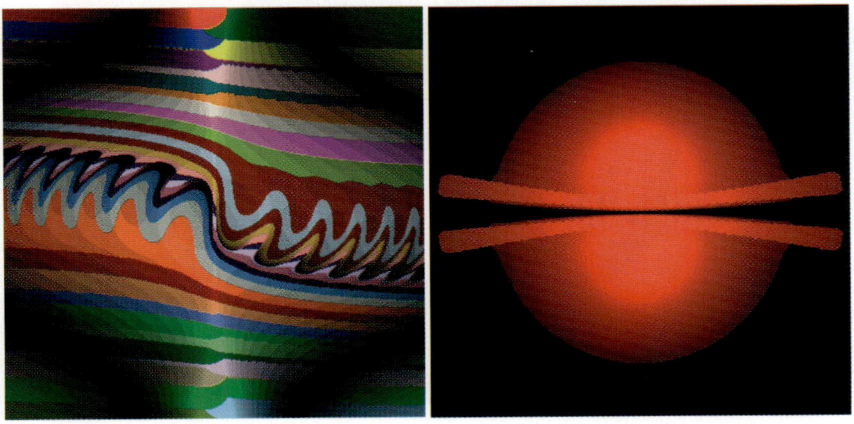

Fig. 19.8 Untitled images.

Fig. 19.9 Two polynomiographs corresponding to the same degree-36 polynomial.

19.1.3 *Polynomiography Based on Levels of Convergence*

Consider the polynomial $cx^{12} - 1$ where c is a complex number.

Fig. 19.10 *Times Square.*

When $c = 1$, the roots form a dozen point on the circumference of a unit circle, placed as the hour marks on a clock. By using different values of c we can create two effects: rotating the points and changing the radius of circle of roots. By multiplying three polynomials of this type, thereby producing a polynomial of degree 36, the author was able to create the polynomiograph in Figure 19.9 (top image). It is interesting that the inspiration behind this polynomiograph was a Persian carpet. In turn this polynomiograph turned into a high-quality hand-woven Persian carpet. Polynomiography can give the blueprint for carpet designs yet to be woven, designs that would not have been possible even for the most experienced of designers.

The bottom image in Figure 19.9 is a polynomiograph whose underlying polynomial is precisely the same polynomial as the top figure in Figure 19.9. What has resulted in the contrast between the two is the coloring and different use of iteration functions. These were purposely done to demonstrate the diversity and choice in polynomiography. This polynomiograph too has been turned into a beautiful carpet.

As with fractal images, one has the ability to zoom in and discover unexpected beauty and complexity that can be used to create yet different types of images. For instance, Figure 19.10, *Times Square*, was created by enlarging a small portion of a polynomiograph of the category that uses Levels-of-Convergence polynomiographs, then using a commercial software and its filters to accentuate certain paths or levels.

These images and remaining images in the section represent variety and richness of polynomiography as a medium of art and as an art form. Figure 19.14 is from an actual exhibition of some of these images as framed artwork. In particular, these images contrast with general fractal images. The optimal size for display of some of the images is poster size. This is because at that size we begin to see and appreciate the real complexity and detail in the images.

Each of the images is based on a single polynomiograph. Even a single polynomial equation and a single polynomiograph could result in the discovery or creation of several different and interesting images much of which depends on the creativity of the individual polynomiographer. The capabilities and sophistication of the underlying polynomiography software is of course another significant parameter in allowing creativity and diversity.

Fig. 19.11 *Cathedral.*

Fig. 19.12 *Life and Death*, two views of the same polynomial.

Fig. 19.13 *Party on the Brooklyn Bridge.*

Fig. 19.14 A Polynomiography exhibition at Rutgers Art Library, 2007.

Fig. 19.15 *Butterfly.*

Fig. 19.16 *Butterfly.*

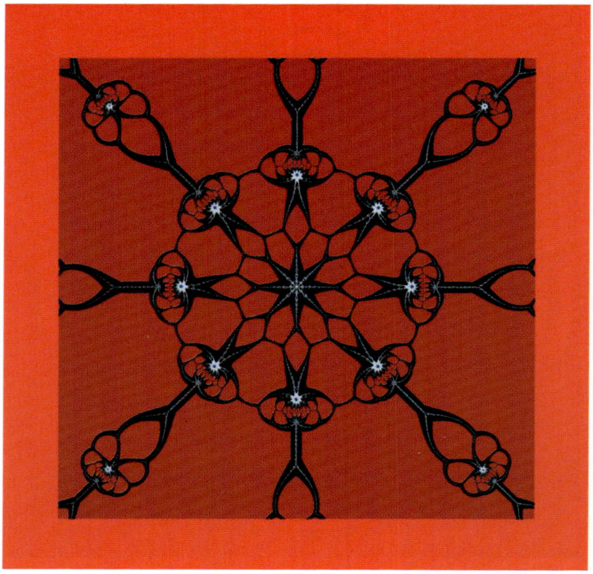

Fig. 19.17 *Acrobats.*

Fig. 19.18 A virtual sculptor inspired by *Acrobats.*

19.1.4 *Symmetric Designs from Polynomiography*

Many symmetric designs can be derived from polynomiography. The significance of symmetry in mathematics, science, and art or design, life and culture is well known. Several interesting books on the subject of symmetry and its significance and origin include Ash and Gross (2006), Lederman *et al.* (2004), and Conway *et al.* (2008). Polynomiography provides a medium for creating symmetric designs and for conveying ideas behind the subject of symmetry at the visual level or mathematical level. It is interesting that while artists and designers appreciate symmetry, often times they would also like to break away from symmetry.

One way to create symmetric designs through polynomiography is by considering polynomials whose roots have symmetric patterns. One can render interesting images by simply considering polynomials whose roots are the roots of unity or more generally n-th root of a real number r. By multiplication of these polynomials, as well as rotation of the roots one can obtain interesting basic designs that can subsequently be turned into intricate designs or paintings. Some sample symmetric images can be seen here. In fact any polynomial with real coefficients will exhibit symmetry with respect to the horizontal axis, that is if polynomiograph window is chosen with y-coordinate within an interval $[-a, a]$. Some example of these is considered next.

19.1.5 *Polynomiography of Numbers*

One interesting application of polynomiography is in encryption of numbers, e.g. ID numbers or credit card numbers, a birthdate, or any other number into a two-dimensional image that resembles a fingerprint. Different numbers will exhibit different fingerprints. Indeed this is a great source for teaching many concepts about numbers and at many levels.

One way to visualize numbers as polynomiographs is to represent them as polynomials. For instance a hypothetical social security number $a_8 a_7 \cdots a_0$ can be identified with the polynomial $P(z) = a_8 z^8 + \cdots + a_1 z + a_0$.

Now we can apply any of the techniques of polynomiography. A particularly interesting visualization results when we makes collective use of the Basic Family, i.e. the Basic Sequence. Figure 19.19 gives several examples. The reader may notice that all the images are distinct except possibly the two lower rightmost images. But that is because they are consecutive numbers. Upon closer look, Figure 19.20, their differences can be noticed immediately.

Given such polynomiograph for a number it should be possible to build scanners that can convert the image back into the original number. The conversion requires the recognition of the roots and the recovery of the corresponding polynomial coefficients.

Fig. 19.19 Polynomiographs of six different nine-digit numbers.

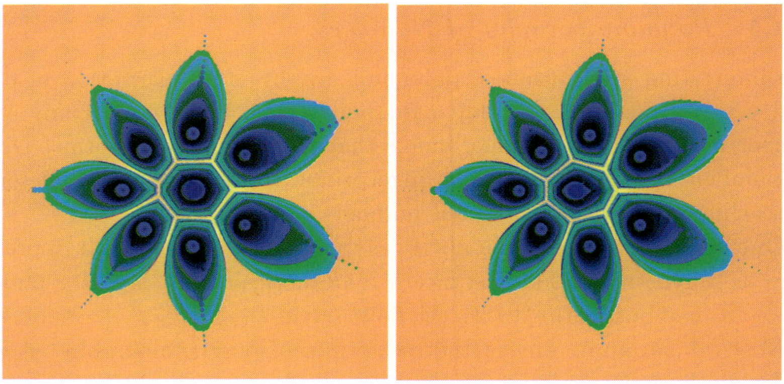

Fig. 19.20 Polynomiographs of the numbers 672123450 and 672123451.

19.1.6 *Some Extensions of Polynomiography*

While we have restricted the attention to 2D images coming from polynomiography, a polynomiograph could inspire or provoke new art form whether digital art, painting, animation, or even sculpture.

Lillian Schwartz, a pioneer in the field of computer art and an accomplished artist, using her own techniques, modified a base polynomiograph to create a new image that would exhibit 3D effect under viewing with 3D glasses (see Kalantari (2005c)). The 3D effect intrinsic in some polynomiographs in turn could suggest 3D architectural structures. In this sense polynomiography provides an alternative means of producing mathematically-inspired sculptors. For instance, while the field of topology has been a major source for such 3D art, e.g. Möbius bands (see Bruter (2007b)), the use of polynomiography offers a set of whole new ideas. As an example in a student project for an undergraduate course on polynomiography at Rutgers, Adrian Sinclair used a 3D software to project *Acrobats* onto a semi-sphere to produce the virtual sculptor shown in Figure 19.18.

Animation with polynomiography offers yet another direction with numerous applications, some educational examples of animation will be considered later in the chapter.

Finally, as with automatically-generated fractals, see Sprott and Pickover (1995), it is easy to produce attractive, automatically-generated polynomiographs. First of all, the k-th digits of any random number, written in ordinary base ten or in binary, can be interpreted as the coefficient of x^k of a polynomial. Then just for this one polynomial alone an infinite number of polynomiographs can be generated where several parameters and coloring schemes can be selected randomly. But even from a given polynomial equation, given as equation or by click of a mouse, one can automatically generate many polynomiographs using randomly generated parameters.

For many images based on polynomiography the reader may visit www.polynomiography.com. This website will continue to release more images and information on polynomiography.

19.1.7 *Glossary of Terms*

We offer here a glossary of terms useful to a non-technical reader.

Complex Number - An algebraic description of the point (a, b) in the Euclidean plane, written as $a + bi$, where $i^2 = -1$.

Polynomial - An expression of the form

$$p(z) = a_n z^n + a_{n-1} z^{n-1} + \cdots + a_1 z + a_0,$$

where n is a given counting number, the coefficients $a_0, \cdots, a_n$ are given complex numbers, and z a variable.

Polynomial Equation - The equation $p(z) = 0$.

Root of a Polynomial- A complex number r such that $p(r) = 0$.

Fundamental Theorem of Algebra (FTA)- The $p(z)$ has n roots. Equivalently, there are n roots $r_1, \cdots, r_n$ such that $p(z) = 0 = (z - r_1)(z - r_2) \cdots (z - r_n)$ (this implies that a polynomial equation is an algebraic description of its roots and conversely, any arbitrary set of n points $r_1, \cdots, r_n$ gives rise to a polynomial equation).

Iteration Function - A recipe that, given any approximation to a polynomial root, however coarse it may be, provides yet another approximation, thereby allowing repetition of the process. In a neighborhood (possibly small) of any root the iterates will converge to that root.

Basin of Attraction of a Root - The set of all initial points such that the corresponding iterates of an iteration function will converge to that root.

Voronoi Regions- A partition of the points in the Euclidean plane into regions based on their proximity to a given finite set S of points. Specifically, the Voronoi region of a point s in a set S is the set of all points in the plane that are closer s than to any other point of S.

Voronoi Coloring- The approximation and coloring of the Voronoi regions of a set of points, whether it is done by hand or by computer.

Newton's Method - The particular iteration function $N(z) = z - p(z)/p'(z)$, where $p'(z)$ is the derivative $p(z)$.

Polynomiography - The computer visualization of polynomial equations under the behavior of iteration functions.

Polynomiograph - An individual polynomiography image viewed at a selected rectangular region in the Euclidean plane.

Julia Set - The boundary of a basin of attraction of polynomial root. (A Julia Set can be defined more generally for other iterative methods).

19.2 Polynomiography in Education

On the one hand polynomials are one of the most important building blocks of mathematics, science and engineering. They have many applications. Polynomials help approximate functions which in turn approximate science and modeling which in turn approximate life and motion. In education as well, polynomials help students get introduced to functions, graphs, solving equations, calculus and much more. Thus polynomials are indispensable abstract objects.

Polynomiography can help young students who are always in the need for visual stimulation to connect to mathematics through playful learning and creativity. This in turn will help them learn complex math. Polynomiography is a medium that helps students to play, to express themselves, to enjoy, and while doing so pick up, easy mathematics, medium mathematics, and even sophisticated mathematics and to reach new frontiers in math and science. This in turn has profound consequence in science and culture. Additionally, through national and international presentations, exhibitions and media attention, polynomiography has received enthusiastic support from artists, engineers, mathematicians, scientists, and the general public.

The spectacular and diverse artworks produced through polynomiography induce striking appreciation of connections between the creativity in art and the intrinsic beauty of mathematics. Polynomiography, as a medium for expression, art, education, discovery or play, can be appreciated by many and without the need for understanding its underlying sophisticated mathematics.

Not only can polynomiography bring art and design into mathematics' and sciences' curricula and education, it can bring mathematics and computer technology to artists who may normally not use mathematics. Thus offering new creative possibilities for artists. Polynomials, some of the most fundamental objects of sciences and math, through polynomiography could suddenly find wider and deeper appreciation by the population at large.

19.2.1 *Polynomiography for Encouraging Creativity in Education*

Well-planned projects using polynomiography will help bring creativity in a very unique fashion in education, both in teaching and learning, from K-12, to college level and beyond. This claim is supported by the reasonably large evidence coming from activities related to polynomiography as art, as a medium for artistry, and as a medium for learning and discovering math while engaged in the creative act of producing aesthetically pleasing images and designs. The most formidable evidence is the extraordinary level of enthusiasm and attention that polynomiography has received after conducting presentations and interacting with wide-ranging audiences. In fact, there is good evidence that even K-12 students exposed to polynomiography have consequently learned of polynomials and shown interest to learn about them. This group of polynomiography enthusiasts consists not only of K-12 and college students, but also of artists, photographers, mathematicians, scientists, and the general public. This medium could actually help students visualize math concepts and can transform a lifeless equation, such as the high school quadratic equation, into beautiful images. In doing so, this can actually help bridge the gap between art and math by bringing polynomial equations to middle and high schools, and more broadly to K-12 education and far beyond.

In what follows we present evidence to the value and significance of polynomiography through sample activities. Specifically we give teacher as well as student survey on the use of polynomiography with respect to creativity in education.

We believe workshops can be designed to introduce teachers to polynomiography and how they can utilize it to inspire creativity in their students through the teaching of lesson plans that can be developed via polynomiography. These lesson plans would get them interested in art-math relations, going from art to science and math, and from science and math to art. Polynomiography is a fantastic medium with proven appeal to teachers and students alike. Through polynomiography the teachers will get encouraged to use their own creativity to develop material that would make the teaching of their subjects more interesting to their students. Such has been the feeling of K-12 teachers in a workshop that was held at Rutgers and in collaboration with mathematicians Iraj Kalantari and Feodor Andreev. The attending teachers could already envision applications of polynomiography in their classrooms and some agreed to develop lesson plans, using a demo software.

Polynomiography can be used in higher education. From courses such as calculus, numerical analysis, dynamical systems, to courses in engineering design, to courses in art can make use of polynomiography in one form or another. Indeed, even a pure mathematician can utilize visualizations through polynomiography to gain deeper insights. In what follows we present feedback from a group of about 30 educators in New Jersey who attended a one-day polynomiography Workshop at Rutgers University in May 2007. The majority of participants were math teachers in middle and of high schools, but also K-12 educators affiliated with museums, or those interested in art-math education, even teachers of astronomy and physics. The feedback was overwhelmingly positive. The teachers were introduced to the subject of polynomiography, its potential applications in education, as well as a demo polynomiography software. Moreover, later in the day they were taken to a computer lab where they each had the opportunity to carry out their own experiment with the demo software. The following is a summary of a written survey.

19.2.2 *Teacher Survey*

I learned new ways to motivate my student – Hands on activities with software and explanation of theories/ideas/concepts – I am trying to create a math/art class and this is helpful to show art in the form of mathematics – I feel I can use this software in my classroom – Fun! New Perspectives - another tool to make mathematics concrete and visual – I am very interested in the intersection of art and math – It directly applies to the subjects I teach – I have been looking for an answer to "why do we need the complex plane." Also, I do the trig using roots of unity. The pictures are much better than just points – Gives me ideas to pursue for Discrete Math Curriculum! – New ideas for interdisciplinary math and art learning experiences – Introduction to the world of [math] and visual design. – Great workshop! – Informative and enjoyable - I would like to see a workshop with artists at some point – A product like this would be a good tool to motivate students – Great job! – Best part was working on software – I indeed had fun trying out the software – I need now to spend more time trying to develop meaningful lessons utilizing the software – I really enjoyed the workshop – I hope to find a use for these ideas and activities in the museum (exhibits/labs/etc.). It was a wonderful workshop! I hope to incorporate it in my classroom, but I need to enhance my knowledge of the computer first – I'd like a guided exploration or at least a few interesting equations to start with, then basic instructions on manipulating the images.

Over the past several years and throughout the development of polynomiography the author has delivered numerous presentations at middle and high schools in New Jersey, college students at Rutgers, other universities in U.S. and abroad. Students have always been impressed with polynomiography asking profound questions and clearly excited about the possibilities. Additionally, collaborators have given presentations/demostrations to their students. As an example polynomiography was tested at Girls Plus Math Camp, held for ages 11-13 at Western Illinois University in July 2007, in collaborations with mathematicians Iraj Kalantari and Feodor Andreev and K-12 camp teachers. . The following is sample student feedback:

19.2.3 *Student Survey*

I like creating my own pictures – It was fun, they looked cool, and I learned a lot and it's useful – It was interesting to see how the numbers and exponents affected the polynomials – It was interesting we could watch an image being made and then change to make another one. We also learned by experimenting and seeing what we got – It was really cool, and you got to play around with it and use your name to get an image – It was interesting – You see a totally different side of something you see everyday – It is a fun way to learn – I have never seen anything like it before and it's very fun to do – It was fascinating and you experimented with numbers to get results – It was fun and a cool math idea to learn – It's cool – Interesting – It was cool to make pictures out of everyday words – It was pretty easy – It seems difficult, but I would like to make a picture I can control – It is fascinating and there is so much you could do with it – I want to know how all those numbers could make such cool pictures. It seems more interesting now – I was kind of confused by it, but now I think it's cool – Specifically I like to know exactly how images are made and the thoughts and reasons behind it – I would like to know more about how they work to understand them better – It would be very interesting to learn more about how it got started – Learning about polynomials was much more fun than I thought and I love learning – If you can do that, what else can you do? – I didn't get it very much but what I got was neat – I like the pictures – I like how they look, and I like how all the numbers show up when you make a complicated one, and it's fun – I like seeing the final product after the equation – Being able to turn numbers into pictures was fun – I liked most of it. It was confusing at times – I liked being on the computer with it – I think the entire concept is fascinating – I loved creating them, but the codes are a little hard – I liked the different shapes of the polynomials.

In addition to such activities exhibition of polynomiography artworks, either in international art-math group exhibitions, or as solo exhibitions, has always aroused the interest of the general public as well as specialized viewers. It is foreseeable that such exhibitions, of 2D and 3D polynomiography-related artworks and possibly interactive media not only would be appreciated by the general public as art, but would equally benefit education and encourage creativity among students and teachers alike. Such exhibitions could complement the workshops, whether for K-12 educators and their students, or for college professors and college students. Such exhibitions could guide visitors, children and adults, in creating their own artwork using an ordinary computer, hence making them appreciate and enjoy the connections between art, math, and technology. These would enhance cultural richness in local institutions and beyond in a unique and significant manner. Such activities could also engage artists. Artists' experiences could result in appropriate features to polynomiography software in order to create a robust medium for graphics, animation, sculpture, 3D and installations and interactive media. This would provoke new directions in art. From the artistic viewpoint these activities also provide an increased vocabulary for other artists to use in extending polynomiography.

19.2.4 *Developing Seminars and Courses Based on Polynomiography*

It is possible to design seminars and courses that are primarily based on the subject of polynomiography, or will use it as a major tool. For instance, it is possible to use polynomiography to teach about polynomials alone, their zeros, iteration functions, Newton's methods, fractals, and much more. These could be fine-tuned with respect to student levels and backgrounds, from K-12 students, to first-year students at college, to advanced undergraduate students, to art students, and more. The author's personal experience with such activities include from First-Year Seminar at Rutgers, to a course for computer science students, to an honors course, and independent studies. The development of polynomiography as an educational medium could advance greatly though related online material and lesson plans. This could involve national and international contributions.

Polynomiography allows visualizations associated with mathematical properties, not necessarily related to polynomials. These not only would help convey important concepts trivially, but could be enjoyed by an artist and the general public. For instance polynomiography animations can be used at various levels: not only at high schools and middle schools, but as a tool that can even make children interested in mathematics and polynomials. On the other hand, it is a sophisticated tool that can be used at college level by students or teachers. Indeed many scientists too should find animation with polynomiography to be a useful tool. Artistic applications of animation with polynomiography are undoubtedly a field worthy of serious attention. Here we give a glimpse of what may be possible through polynomiography animation with a basic polynomiography software. The animation potentials in polynomiography are indeed vast. In achieving some of the advanced educational or artistic applications, it is desirable to develop more sophisticated polynomiography software, at the same time software that is user-friendly for the particular audience or purpose being considered. In what follows we will demonstrate a few examples that capture the applications of polynomiography in animation for mathematical visualization (link to the actual animations may be found at www.polynomiography.com). More explanations about the animations is given in Kalantari *et al.* (2004).

Example 19.1. (Animation of Approximate Voronoi Regions) The Voronoi region of a root θ of $p(z)$ is a convex polygon defined by the locus of points which are closer to this root than to any other root. The boundary of Voronoi regions are defined by the set of points that are equidistant to two distinct roots, i.e. a straight line perpendicular to the line segment joining the two roots, at their midpoint. Any finite set of points in the plane corresponds to the set of roots of a polynomial equation, and conversely. The Basic Family has the property that for large m the basins of attraction provide good approximation of the Voronoi regions. Although the Voronoi region of a given set of points can be computed very efficiently using computational geometry techniques, if the set of points is given as a polynomial equation, then polynomiography provides a direct approach for computing the Voronoi regions without the need to compute the roots in advance. In particular, this becomes a desirable feature when the polynomial is sparse having a few nonzero coefficients. Polynomiography is a convenient medium to demonstrate or discover Voronoi regions as demonstrated in Figures 19.21 first and second rows.

Example 19.2. (Animation of Root Sensitivity) It is well known that the roots of polynomials maybe sensitive to small changes in their coefficients. Classical example is the polynomial $p(z) = (z-1)(z-2)\cdots(z-n)$. For instance, for $n = 7$ the coefficient of z^6 is -28 and even decreasing it to -28.002 causes somewhat large change in the roots, indeed some become complex. This phenomenon can be visualized via polynomiography. Figure 19.14 third row shows a few instances corresponding to gradual changes in the coefficient of z^6.

Example 19.3. (Animation of Multiplication of Two Complex Numbers and Its Meaning as Rotation) Indeed polynomiography is a good medium for teaching properties of complex numbers. Consider the polynomial $p(z) = c(z-z_1)\cdots(z-z_n)$. Let δ be a complex number. The roots of $p(\delta z)$ are those of $p(z)$ multiplied by the complex number $1/\delta$. Figures 19.21, last row, gives two polynomiographies for $p(z) = z^4 - 1$ and $p(e^{i\pi/3}z)$.

19.3 Polynomiography in Mathematics and Science

Aside from the educational or artistic applications mentioned previously, polynomiography will undoubtedly find many applications in mathematics and sciences. Polynomials are most certainly foundational to mathematics and science.

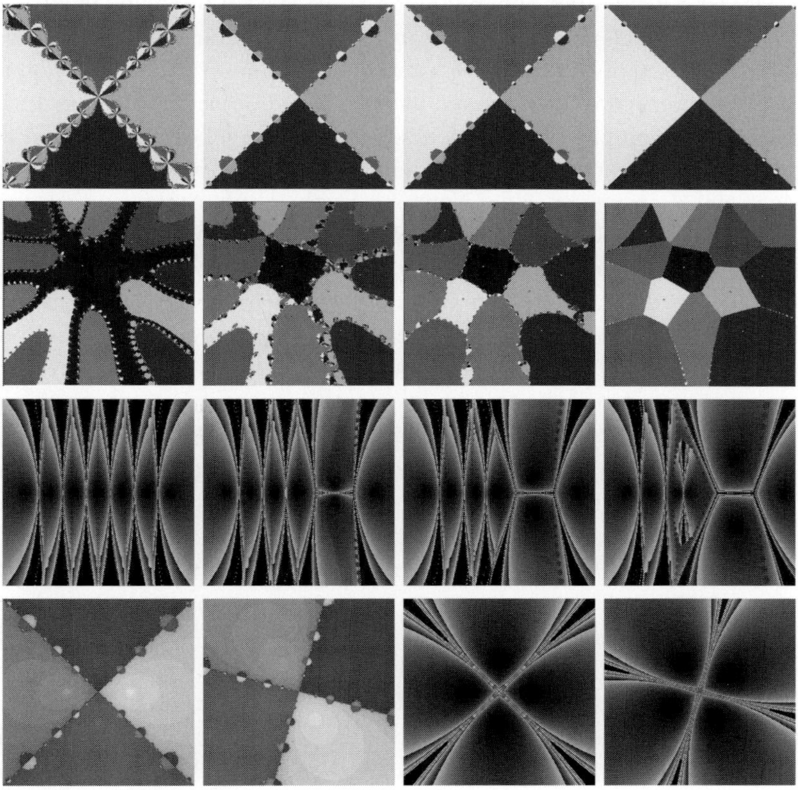

Fig. 19.21 Evolution of basins of attraction to Voronoi regions via $B_m(z)$: $p(z) = z^4 - 1$ (top row); the case of random points (second row); root sensitivity (third row); and animation of rotation (fourth row). For the actual animations visit www.cs.rutgers.edu/ kalantar/Animation.

They are the building blocks of mathematics and science. The need to understand them is not merely a matter of applying them to a particular situation at hand, it is a matter of appreciating the very essence of science and math. It is to understand what numbers are and what mathematician of every era has pondered about. Polynomiography opens the way to new understandings and usage of polynomials and brings their appreciation and utility to an audience never before conceived of.

Throughout the history of mathematics and science numerous special polynomials have been discovered or invented with special applications. But we may ask, do we fully understanding these special polynomials? Clearly the answer depends on what we mean by understanding them and soon we realize that we really do not have a good visual feeling of these polynomials. Real polynomials are often viewed as real-valued objects even though their roots could be complex. This of course is because we may be concerned with them in the context of real polynomials. But this is not always justified. Does it make sense to speak of Newton's method only in the context of real roots? Certainly this view is limited if we are trying to find a root of a real polynomial which may not have any real roots. Whether or not directly noticeable, in this case the dynamics of Newton's method on the real line is influenced by the existence of complex roots. The second dimension in real polynomials, that is the imaginary dimension, is present whether or not we may need to acknowledge it. In this sense polynomiography will bring more insight to the nature of real polynomials. The animation of the phenomenon sensitivity considered in the previous section brings to view quite vividly the connection between real and complex roots as a result of minor changes in the coefficients.

An obvious application of polynomiography is that it will benefit the study of dynamical systems and root-finding algorithms, fractals and much more.

On the one hand, numerous questions arise with respect to the dynamics of iteration functions and the Basic Family and its variants. On the other hand, polynomiography of larger and larger degree classes of polynomials should give rise to new applications. It should give rise to new theoretical and computational challenges such as the development of sophisticated polynomiography software to handle very large degree polynomials. This would require the use of sophisticated and efficient algorithms, from parallel algorithms to Fast Fourier Transform (FFT). The evaluation of pi to larger and larger number of decimal or binary digits is the sort of task that not only has called for such sophistication in algorithms and technology, but has inspired new algorithms and mathematics. In the case of polynomials it is very likely that it gives rise to new designs and artwork. Perhaps designs that would also inspire or include 3D sculptors and/or architecture.

The above-mentioned applications are by no means exhaustive and undoubtedly many novel applications would be discovered. For instance, polynomiography of natural numbers could someday give rise to new encryption algorithms, if not as alternative to the well-known and universal algorithms, such as RSA, but simply by virtue of the fact that it encrypts a number as a visually pleasing image, not just a bar-code. In the author's experiences with respect to the educational potentials of polynomiography, a noticeable excitement among the youth comes from an inherent visualization that can be associated with a natural number. For instance the conversion of such numbers as their birthdate or cell phones into polynomiographs. They immediately observe symmetry and the connection between the number of polynomial roots and the degree of the polynomial so that the fundamental theorem of algebra suddenly becomes an interesting matter of fact, rather than a typically uninteresting and forgettable theorem of mathematics. In what follows we give a specific mathematical application.

19.3.1 *Polynomiography for Measuring the Average Performance of Root-finding Algorithms*

Here we will consider the use of polynomiography in measuring the performance of root-finding algorithms. In particular, we test the efficiency of the first few members of the Basic Family and the Euler-Schröder Family, as well as some individual iteration functions. Some comparison of the Basic Family and its multipoint versions for computing real roots was presented in Chapter 12. For instance, empirically it was shown that for the application mentioned therein B_3 or B_4 is generally faster than the Newton's method. The question of "practical" performance of the Basic Family is certainly of interest and worthy of extensive experimentation because, in particular, it is not enough to consider their local performance only. The word "practical" here in itself is dependent upon the particular usage: computing a single root, computing all roots, polynomiography and specific usages discussed in the chapter.

Measuring the performance of two iteration functions against each other depends on criteria that may not be universal, but dependent on the particular usage. From the practical or scientific point of view, efficiency is the most significant criteria. But even that measure may depend on the particular polynomial and should be averaged in some respects. We also know that no generally convergent rational iteration function exists for all polynomials. Thus in measuring two different iteration functions it would not make sense to pick a particular initial point and apply the two iteration functions for approximation of a root of a given polynomial. While the orbit with respect to one iteration function may converge to a root, the orbit with respect to a second iteration function may converge to another root, or it may not converge at all. Thus instead of measuring the time required for the convergence of an algorithm applied to a certain initial point, we compute the time needed to iterate all of the points from a certain rectangular region. That is to say we examine the global performance instead of the local performance. The time needed to produce the corresponding polynomiograph serves as an indicator for the effectiveness of the global performance of the corresponding algorithm. Of course because an orbit may not converge, we must decide on termination of iterations at some upper bound.

From the implementational point of view, to measure the performance of root-finding algorithms over a certain rectangular area R we measure the time needed to iterate every point from discretized R. Iterations stop if we get *close* to a root of $p(z)$ or when a certain maximum number of iterations is reached (in the latter case we say the algorithm fails to converge).

The algorithms we have explored for the intended measurement can be used to visualize areas of convergence through a corresponding polynomiograph as described in Chapter 17. In what follows we next make a comparison of some iteration functions visually as well as with respect to computed times. For the experiment we considered the first four members of the Basic Family, B_2, B_3, B_4, and B_5, as well as the first four members of Euler-Schöder family, E_2, E_3, E_4, and E_5. These are specifically given at the end.

Our experiments using this approach in Andreev *et al.* (2005) indicate measurable superiority of the Basic Family over the Euler-Schröder counterparts and in terms of speed of convergence. Indeed in some cases, such as that shown in Figure 19.22, even visually one can conclude that one iteration function is superior over another. Moreover, the high order members of the Basic Family, B_3, B_4, and to a certain extent B_5, outperform Newton's method for corresponding polynomials. It is thus not only pedagogically important to consider higher order members of the Basic Family, but from practical point of view as well since they are significant alternatives to Newton's method which is often considered to be the method of choice. In Figure 19.23 we have compared the polynomiography of $z^3 - 2z + 2$, a polynomial under which Newton's method fails to be generally convergent, with three different iteration functions, B_3, E_3, and McMullen's generally convergent algorithms for cubics. In this case B_3 not only outperforms McMullen's in the sense of time, but in terms of allocating regions of convergence more equally than does McMullen's for this polynomial which seems to exhibit a much larger basin of attraction to the real root.

Fig. 19.22 Polynomiographs of $z^3 - 1$: B_m versus E_m for $m = 3, 4, 5$; rows 1, 2 and 3, respectively.

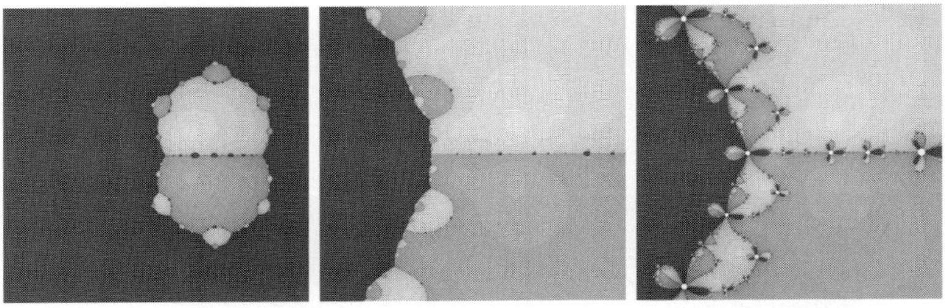

Fig. 19.23 Polynomiographs of $z^3 - 2z + 2$ under McMullen's iteration function, B_3, and E_3.

$$
\begin{aligned}
B_2(z) &= z - \frac{p}{p'} \\
B_3(z) &= z - \frac{2p'p}{2(p')^2 - p''p} \\
B_4(z) &= z - \frac{6(p')^2p - 3p''p^2}{p'''p^2 + 6(p')^3 - 6p''p'p} \\
B_5(z) &= z - 4p\frac{6(p')^3 - 6p''p'p + p'''p^2}{24(p')^4 - 36p''(p')^2p + 8p'''p'p^2 + 6(p'')^2p^2 - p^{(4)}p^3} \\
E_2(z) &= z - \frac{p}{p'} \\
E_3(z) &= z - \frac{p}{p'} - \left(\frac{p}{p'}\right)^2 \frac{p''}{2p'} \\
E_4(z) &= E_3(z) + \left(\frac{p}{p'}\right)^3 \left(\frac{p'''}{6p'} - \frac{p''}{2p'^2}\right) \\
E_5(z) &= E_4(z) - \left(\frac{p}{p'}\right)^4 \left(\frac{p^{(4)}}{4!p'} - \frac{5p''p'''}{12p'^2} + \frac{5p''^3}{8p'^3}\right).
\end{aligned}
$$

19.4 Conclusions

In summary, there is no doubt that polynomiography has tremendous applications in art, science, mathematics and education. There are numerous possibilities for explorations which we hope to undertake and to bring collaborations and support of the others to carry out specific projects.

Chapter 20

Approximation of Square-Roots Revisited: Basic Family, Continued Fractions and Factorization

Approximation of square-roots has had deep impact in the development of science and mathematics, from the intellectual point of view, as well as both theoretical and practical ones. Well-known examples are the discovery of irrational numbers, the complex numbers, the concept of iteration, to generalizations such as the approximations of roots of unity. Indeed the approximation of cube roots of unity can be considered as a starting point in the study of dynamical systems associated with iterations of rational functions, hence Fatou and Julia sets, fractals, and much more. To this author the problem of approximation of square roots was the genesis of the book, and the field of polynomiography (see e.g. Kalantari (2004a)). In this brief chapter we explore some interesting connections between the continued fractions and the Basic Family. In particular, we consider continued fractions for the special case of approximation of square roots and its connections with factorization algorithms. In a sense the Basic Sequence provides an alternative to continued fractions.

20.1 Regular Continued Fractions and the Basic Family

The theory of continued fractions is quite rich and interesting with numerous practical applications, one of which is devising one of the first successful algorithms for factorization of integers. We will give a brief description of continued fractions, then make a comparison with the Basic Family and Basic Sequence.

Let r be a positive real number. The regular continued fraction expansion of r is defined as follows.

Set $r_0 = r$, then for $i \geq 0$ define

$$b_i = \lfloor r_i \rfloor, \quad r_{i+1} = \frac{1}{r_i - b_i}.$$

The m-th *convergent* is the rational number $\mathbf{A}_m/\mathbf{B}_m$ defined as

$$\frac{\mathbf{A}_m}{\mathbf{B}_m} = b_0 + \cfrac{1}{b_1 + \cfrac{1}{b_2 + \cfrac{1}{b_3 + \cdots + \cfrac{1}{b_m}}}}$$

It can be shown (see e.g. Riesel (1994)) $\mathbf{A}_m$ and $\mathbf{B}_m$ satisfy the following:

Theorem 20.1.

$$\begin{cases} \mathbf{A}_m = \mathbf{B}_m \mathbf{A}_{m-1} + \mathbf{A}_{m-2} \\ \mathbf{B}_m = \mathbf{B}_m \mathbf{B}_{m-1} + \mathbf{B}_{m-2} \end{cases},$$

where the initial conditions are

$$\mathbf{A}_{-1} = 0, \quad \mathbf{B}_{-1} = 0, \quad \mathbf{A}_0 = b_0, \quad \mathbf{B}_0 = 1.$$

Moreover,

$$\lim_{m\to\infty} \frac{\mathbf{A}_m}{\mathbf{B}_m} = r.$$

□

For a natural number it can be shown, see e.g Riesel (1994), that:

Theorem 20.2. *Continued fraction convergents corresponding to a natural number N satisfy the bound*

$$\left|\frac{\mathbf{A}_m}{\mathbf{B}_m} - \sqrt{N}\right| < \frac{1}{\mathbf{B}_m^2\sqrt{5}}.$$

□

Consider now $p(z) = z^2 - N$, where N is a natural number. To approximate $\sqrt{N}$ using the Basic Sequence $\{B_m(z), m \geq 2\}$ and its general formula we have:

$$D_m(z) = det \begin{pmatrix} 2z & 1 & \dots & 0 & 0 \\ z^2 - N & 2z & \ddots & \ddots & 0 \\ 0 & z^2 - N & \ddots & \ddots & \vdots \\ \vdots & \vdots & \ddots & \ddots & 1 \\ 0 & 0 & \dots & z^2 - N & 2z \end{pmatrix}.$$

Also, it turns out that independent of m, z, or N we have

$$\widehat{D}_{m,m+1}(z) = det \begin{pmatrix} 1 & 0 & \dots & 0 & 0 \\ 2z & 1 & \ddots & \ddots & 0 \\ z^2 - N & 2z & \ddots & \ddots & \vdots \\ \vdots & \vdots & \ddots & \ddots & 0 \\ 0 & 0 & \dots & 2z & 1 \end{pmatrix} = 1.$$

For any complex input z_0 we have

$$B_m(z_0) = z_0 - p(z_0)\frac{D_{m-2}(z_0)}{D_{m-1}(z_0)} = \frac{z_0 D_{m-1}(z_0) - p(z_0)D_{m-2}(z_0)}{D_{m-1}(z_0)}.$$

We define

$$\mathcal{A}_m(z_0) = z_0 D_{m-1}(z_0) - p(z_0)D_{m-2}(z_0),$$
$$\mathcal{B}_m(z_0) = D_{m-1}(z_0).$$

Thus

$$B_m(z_0) = \frac{\mathcal{A}_m(z_0)}{\mathcal{B}_m(z_0)}.$$

We can think of the Basic Sequence as alternative convergents to the regular continued fraction convergents. In fact from analysis in Chapter 1 we have:

Theorem 20.3. *For any complex number z_0 with positive real part we have*

$$B_m(z_0) - \sqrt{N} = \frac{\mathcal{A}_m(z_0)}{\mathcal{B}_m(z_0)} - \sqrt{N} = (-1)^m \frac{(\sqrt{N} - z_0)^m}{D_{m-1}(z_0)}.$$

Moreover,

$$\lim_{m\to\infty} \frac{\mathcal{A}_m(z_0)}{\mathcal{B}_m(z_0)} = \sqrt{N}.$$ □

We may thus refer to $\{B_m(z_0)\}$ as *Basic Sequence convergents.*

Theorem 20.4. *Let $z_0 = \lfloor \sqrt{N} \rfloor$ or $z_0 = \lceil \sqrt{N} \rceil$. Then*

$$\left| \frac{\mathcal{A}_m(z_0)}{\mathcal{B}_m(z_0)} - \sqrt{N} \right| \leq \frac{2\sqrt{N}|\sqrt{N} - z_0|^m}{(z_0 + \sqrt{N})^{m+1} - 1}.$$

Proof. An alternative formula for $D_m(z_0)$ is

$$D_m(z_0) = \frac{(z_0 + \sqrt{N})^{m+1} - (z_0 - \sqrt{N})^{m+1}}{2\sqrt{N}}.$$

The formula can be argued in several different ways, for instance from the recurrence relation and representation in terms of roots of $p(z_0) = z_0^2 - \sqrt{N}$. Since $|z_0 - \sqrt{N}| < 1$ we conclude

$$D_m(z_0) \geq \frac{(z_0 + \sqrt{N})^{m+1} - 1}{2\sqrt{N}}.$$

Replacing this in Theorem 20.3 we get the desired results. □

20.2 Regular Continued Fraction Convergents Versus Basic Sequence Convergents

We make a comparison of continued fraction convergents and the Basic Sequence convergents for approximation of $\sqrt{N}$. In contrast to the continued fraction approximation, using the Basic Sequence convergents any positive z_0 can be used as initial seed. In particular, we may choose

$$z_0 = \lfloor \sqrt{N} \rfloor, \quad \text{or} \quad z_0 = \lceil \sqrt{N} \rceil.$$

While the error in approximation via the m-th convergent of the continued fraction is based on the knowledge of the m-th partial denominator $\mathbf{B}_m$, the error in the corresponding m-th Basic Sequence convergent is defined in terms of m and can be approximated in terms of floor or ceiling of $\sqrt{N}$. In fact the analysis in Theorem 20.4 says we can use as the initial input $z_0 = \lceil \sqrt{N} \rceil$.

In what follows we give a computational experimentation using $z_0 = \lfloor \sqrt{N} \rfloor$.

For $N = 2, 3, 5, 6$, using a fixed value, say $m = 8$ in Mathematica software we get:

$$e_m = \left| \frac{\mathbf{A}_m}{\mathbf{B}_m} - \sqrt{N} \right| = \left| \frac{\mathcal{A}_m(\lfloor \sqrt{N} \rfloor)}{\mathcal{B}_m(\lfloor \sqrt{N} \rfloor)} - \sqrt{N} \right| = e'_m.$$

However for $N = 7$ we get

$$\left| \frac{\mathbf{A}_m}{\mathbf{B}_m} - \sqrt{7} \right| = \left| \frac{127}{48} - \sqrt{7} \right| = 0.00008202226874274283171 8.$$

$$\left| \frac{\mathcal{A}_m(\lfloor \sqrt{N} \rfloor)}{\mathcal{B}_m(\lfloor \sqrt{N}) \rfloor)} - \sqrt{7} \right| = \left| \frac{108497}{41008} - \sqrt{7} \right| = 7.373162131551089968 1 \times 10^{-7}.$$

$$e_m - e'_m = 0.000081284952529587722721.$$

This particular comparison indicates two facts: First, continued fractions are not identical with Basic Sequence. Second, for $N = 7$, $m = 8$, the Basic Sequence outperforms the continued fractions and give better error.

The size of the numerators and denominators in continued fractions is superior to the corresponding ones in the Basic Sequence convergents. This follows from the fact that continued fraction convergents to an irrational number α is the *best approximations* in the following sense:

Suppose that for an irrational number ξ we set $||\xi||$ to be the smaller of the two distances between ξ and $\lfloor \xi \rfloor$ or $\lceil \xi \rceil$.

Now if $\mathbf{A}_m/\mathbf{B}_m$ is the m-th continued fraction convergents, then

$$\|\mathbf{B}_m\alpha\| = |\mathbf{A}_m\alpha - \mathbf{B}_m|.$$

Moreover, for any rational number A'/B', with $1 \leq B' < \mathbf{B}_m$ we have

$$\|B'\alpha\| > \|\mathbf{B}_m\alpha\|.$$

For the proof of this property see e.g. Lang (1995). In other words

$$|\mathbf{A}_m - \mathbf{B}_m\alpha| < |A' - B'\alpha|.$$

Dividing the above by $\mathbf{B}_m$ we may write

$$\left|\frac{\mathbf{A}_m}{\mathbf{B}_m} - \alpha\right| < \frac{B'}{\mathbf{B}_m}\left|\frac{A'}{B'} - \alpha\right| < \left|\frac{A'}{B'} - \alpha\right|.$$

In summary, with a rational number having smaller denominator than $\mathbf{B}_m$ we could not get a better approximation.

Since the m-th Basic Sequence Concergent gives a better error to $\sqrt{N}$ than the m-th continued fraction convergent, the above implies that $\mathcal{A}_m$ and $\mathcal{B}_m$ have larger magnitude than $\mathbf{A}_m$ and $\mathbf{B}_m$, respectively. However, the notion of "best approximation" does not necessarily imply overall superiority and there is a trade-off between the use of continued fractions and Basic Sequence convergents. In Figures 20.1 and 20.2 we give a comparison of the errors between these two approximations for N up to 2000. Clearly, in the sense of error, the Basic Sequence out-performs continued convergents.

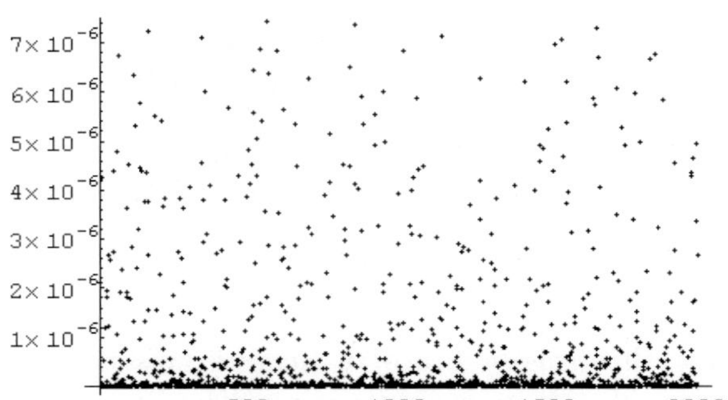

Fig. 20.1 Continued fraction error $\frac{\mathbf{A}_m}{\mathbf{B}_m} - \sqrt{N}$, $N = 1, \ldots, 2000$, $m = 7$.

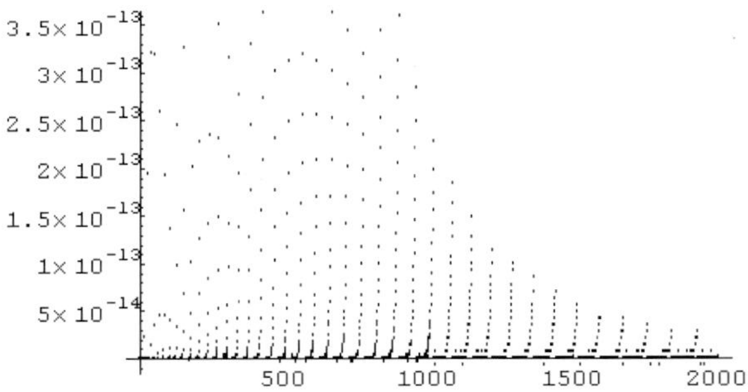

Fig. 20.2 Basic Sequence error $\frac{\mathcal{A}_m}{\mathcal{B}_m} - \sqrt{N}$, $N = 1, \ldots, 2000$, $m = 7$.

20.3 Applications of Continued Fractions and Basic Sequence in Factorization

Factorization of integers has been one of the most fascinating and challenging problems in the theory of numbers, theory of algorithms, and theoretical computer science with such important practical applications as cryptography, see e.g. Cohen (1993).

The goal of this section is to give a brief introduction to classical and significant connections between continued fractions and factorization of integers. Then, exploring the connections of the Basic Family in that light. We first give some definitions.

Definition 20.1. If a, b and n are integers such that $a - b$ is divisible by n, we write $a \equiv b \mod n$ (reads, "a is congruent to b modulo n").

The Greatest Common Divisor of two integers a, n, written as $GCD(a, b)$, can be efficiently computed via Euclid's algorithm.

Definition 20.2. If $GCD(a, n) = 1$ and if the congruence

$$x^2 \equiv a \mod n$$

has a solution x, then a is said to be a *quadratic residue of* n. If there is no solution, a is said to be a *quadratic non-residue.*

For the case when n is a prime, in Legendre's notation for the above two cases we write

$$(\frac{a}{n}) = 1, \quad (\frac{a}{n}) = -1.$$

Jacobi extended the definition to composite n. Fundamental properties of residue are summarized in the following theorem

Theorem 20.5. *Let m and n be odd naturals and $GCD(a,n) = GCD(b,n) = 1$. Then*

$$(\frac{a}{n})(\frac{b}{n}) = (\frac{ab}{n})$$

$$(\frac{-1}{n}) = (-1)^{\frac{1}{2}(n-1)}, \quad (\frac{2}{n}) = (-1)^{\frac{1}{8}(n^2-1)}$$

$$(\frac{m}{n}) = (\frac{n}{m})(-1)^{\frac{1}{2}(n-1)\frac{1}{2}(m-1)}.$$

□

The first equation shows that the product of two quadratic residues, or two quadratic non-residues is a quadratic residue, while the product of a quadratic residue and a quadratic non-residue is a quadratic non-residue. The next two special cases allow extensions to more general cases of n and m. The third equation is Gauss's quadratic reciprocity law, considered to be one of the highlight of number theory.

We now describe Legendre's congruence and how it could lead to finding an integer factor of a given natural number N. It is a more powerful version of Fermat's method for factoring which is based on finding x and y such that $x^2 - y^2 = N$. The congruence

$$x^2 \equiv y^2 \mod N$$

has solutions $x = \pm y$, and these are called the trivial solutions. When N is a composite number, the congruence has nontrivial solutions and these can be used to find a factor of N. Since a nontrivial solution gives the following

$$(x-y)(x+y) \equiv 0 \mod N,$$

to find a factor of N it suffices to compute the GCD of $x \pm y$ and N. A prime factor p in this GCD must necessarily be a nontrivial factor of N since p cannot divide both $x+y$ and $x-y$, otherwise both x and y are zero mod N, implying that they are a trivial pair of solutions. The use of Legendre's congruence for factorization serves as the root of several significant algorithms for integer factorization. These include old and modern applications. What remains to be done in using Legendre's congruence is to find a way in which nontrivial solutions can be found.

Continued fractions provide a powerful method that gives rise to solutions. Indeed some of the first successful algorithms for factorization that made use of a computer utilized continued fractions. These include Shanks' "Square Forms Factorization" method, SQUFOF, and the Morrison-Brillhart's "Continued Fraction Method" method, CFRAC, see e.g. Riesel (1994) and Cohen (1993) for these and other factorization methods.

The regular continued fractions for $\sqrt{N}$ can be shown to satisfy the following (see Riesel (1994)):

$$\mathbf{A}_m^2 - \mathbf{B}_m^2 N = (-1)^m \mathbf{Q}_m,$$

where $\mathbf{Q}_m$ is a natural number. It thus follows that

$$\mathbf{A}_m^2 \equiv (-1)^m \mathbf{Q}_m \bmod N.$$

If for the case when m is even, $(-1)^m \mathbf{Q}_m$ happens to be a perfect square, say R^2, then

$$(\mathbf{A}_m + R)(\mathbf{A}_m - R) \equiv 0 \bmod N.$$

In fact $\mathbf{Q}_m$ turns out be computable without the knowledge of $\mathbf{A}_m$. Then using Euclid's algorithm we can then check if $\mathbf{A}_m \pm R$ and N have a nontrivial factor p.

Remark 20.1. To be consistent with continued fraction literature our $\mathbf{Q}_m$ above should have been written as $\mathbf{Q}_{m+1}$. However, for notational convenience and since we do not derive it explicitly we have chosen the index to be m as opposed to the traditional $m+1$.

The above properties of continued fractions together with the laws on quadratic residues, such as quadratic reciprocity (Theorem 20.5) allow efficient techniques for testing if $(-1)^m \mathbf{Q}_m$ is a quadratic residue mod N or to form new ones from those that are quadratic non-residue mod N which in a way is the basis of Morrison and Brillhart's algorithm. While Shanks' algorithm has to wait before generating an even m such that $\mathbf{Q}_m$ is a perfect square, Morrison and Brillhart's algorithm tries to use combinations which yield squares by multiplication of quadratic residues generated so as to get a perfect quare. We describe this next.

In Shanks' algorithm it is shown that for all m, $\mathbf{Q}_m < 2\sqrt{N}$. Since $\sqrt{N}$ is reasonably smaller than N, for certain values of m we can try to compute the prime factorization of $\mathbf{Q}_m$. To decide on what values of m to select, a prime number $p*$ is chosen as an upper bound. Suppose that we have been able to generate a set of m values, say $m_1, \ldots, m_t$ such that for each

$i = 1, \ldots, t$ the complete prime factorization of each $\mathbf{Q}_{m_i}$ is known. If the set of primes up to $p*$ is $p_1, \ldots, p_r$, then for $i = 1, \ldots, t$

$$(-1)^m \mathbf{Q}_{m_i} = (-1)^m \prod_{j=1}^{r} p_j^{k_{ij}},$$

for some nonnegative set of integers k_{ij}. To explain the main idea behind Morrison and Brillhart's CFRAC algorithm, consider a vector $x = (x_1, \ldots, x_t)$ of zero-one variables x_i, $i = 1, \ldots, t$. Then multiplying the above expansions of $\mathbf{Q}$'s corresponding to $x_i = 1$, we may write

$$\prod_{i=1}^{t} ((-1)^{m_i} \mathbf{Q}_{m_i})^{x_i} = (-1)^{K_0(x)} \prod_{j=1}^{r} p_j^{K_j(x)}$$

where

$$K_0(x) = \sum_{i=1}^{t} m_i x_i,$$

and for $j = 1, \ldots, r$,

$$K_j(x) = \sum_{i=1}^{t} k_{ij} x_i.$$

The idea of CFRAC algorithm is to find zero-one vector $x = (x_1, \ldots, x_t)$ such that $K_j(x)$ is even for all $j = 0, 1, \ldots, r$. Assuming such an x vector is found, we have

$$\prod_{i=1}^{t} (\mathbf{A}_{m_i}^2)^{x_i} \equiv \prod_{i=1}^{t} (\mathbf{Q}_{m_i})^{x_i} = \prod_{j=1}^{r} p_j^{K_{j(x)}}.$$

Thus if we set

$$\mathbf{A}^2(x) = \prod_{i=1}^{t} (\mathbf{A}_{m_i})^{x_i}, \quad \mathbf{Q}(x) = \prod_{i=1}^{t} (\mathbf{Q}_{m_i})^{x_i},$$

then we

$$\mathbf{A}(x) \equiv \mathbf{Q}(x) \mod N,$$

and $\mathbf{Q}(x) = R^2(x)$, a perfect square with

$$R(x) = \prod_{j=1}^{r} p_j^{K_{j(x)/2}}.$$

Thus we can proceed as in Legendre's congruence. As an example suppose that for some N we have

$$\mathbf{A}_1^2 \equiv (-1)2^3 \times 7^2 \times 11 \mod N,$$

$$\mathbf{A}_7^2 \equiv (-1)2^9 \times 5 \mod N,$$

$$\mathbf{A}_{12}^2 \equiv 5 \times 9 \mod N.$$

Then

$$(\mathbf{A}_1\mathbf{A}_7\mathbf{A}_{12})^2 \equiv 2^{12} \times 5^2 \times 7^2 \times 11^2 \mod N \equiv (2^6 \times 5 \times 7 \times 11)^2 \mod N.$$

In Shanks' algorithm the partial convergents $\mathbf{A}_m$, are not computed explicitly while Morrison-Brillhart's algorithm requires their values. The advantage in the latter algorithm is clear because it allows forming quadratic residues in a variety of ways. From the implementation point of view there are more technical issues we omit from the description here.

The Basic Sequence convergents, $\mathcal{A}_m(\lfloor\sqrt{N}\rfloor)/\mathcal{B}_m(\lfloor\sqrt{N}\rfloor)$, can also be shown to satisfy the relation

$$(\mathcal{A}_m(\lfloor\sqrt{N}\rfloor))^2 \equiv (-1)^m \mathcal{Q}_m \mod N,$$

for some integer $\mathcal{Q}_m$. In fact we get an explicit formula for $\mathcal{Q}_m$.

Theorem 20.6. *Let $z_0 = \lfloor\sqrt{N}\rfloor$, or $z_0 = \lceil\sqrt{N}\rceil$. Then for all m*

$$\mathcal{A}_m^2(z_0) \equiv z_0^{2m} \mod N.$$

In particular, if we set $R = z_0^m$, then we get the Legendre congruence

$$(\mathcal{A}_m(z_0) - z_0^m)(\mathcal{A}_m(z_0) + z_0^m) \equiv 0 \mod N.$$

Proof. Suppressing z_0, from the expansion of $B_m(z)$ and upon multiplying it by $D_{m-1} = \mathcal{B}_m$ we get

$$\mathcal{A}_m - \sqrt{N}\mathcal{B}_m = (-1)^m(\sqrt{N} - z_0)^m.$$

We may assume N is not a perfect square since otherwise from the above the theorem is obvious. Squaring both sides we get

$$\mathcal{A}_m^2 - 2\sqrt{N}\mathcal{A}_m\mathcal{B}_m + N\mathcal{B}_m^2 = (\sqrt{N} - z_0)^{2m}.$$

Each side of the equation can be simplified in the form

$$u + v\sqrt{N}, \quad u, v \in \mathbb{N}.$$

The right-hand side can be written as

$$(\sqrt{N} - z_0)^{2m} = \sum_{i=0}^{2m} (-1)^i \binom{2m}{i} \sqrt{N}^{\,i} z_0^{2m-i}$$

$$= \sum_{i \text{ even}} (-1)^i \binom{2m}{i} \sqrt{N}^{\,i} z_0^{2m-i} + \sum_{i= \text{ odd}} (-1)^i \binom{2m}{i} \sqrt{N}^{\,i} z_0^{2m-i}.$$

The first sum in the right-hand side of the last equality is an integer u, while the second sum is of the form $v\sqrt{N}$. By equating the corresponding parts of the left-hand side and the right-hand side of the main equation and moreover, noting that $u \equiv z_0^{2m} \mod N$ the proof is immediate. □

Remark 20.2. The above theorem implies that in using the Basic Sequence convergents for the factorization of an integer N, the quantity $z_0^m \mod N$ can be computed efficiently. Thus whether used as in Shanks style, or Morrison-Brillhart style algorithm, setting $R_m = z_0^m \mod N$, for all m even we have

$$(\mathcal{A}_m - R_m)(\mathcal{A}_m + R_m) \equiv 0 \mod N.$$

Thus Legendre congruence is satisfied and we may test for factors of N. Whether or not this happens to produce good factors is worthy of scrutiny and future experimentation. It may happen that the Basic Sequence convergents coincide with those of the continued fractions. This raises several interesting questions, such as when do the two coincide? Moreover, the fact that they are unequal suggests the possibility of using the combination of the two convergents in producing even newer perfect squares, that is to say multiplication of the congruences of the type $\mathbf{A}_i^2 \equiv (-1)^i \mathbf{Q}_i \mod N$ with those of the type $\mathcal{A}_j^2 \equiv (-1)^j \mathcal{Q}_j \mod N$.

20.4 Basic Sequence for Approximation of Higher Roots of a Number and its Factorization

In this section we consider the Basic Sequence for approximation of general roots of a positive integer N, say for the approximation of $\theta = N^{1/t}$, where t is a natural number. We assume θ is not a natural number. In this general case we construct the Basic Sequence with respect to the polynomial.

$$p(z) = z^t - N.$$

Analogous to the case of $\sqrt{N}$ we can use $z_0 = \lfloor N^{1/t} \rfloor$ or $z_0 = \lceil N^{1/t} \rceil$. The higher value of t however would result in an expansion for $B_m(z_0)$ with more than one term. For instance, suppose that $t = 3$. Then we get

$$B_m(z) - \theta = z - p(z)\frac{D_{m-2}(z)}{D_{m-1}(z)}$$

$$= (-1)^m \frac{\widehat{D}_{m-1,m}(z)}{D_{m-1}(z)}(z-\theta)^m + (-1)^m \frac{\widehat{D}_{m-1,m+1}(z)}{D_{m-1}(z)}(z-\theta)^{m+1},$$

where

$$D_m(z) = det \begin{pmatrix} 3z^2 & 3z & 1 & \dots & 0 & 0 \\ z^3-N & 3z^2 & 3z & \ddots & \ddots & 0 \\ 0 & z^3-N & 3z^2 & \ddots & \ddots & \vdots \\ \vdots & \vdots & \vdots & \ddots & \ddots & 3z \\ 0 & 0 & 0 & \dots & z^3-N & 3z^2 \end{pmatrix},$$

$$\widehat{D}_{m,m+1}(z) = det \begin{pmatrix} 3z & 1 & 0 & \dots & 0 & 0 \\ 3z^2 & 3z & 1 & \ddots & \ddots & 0 \\ z^3-N & 3z^2 & 3z^2 & \ddots & \ddots & \vdots \\ \vdots & \vdots & \ddots & \ddots & \ddots & 1 \\ 0 & 0 & 0 & \dots & 3z^2 & 3z \end{pmatrix},$$

$$\widehat{D}_{m,m+2}(z) = det \begin{pmatrix} 3z & 1 & 0 & \dots & 0 & 0 \\ 3z^2 & 3z & 1 & \ddots & \ddots & 0 \\ z^3-N & 3z^2 & 3z^2 & \ddots & \ddots & \vdots \\ \vdots & \vdots & \ddots & \ddots & \ddots & 0 \\ 0 & 0 & 0 & \dots & 3z^2 & 1 \end{pmatrix}.$$

Expanding the determinant of $\widehat{D}_{m,m+2}(z)$ along the last column, it follows that

$$\widehat{D}_{m,m+2}(z) = \widehat{D}_{m-1,m}(z).$$

These allow bounding the error with respect to the general Basic Sequence convergents. We avoid computing explicit bounds such as the case of $t = 2$ derived earlier in the chapter. However, it would certainly be interesting to make a detailed comparison of the Basic Sequence convergents for $N^{1/t}$ and the corresponding continued fraction convergents.

We conclude this chapter by considering the corresponding Basic Sequence convergents arising in approximating $N^{1/t}$, say using z_0 as the floor or ceiling of $N^{1/t}$, and suggest their application with respect to factorization of N. These convergents analogously could be written as

$$\frac{\mathcal{A}_m}{\mathcal{B}_m},$$

where naturally, they depend also on t. It can be shown that

Theorem 20.7.

$$\mathcal{A}_m^t \equiv (-1)^m \mathcal{Q}_m \mod N.$$

□

Suppose that m is even and $\mathcal{Q}_m = R^t$, for some integer R. Then, using the identity $(a^t - b^t) = (a^{t-1} + a^{t-2}b + \cdots + b^{t-1})$ we have

$$(\mathcal{A}_m - R)(\mathcal{A}_m^{t-1} + \cdots + R^{t-1}) \equiv 0 \mod N.$$

But this implies we can search for a prime factor of N by computing the GCD of N and the factors above.

In summary, beyond square roots, the higher roots of an integer N can be used to get new Basic Sequence convergents, thus potentially increasing the chances of factorization. This is at the cost of new computations, but many interesting questions arise with respect to convergents from Basic Sequence in approximation of t-th root of N.

Problem 1. For what values of N are the corresponding continued fraction convergents and Basic Sequence Convergents identical?

Problem 2. Derive a bound on the error $|B_m(z_0) - \theta|$ where $\theta = N^{1/t}$, $t \geq 3$, N a natural number, and z_0 is either the floor of ceiling of $N^{1/t}$.

Problem 3. Make a theoretical or thorough computational comparison between the Basic Sequence convergents error $|B_m(z_0) - \theta|$ and the error in the regular continued fraction approximation of $\theta = N^{1/t}$, where z_0 is floor of ceiling of $N^{1/t}$.

Problem 4. Make a theoretical or thorough computational investigation of the use of the Basic Sequence convergents in the factorization of an integer N, using the approximation of $\theta = N^{1/t}$, for $t \geq 2$, allowing also z_0 to take on other values than the floor or ceiling of θ.

Problem 5. More generally, suppose θ is an algebraic number whose minimal polynomial is $p(z)$. Make a theoretical comparison between the error $|B_m(z_0) - \theta|$ and the continued fraction approximation of θ, where z_0 is floor of ceiling of θ, or other integer values for which the Basic Sequence is guaranteed to converge.

Chapter 21

Further Applications and Extensions of the Basic Family and Polynomiography

In this chapter we will informally describe some extensions of the Basic Family and polynomiography to the more general case of analytic functions, as well as to higher dimensions, other domains, and more. We then propose means by which polynomiography can become widely utilized through a digital online media and collaborations.

21.0.1 *Extensions to Analytic Functions*

In order to find a zero of an analytic function $f(z)$ over $\mathbb{C}$ via iteration functions developed for a polynomial $p(z)$, we simply replace in the formulas p with f. This implies that each iteration function considered in this book, including the Basic Family and any of its variants can all be applied to approximate zeros of $f(z)$. In particular for each $B_m(z)$ we can generate visualizations based on the dynamics of the iterates.

How should one address the corresponding image? Although the function $f(z)$ may not be a polynomial, it may still make sense to address the image as a "polynomiograph." This is justifiable because one could argue that any evaluation of the underlying analytic function can be considered as an evaluation with respect to a polynomial approximation to $f(z)$.

For instance, in Figure 21.1 we apply Newton's method to give polynomiographs of the first few Taylor polynomial approximations from the series expansion:

$$\cos(z) = 1 - \frac{z^2}{2!} + \frac{z^4}{4!} - \frac{z^6}{6!} + - \cdots,$$

as well as the "polynomiographs of $\cos(z)$ under Newton's method." The figure consists of the polynomiograph in a rectangle centered at the origin, and a closeup near the origin. The polynomiographs for $\cos(z)$ are given as the right pair of images in the last row of the figure. As we see the sequence of polynomiographs appear "convergent" to the images of $\cos(z)$. This justifies addressing the image as a polynomiograph.

From the visual point of view, the world of polynomials is already infinitely rich. Moreover, polynomials form a very significant class, as also seen and justified in this book. However, polynomiography of certain analytic functions such as trigonometric or rational functions could not only turn out to be interesting both from the point of view of dynamical systems and theory, but from the point of view of all the other applications we have discussed previously in this book.

The reader may notice that the sequence of polynomiographs of Taylor polynomials quite nicely conveys the notion of convergence. An animation of this effect by taking higher and higher degree polynomials would be a very interesting demonstration of convergence and will be developed elsewhere.

From the computational point of view when $f(z)$ is an analytic function, e.g. a rational function, the efficiency of the corresponding visualizations could dramatically decrease as we make use of higher order members of the Basic Family. While applying low order members of the Basic Family such as Newton's and Halley's would be computationally feasible, the higher order members may require higher derivatives of a rational function which may be computationally intensive.

As we have seen for polynomials the Basic Sequence has the property that for any point in a Voronoi region of a root it converges to that root. Such sequence can also be defined for an arbitrary analytic function but the question of convergence remains to be answered. The reader may recall that in Chapter 14 formulas were derived for π based on the convergence of sequences that can be viewed as Basic Sequences corresponding to appropriately selected trigonometric functions, and appropriately selected inputs.

For instance, if we consider $f(z) = \sin(z) - .5$, the corresponding Basic Sequence at 0 was proved to converge to $\pi/6$. This would intuitively imply that in a neighborhood of the origin the corresponding Basic Sequence would be convergent to $\pi/6$ as well. In fact in this case the Basic Sequence itself is efficiently computable because of periodicity of the derivatives and the nature of the formulas involved. However, since $f(z)$ has infinitely many zeros, the notion of Voronoi region is ambiguous and needs to be defined appropriately. Furthermore, for such function the behavior and convergence of the Basic Sequence for a general input needs to be examined.

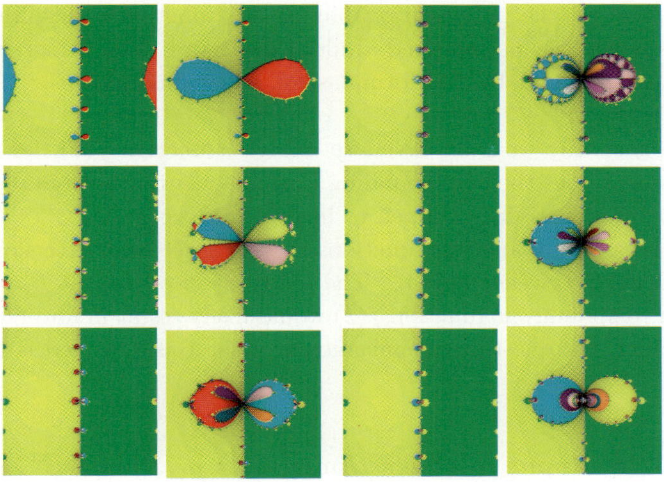

Fig. 21.1 Polynomiographs of $P_4(z), P_6(z), P_8(z)$ and their closeups near the origin - the first two columns and top to bottom; and polynomiographs of $P_{10}(z), P_{12}(z)$ and $\cos(z)$ and their closeups - the last two columns.

Fig. 21.2 "Masked Queen" a polynomiograph based on approximation of π.

Nevertheless, the polynomiography of this function as inspired by the Basic Sequence can be rendered. Indeed in Figure 21.2, "Masked Queen" is an image inspired by the formulas for approximation of π (Kalantari (2000b)).

More generally we ask: Given an analytic function $f(z)$, for what set of points in $\mathbb{C}$ does the corresponding Basic Sequence converge to a zero of $f(z)$, assuming $f(z)$ does happen to have a finite zero.

The dynamics of analytic functions has of course been extensively studied, see e.g. Bergweiler (1993). The study of iterations of $B_m(z)$ corresponding to an analytic function $f(z)$ would likely result in novel questions. In particular, the question of convergence of the Basic Sequence at various inputs.

21.0.2 *Extensions to Other Dimensions or Domains*

Several natural questions arise with respect to generalizations to polynomials over several variables, or higher dimensions, or over restriction to complex domain. These require further research beyond the present book but in some cases there are suitable extensions using the existing methodologies given here.

For instance, given a complex polynomial in two complex variables $p(z, w)$, we can fix one variable, say w, and then do polynomiography for the resulting complex polynomial. This procedure, once repeated for distinct values of w, results in a cross-sectional polynomiography that can be displayed as a sequence of images or in the form of animation.

Another possible extension of polynomial root-finding or polynomiography is the case of finite fields. Over such fields factorization and polynomial zeros are of numerous applications. Yet another extension is to quaternions and more general extensions of complex numbers. For quaternions and extension see e.g. Conway and Smith (2003). Other possibilities worthy of exploration are the case of fields where polynomials still obey the fundamental theorem of algebra, see e.g. Bruter (2007a).

21.0.3 *Polynomiography for Designing Shapes*

A natural yet interesting question is whether a particular shape can be realized as a polynomiograph for an appropriately selected polynomial equation. Clearly one would not expect an arbitrary shape to be described as polynomiograph. Nevertheless, knowing the power of just a few points in giving rise to magnificent images through polynomiography, it is plausible that many geometric shapes would be producible as a polynomiograph. It would be conceivable that one could catalogue a collection of some basic geometric shapes that could be used as basic building blocks for different applications. The educational or artistic applications would be enormous.

21.1 Toward a Digital Media Based on Polynomiography

Throughout the book we have spoken of many potential applications of polynomiography. In this final section we will summarize these and emphasize the possibility that polynomiography as introduced and examined in this book, could help lay the foundation to a new multidisciplinary field. Polynomiography, although based on sophisticated algorithmic visualizations of one of the most basic and fundamental tasks in sciences and math - solving a polynomial equation - bridges art and math in a significant and unique way which could open up tremendous artistic and educational possibilities for a wide range of potential users, from children to adults and experts of different kinds. As a medium for art, education, discovery or play, it can be appreciated by many, even without the need for understanding its underlying mathematics. In particular, it could help young students to connect to mathematics and algorithms through playful learning and creativity and help them learn difficult concepts and reach new frontiers in math and science.

Polynomiography can thus be used as the basis of a technology for the encouragement of creativity in multidisciplinary teaching and learning experiences, and to develop curricula for a wide range of educational courses. This claim is supported both through student surveys at middle and high schools, university students in different fields, as well as teacher surveys. Although these have been based on limited access to polynomiography software, much interest has been demonstrated by both groups. In higher education too, students and teachers from courses such as calculus, numerical analysis, dynamical systems, to courses in engineering design, to courses in art, can make use of polynomiography in one form or another. From undergraduate and graduate students to scientists and pure mathematicians, they could all find visualizations through polynomiography not only revealing, but also capable of provoking new ideas and discoveries. Through national and international presentations, exhibitions and media attention, it has been determined that there is sufficient interest in polynomiography from among artists, engineers, mathematicians, scientists who foresee applications to their particular fields of interest.

The time is most apt to explore this wide and varied range of interests to propel the medium of polynomiography for applications. In doing so there is the need to develop, refine, expand, specialize, and fine-tune our existing prototype software in order to adapt the application to specific objectives and environments. This is a future goal we hope to accomplish. Once polynomiography software is widely available, national and international experts with specialties in a wide range of fields such as mathematics, mathematics education, engineering, sciences, K-12 education, computer art and fine art could help create and maintain an international digital network around polynomiography and develop multi-disciplinary teaching and learning curricula, possibly even in different languages. International interactions and collaborations could further enhance this task and could in turn have consequences for improvement of the teaching of mathematics and art through polynomiography.

Polynomiography could bring art and design into mathematics, sciences, and education, it can also bring mathematics and computer technology to artists who may normally not use mathematics. Thus offering new creative possibilities. Polynomials could even receive appreciation by the population at large. Certainly, the development of such digital media and execution of its goals not only demands international desire, but the necessary support. However, whether or not these goals would ever become fully realized, polynomiography seems to be moving forward, promising to find its way into education and beyond.

Bibliography

Ahlfors, L. (1979). *Complex Analysis*, 3rd edn. (McGraw-Hill Book Co., New York).

Alefeld, G. (1981). On the convergence of Halley's method, *Amer. Math. Monthly* **88**, pp. 530–536.

Andreev, F., Kalantari, B. and Kalantari, I. (2005). Measuring the average performance of root-finding algorithms and imaging it through polynomiography, in *Proceedings of 17th IMACS World Congress, Scientific Computation, Applied Mathematics and Simulation* (Paris, France).

Ash, A. and Gross, R. (2006). *Fearless Symmetry: Exposing the Hidden Patterns of Numbers* (Princeton University Press, Princeton, NJ).

Atkinson, K. E. (1989). *An Introduction to Numerical Analysis*, 2nd edn. (John Wiley & Sons, Inc.).

Bailey, D. F. (1989). Historical survey of solution by functional iteration, *Mathematics Magazine* **62**, 3, pp. 155–166.

Bailey, D. H. (1988). The computation of Pi to 29,360,000 decimal digits using Borweins' quadratically convergent algorithm, *Math. of Computation* **50**, 181, pp. 283–296.

Bailey, D. H., Borwein, J. M., Borwein, P. B. and Plouffe, S. (1997). The quest for pi, *Math. Intelligencer* **19**, pp. 50–57.

Bak, J., Ding, P. and Newman, D. (2007). Extremal points, critical points, and saddle points of analytic functions, *American Mathematical Monthly* **114**, 6, pp. 540–546.

Bak, J. and Newman, D. J. (1997). *Complex Analysis*, 2nd edn. (Springer-Verlag, New York).

Baker, G. A. and Graves-Morris, P. R. (1996). *Padé Approximants, Encyclopedia of Mathematics and its Applications*, Vol. 59, 2nd edn. (Cambridge University Press).

Barbeau, E. J. (1989). *Polynomials* (Problem Books in Mathematics, Springer-Verlag, New York).

Barna, B. (1956). Über die divergenzpunkte des newtonsches verfahrens zur bestimmung von wurzeln algebraischen gleichungen. ii, *Publ. Math. Debrecen* **4**, pp. 384–397.

Barnsley, M. (1988). *Fractals Everywhere* (Academic Press, Boston, MA).

Bateman, H. (1938). Halley's method for solving equations, *Amer. Math. Monthly* **45**, pp. 11–17.

Beardon, A. F. (1991). *Iteration of Rational Functions: Complex Analytic Dynamical Systems* (Springer-Verlag, New York).

Beckmann, P. (1971). *History of* π (St. Martin's Press, New York).

Ben-Israel, A. (1969). Theorems of the alternative for complex linear inequalities, *Israel J. Math.* **7**, pp. 121–136.

Berggren, L., Borwein, J. and Borwein, P. (1997). *Pi: A Source Book* (Springer-Verlag, New York).

Bergum, G. E., Philippou, A. N. and Hordam, A. F. (1992). *Applications of Fibonacci Numbers*, Vol. 5 (Kluwer Academic Publishers).

Bergweiler, W. (1993). Iterations of meromorphic functions, *Bull. AMS* **29**, pp. 151–188.

Bini, D. and Pan, V. Y. (1994). *Polynomials and Matrix Computations, Vol. 1: Fundamental Algorithms* (Birkhäuser, Boston, Cambridge, MA).

Blanchard, P. (1984). Complex analytic dynamics on the Riemann sphere, *Bull. AMS* **11**, pp. 85–141.

Blum, L., Cucker, F., Shub, M. and Smale, S. (1998). *Complexity and Real Computation* (Springer-Verlag, New York).

Bodewig, E. (1949). On types of convergence and on the behavior of approximations in the neighborhood of a multiple root of an equation, *Quart. App. Math.* **7**, pp. 325–333.

Borwein, J. M., Borwein, P. and Bailey, H. (1989). Ramanujan, modular equations, and approximation to pi, or how to compute one billion digits of pi, *Amer. Math. Monthly* **96**, 3, pp. 201–219.

Borwein, P. and Erdélyi, T. (1995). *Polynomials and Polynomial Inequalities*, Vol. 161 (Springer-Verlag, New York).

Brent, R. P. (1976). Fast multiple-precision evaluation of elementary functions, *JACM* **23**, pp. 242–251.

Brezinski, C. (1990). *History of Continued Fractions and Padé Approximants* (Springer, Berlin).

Bruter, C. P. (2007a). Du nouveau du côtá des nombres? *Quadrature* **66**, pp. 8–14.

Bruter, C. P. (ed.) (2007b). *Mathematics and Art: Mathematical Visualization in Art and Education* (Springer, New York).

Buff, X. and Henriksen, C. (2003). On König's root-finding algorithms, *Nonlinearity* **16**, pp. 989–1015.

Cayley, A. (1897). The Newton-Fourier imaginary problem, *American Journal of Mathematics* **2**, p. 97.

Chan, R. H. and Ng, M. K. (1996). Conjugate gradient methods for Toeplitz systems, *SIAM Rev.* **38**, pp. 427–482.

Chudnovsky, D. V. and Chudnovsky, G. V. (1987). Approximations and complex multiplication according to Ramanujan, in *Ramanujan Revisited: Proceedings of the Centenary Conference, University of Illinois at Urbana-Champaign* (Academic Press, Boston, MA), pp. 375–472.

Cohen, H. (1993). *A Course in Computational Algebraic Number Theory* (Springer-Verlag, New York).

Conway, J. H., Burgiel, H. and Goodman-Strauss, C. (2008). *The Symmetries of Things* (A K Peters, Wellesley, MA).

Conway, J. H. and Smith, D. (2003). *On Quaternions and Octonions: Their Geometry, Arithmetic And Symmetry* (A K Peters, Wellesley, MA).

Crass, S. and Doyle, P. (1997). Solving the sextic by iteration, *Internat. Math. Res. Notices* **163**, pp. 83–99.

Curry, J., Garnett, L. and Sullivan, D. (1983). On the iteration of rational functions: Computer experiment with Newton's method, *Comm. Math. Phys.* **91**, pp. 267–277.

Cuyt, A., Peterson, V. B., Verdonk, B., Waadeland, H. and Jones, W. B. (2008). *Handbook of Continued Fractions for Special Functions* (Springer, New York).

Dahlquist, G. and Björck, Å. (1974). *Numerical Methods* (Prentice-Hall, Englewood Cliffs, New Jersey).

Dediue, J. P. (1997). Estimation for the separation number of a polynomial system, *J. Symbolic Computation* **24**, pp. 683–693.

Devaney, R. L. (1986). *Introduction to Chaotic Dynamic Systems* (Benjamin Cummings).

Devaney, R. L. (1992). *A First Course in Chaotic Dynamic Systems Theory and EXPERIMENT* (ABP).

Devaney, R. L. (ed.) (1994). *Complex Dynamical Systems, The Mathematics Behind the Mandelbrot and Julia Set*, Vol. 49 (American Mathematical Society Lecture Note Series).

Dimitrov, D. K. (1998). A refinement of the Gauss-Lucas theorem, *Proceedings of the American Mathematical Society* **126**, 7, pp. 2065–2070.

Douady, A. and Hubbard, J. (1985). On the dynamics of polynomial-like mappings, *Ann. Sci. Ecole Norm Sup.* **18**, pp. 287–344.

Doyle, P. and McMullen, C. (1989). Solving the quintic by iteration, *Acta Math.* **163**, pp. 151–180.

Drakopoulos, V., Argyropoulos, N. and Bohm, A. (1999). Generalized computation of Shroder iteration functions to motivate families of Julia and Mandelbrot-like sets, *SIAM J. Numer. Anal.* **36**, pp. 417–435.

Emmer, M. (ed.) (1993). *The Visual Mind: Art and Mathematics* (MIT Press, Cambridge, USA).

Falconer, K. J. (1990). *Fractal Geometry: Mathematical Foundations and Applications* (John Wiley & Sons Inc.).

Fatou, P. (1919). Sur les équations fonctionelles, *Bull. Soc. Math. France* **47**, pp. 161–271.

Feinberg, M. (1963). Fibonacci-Tribonacci, *Fibonacci Quarterly* **1**, pp. 71–74.

Fiduccia, C. M. (1985). An efficient formula for linear recurrences, *SIAM J. Comput* **14**, pp. 106–112.

Fine, B. and Rosenberger, G. (1997). *The Fundamental Theorem of Algebra* (Springer-Verlag, New York).

Ford, W. F. and Pennline, J. A. (1996). Accelerated convergence in Newton's method, *SIAM Review* **38**, pp. 658–659.

Frame, J. S. (1944). A variation of Newton's method, *Amer. Math. Monthly* **51**, pp. 36–38.

Frame, J. S. (1945). Remarks on a variation of Newton's method, *Amer. Math. Monthly* **52**, pp. 212–214.

Frame, J. S. (1953). The solution of equations by continued fraction, *Amer. Math. Monthly* **60**, pp. 293–305.

Gander, W. (1985). On Halley's iteration method, *Amer. Math. Monthly* **92**, pp. 131–134.

Gerlach, J. (1994). Accelerated convergence in Newton's method, *SIAM Review* **36**, pp. 272–276.

Glick, J. (1988). *Chaos: Making a New Science* (Penguin Books).

Golub, G. and Loan, C. V. (1996). *Matrix Computations*, 3rd edn. (The John Hopkins University Press, Baltimore, MD).

Goodman, A. W. (1975). Remarks on Gauss-Lucas theorem in higher dimensional space, programming, *Proceedings Of Amer. Math. Soci.* **55**, 1, pp. 97–102.

Gries, D. and Levin, G. (1980). Computing Fibonacci numbers (and similarly defined functions) in log time, *IPL* **11**, pp. 68–69.

Haesler, F. v. and Peitgen, H. O. (1989) (Kluwer Academic Publishers, Dordecht).

Hajja, M. (2006). *Private Communication.*

Halley, E. (1694). A new, exact, and easy method of finding roots of any equations generally, and that without any previous reduction, *Philos. Trans. Roy. Soc. London* **18**, pp. 136–145.

Hamilton, H. J. (1950). A type of variation of Newton's method, *Amer. Math. Monthly* **57**, pp. 517–522.

Hansen, E. and Patrick, M. (1977). A family of root finding methods, *Numer. Math.* **27**, pp. 257–269.

Hawkins, J. (2002). McMullen's root-finding algorithm for cubic polynomials, *Comm. Math. Phys.* **130**, pp. 2583–2592.

Henrici, P. (1974). *Applied and Computational Complex Analysis*, Vol. I (Wiley, New York).

Hildebrand, F. B. (1974). *Introduction to Numerical Analysis*, 2nd edn. (McGraw-Hill, New York).

Householder, A. S. (1970). *The Numerical Treatment of a Single Nonlinear Equation* (McGraw-Hill, New York).

Jamieson, M. J. (1989). Rapidly converging iterative formulae for finding square roots and their computational efficiencies, *Comput. J.* **32**, pp. 93–94.

Jin, Y. (2005a). *Combinatorics of Polynomial Root-Finding*, Ph.D. thesis, Department of Computer Science, Rutgers University, NJ, USA.

Jin, Y. (2005b). *Private Communication.*

Jin, Y. (2006). On efficient computation and asymptotic sharpness of Kalantari's bounds for zeros of polynomials,, *Mathematics of Computation* **75**, pp. 1905–1912.

Jin, Y. and Kalantari, B. (2005a). An algebraic derivation of a variant of basic family for finding multiple roots of a polynomial, in *Proceedings of 17th IMACS World Congress, Scientific Computation, Applied Mathematics and*

Simulation (Paris, France).

Jin, Y. and Kalantari, B. (2005b). Symmetric functions and root-finding algorithms, *Advances in Applied Mathematics* **34**, pp. 156–174.

Jin, Y. and Kalantari, B. (2007). On general convergence of the basic family for extracting radicals, *J. Comp. Appl. Math.* **206**, pp. 832–842.

Jones, W. B. and Thron, W. J. (1980). *Continued Fractions, Encyclopedia of Mathematics and its Applications*, Vol. 11 (Addison Wesley, Reading, M.A).

Julia, G. (1918). Sur les équations fonctionelles, *J. Math. Pure Appl.* **4**, pp. 47–245.

Kalantari, B. (1997). A lower bound on determinants from linear programming, Tech. Rep. DCS-TR-330, Department of Computer Science, Rutgers University, New Brunswick, New Jersey.

Kalantari, B. (1998a). Approximation of polynomial root using a single input and the corresponding derivative values, Tech. Rep. DCS-TR-369, Department of Computer Science, Rutgers University, New Brunswick, New Jersey.

Kalantari, B. (1998b). Halley's method is the first member of an infinite family of cubic order root-finding methods, Tech. Rep. DCS-TR-370, Department of Computer Science, Rutgers University, New Brunswick, New Jersey.

Kalantari, B. (1999). On the order of convergence of a determinantal family of root-finding methods, *BIT* **39**, pp. 96–109.

Kalantari, B. (2000a). Generalization of Taylor's Theorem and Newton's method via a new family of determinantal interpolation formulas and its applications, *J. Comp. Appl. Math.* **126**, pp. 287–318.

Kalantari, B. (2000b). New formulas for approximations of pi and other transcendental numbers, *Numerical Algorithms* **24**, pp. 59–81.

Kalantari, B. (2002a). Can polynomiography be useful in computational geometry? in *DIMACS Workshop on Computational Geometry* (New Brunswick, NJ), `http://dimacs.rutgers.edu/Workshop/CompGeom/abstracts/005.pdf`.

Kalantari, B. (2002b). Polynomiography: A new intersection between mathematics and art, Tech. Rep. DCS-TR-506, Department of Computer Science, Rutgers University, New Brunswick, NJ, `http://www.polynomiography.com/images/artmath.pdf`.

Kalantari, B. (2004a). A new visual art medium: Polynomiography, *ACM SIGGRAPH Computer Graphics Quarterly* **38**, pp. 21–23.

Kalantari, B. (2004b). On homogeneous linear recurrence relations and approximation of zeros of complex polynomials, in M. B. Nathanson (ed.), *Unusual Applications in Number Theory , DIMACS Series in Discrete Mathematics and Theoretical Computer Science*, Vol. 64, pp. 125–143.

Kalantari, B. (2004c). Polynomiography and applications in art, education, and science, *Computers & Graphics* **28**, pp. 417–430.

Kalantari, B. (2004d). Polynomiography in art and design, in *Proceedings of the Fourth International Mathematica & Design Conference*, Vol. 4 (Argentina), pp. 305–311.

Kalantari, B. (2005a). Corrigendum to An infinite family of bounds on zeros of analytic functions and relationship to Smale's bound, *Mathematics of*

Computation **74**, p. 2101.

Kalantari, B. (2005b). An infinite family of bounds on zeros of analytic functions and relationship to Smale's bound, *Mathematics of Computation* **74**, pp. 841–852.

Kalantari, B. (2005c). Polynomiography: From the fundamental theorem of algebra to art, *LEONARDO* **38**, pp. 233–238.

Kalantari, B. (2006). Two and three-dimensional art inspired by polynomiography, *Visual Mathematics* **8**, 1, also in Proceedings of BRIDGES (Mathematical Connections in Art, Music, and Science), Banff, Canada, 321-328, 2005.

Kalantari, B. (2007). Polynomiography in art and design, in *Proceedings of the Fifth International Mathematics & Design Conference*, Vol. 5 (Brazil), pp. 63–70.

Kalantari, B. and Gerlach, J. (2000). Newton's method and generation of a determinantal family of iteration functions, *J. of Comp. and Appl. Math.* **116**, pp. 195–200.

Kalantari, B. and Jin, Y. (2003). On extraneous fixed-points of the basic family of iteration functions, *BIT* **43**, pp. 453–458.

Kalantari, B. and Kalantari, I. (1996). High order iterative methods for approximating square roots, *BIT* **36**, pp. 395–399.

Kalantari, B., Kalantari, I. and Andreev, F. (2004). Animation of mathematical concepts using polynomiography, in *Proceedings of SIGGRAPH, Educator's Program*.

Kalantari, B., Kalantari, I. and Zaare-Nahandi, R. (1997). A basic family of iteration functions for polynomial root finding and its characterizations, *J. of Comp. and Appl. Math.* **80**, pp. 209–226.

Kalantari, B. and Park, S. (2001). A computational comparison of the first nine members of a determinantal family of root-finding methods, *J. of Comp. and Appl. Math.* **130**, pp. 197–204.

Kalantari, B. and Pate, T. H. (2001). A determinantal lower bound, *Linear Algebra and its Application* **326**, pp. 151–159.

Kneisl, K. (2001). Julia sets for the super-Newton method, Cauchy's method and Halley's method, *Chaos* **11**, pp. 359–370.

Knuth, D. E. (1973). *The Art of Computer Programming, Volume I: Fundamental Algorithms*, 2nd edn. (Addison Wesley, Reading, MA).

Koshy, T. (2001). *Fibonacci and Lucas Numbers with Application* (John Wiley & Sons Inc.).

Kung, H. T. (1974). A new upper bound on the complexity of derivative evaluation, *Info. Proc. Lett.* **2**, pp. 146–147.

Kung, H. T. (1976). New algorithms and lower bounds for the parallel evaluation of certain rational expressions and recurrences, *JACM* **23**, pp. 252–261.

Lang, S. (1995). *Introduction to Diophantine Approximations* (Springer-Verlag, New York).

Lederman, L. M., C.T. and Hill (2004). *Symmetry and the Beautiful Universe* (Prometheus Books, Amherst, New York).

Lei, T. (ed.) (2000). *Mandelbrot Set, Themes and Variations*, Vol. 274 (London

Mathematical Society Lecture Note Series).

Mandelbrot, B. B. (1983). *The Fractal Geometry of Nature* (W. F. Freeman, New York).

Mandelbrot, B. B. (1993). Fractals and art for the sake of science, in M. Emmer (ed.), *The Visual Mind: Art and Mathematics* (MIT Press, Cambridge, USA), pp. 11–14.

Marcus, M. and Minc, H. (1964). *A Survey of Matrix Theory and Matrix Inequalities* (Allyn and Bacon, Boston, MA).

Marden, M. (1966). *Geometry of Polynomials* (Mathematical Surveys and Monographs, Amer. Math. Soc.).

Mazur, B. (2003). *Imagining Numbers (particularly the square root of minus fifteen)* (Picador).

McMullen, C. (1987). Families of rational maps and iterative root-finding algorithms, *The Annals of Math.* **125**, 2, pp. 467–493.

McMullen, C. (1988). Braiding of the attractor and the failure of iterative algorithms, *Invent. math.* **91**, pp. 259–272.

McMullen, C. (1994). *Complex Dynamics and Renormalization*, Vol. 135 (Princeton University Press, Princeton, NJ).

McMullen, C. (2004). *Algebra and Dynamics* (Course Notes).

McMullen, C. (2007). *Mandelbrot set is universal* (Online).

McNamee, J. M. (1993). A bibliography on root of polynomials, *J. of Comp. and Appl. Math.* **47**, pp. 391–394.

McNamee, J. M. (2007). *Numerical Methods for Roots of Polynomials: Part I*, Vol. 14 (Elsevier, Amsterdam).

McNamee, J. M. and Olhovsky, M. (2005). A comparison of a priori bounds on (real or complex) roots of polynomials, in *Proceedings of 17th IMACS World Congress, Scientific Computation, Applied Mathematics and Simulation* (Paris, France).

Miles, E. P. (1960). Generalized Fibonacci numbers and associated matrices, *Amer. Math. Monthly* **6**, pp. 745–757.

Milnor, J. (2006). *Dynamics in One Complex Variable*, Vol. 160, 3rd edn. (Princeton University Press, Princeton, NJ).

Mossinghoff, M. J. (1998). Polynomials with small Mahler measure, *Mathematics of Computation* **67**, p. 16971705.

Nahin, P. J. (1998). *An Imaginary Tale: The Story of "i" [the Square Root of Minus One]* (Princeton University Press).

Nehari, Z. (1961). *Introduction to Complex Analysis* (Allyn and Bacon INC., Boston).

Ostrowski, A. M. (1966). *Solution of Equations and System of Equations*, 2nd edn. (Academic Press, New York).

Pan, V. Y. (1994). New techniques for approximating complex polynomial zeros, in *Proc. 5th Ann. ACM-SIAM Symp. on Discrete Algorithms*, pp. 260–270.

Pan, V. Y. (1997). Solving a polynomial equation: some history and recent progress, *SIAM Review* **39**, pp. 187–220.

Peitgen, H. O. and Richter, P. H. (1992). *The Beauty of Fractals* (Springer-Verlag, New York).

Peitgen, H. O., Saupe, D. and Haesler, F. v. (1984). Cayley's problem and Julia sets, *Math. Intelligencer* , pp. 11–20.

Peitgen, H. O., Saupe, D., Jurgens, H. and Yunker, L. (1992). *Chaos and Fractals* (Springer-Verlag, New York).

Peterson, I. (2001). *Fragments of Infinity, A Kaleidoscope of Math and Art* (John Wiley & Sons, Inc.).

Petković, M. and Hereceg, D. (1999). On rediscovered iteration methods for solving equations, *J. of Comp. and Appl. Math.* **107**, pp. 275–284.

Popovski, D. B. (1980). A family of one-point iteration formulae for finding roots, *Internat. J. Computer Math.* **8**, pp. 85–88.

Potra, F.-A. and Ptak, V. (1984). *Nondiscrete induction and iterative processes, Research Notes in Mathematics* (Pitman Advanced Publishing Program).

Prasolov, V. V. (2004). *Polynomials* (Springer-Verlag, Berlin).

Riesel, H. (1994). *Prime Numbers and Computer Methods for Factorization* (Birkhäuser, Boston, Cambridge, MA).

Salamin, E. (1976). Computation of pi using arithmetic-geometric mean, *Math. of Computation* **30**, pp. 565–570.

Scavo, T. R. and Thoo, J. B. (1995). On the geometry of Halley's method, *Amer. Math. Monthly* **102**, pp. 417–426.

Schröder, E. (1870). On infinitely many algorithms for solving equations (German), *Math. Ann.* **2**, pp. 317–365, English translation by G.W. Stewart, TR-92-121, Institute for Advanced Computer Studies, University of Maryland, College Park, MD, 1992.

Sheil-Small, T. (2002). *Complex Polynomials, Cambridge Studies in Advance Mathematics*, Vol. 75 (Cambridge University Press, New York).

Shishikura, M. (1987). On the quasiconformal surgery of rational functions, *Ann. Sci. Ecole Norm. Sup.* **20**, pp. 1–29.

Shishikura, M. (1990). The connectivity of Julia sets and fixed points, *IHES Preprint* .

Shishikura, M. (1994). The boundary of the Mandelbrot set has Hausdorff dimension two, *Astérisue* **222**, pp. 389–405.

Shub, M. and Smale, S. (1985). Computational complexity: On the geometry of polynomials and a theory of cost, part I, *Ann. Sci. Ecole Norm. Sup.* **18**, pp. 107–142.

Shub, M. and Smale, S. (1986). On the existence of generally convergent algorithms, *J. Complexity* **1**, pp. 2–11.

Siegel, C. (1942). Iterations of analytic functions, *Ann. of Math.* **43**, pp. 607–616.

Smale, S. (1985). On the efficiency of algorithms of analysis for solving equations, *Bull. Amer. Math. Soc.* **13**, pp. 87–121.

Smale, S. (1986). Newton's method estimates from data at one point, in R. E. Ewing, K. I. Gross and C. Martin (eds.), *The merging of Disciplines: New Directions in Pure, Applied, and Computational Mathematics*, pp. 185–196.

Sprott, J. and Pickover, C. (1995). Automatic generation of quadratic map basins, *Computers & Graphics* **19**, pp. 309–313.

Stewart, J. K. (1951). Another variation of Newton's method, *Amer. Math. Monthly* **58**, pp. 331–334.

Sullivan, D. (1985). Quasiconformal homemorphisms and dynamics I: Solution of the Fatou-Julia problem on wandering domains, *Ann. of Math.* **122**, pp. 401–418.

Takahashi, D. and Kanada, Y. (1998). Calculation of π to 51.5 billion decimal digits on distributed memory parallel processors, *Trans. Inform. Process. Soc. Japan* **39**, pp. 2074–2083.

Traub, J. F. (1964). *Iterative Methods for the Solution of Equations* (Prentice-Hall, Englewood Cliffs, NJ).

Traub, J. F. (1966). A class of globally convergent iteration functions for the solution of polynomial equations, *Mathematics of Computations* **20**, pp. 113–138.

van der Waerden, B. L. (1970). *Algebra*, Vol. 1 (F. Ungar Publishing Co., New York).

Varona, J. L. (2002). Graphic and numerical comparison between iterative methods, *Math. Intelligencer* **24**, pp. 37–46.

Vrscay, E. R. and Gilbert, W. J. (1988). Extraneous fixed points, basin boundaries and chaotic dynamics for Schröder and König iteration functions, *Numer. Math.* **52**, pp. 1–16.

Wall, H. S. (1948). A modification of Newton's method, *Amer. Math. Monthly* **55**, pp. 90–94.

Wang, X. H. (1994). A summary on continues complexity theory, *Contemporary Mathematics* **163**, pp. 155–170.

Weyl, H. (1924). Randbemerkungen zu Hauptproblemen der Mathematik. II Fundamentalsatz der Algebra und Grundlagen der Mathematik, *Math. Z.* **20**, pp. 142–150.

Wolfram, S. (2004). *Private Communication.*

Yap, C. K. (1999). *Fundamental Problems of Algorithmic Algebra* (Oxford University Press, Oxford).

Yeyios, A. K. (1992). On two sequences of algorithms for approximating square roots, *J. Comput. Appl. Math.* **40**, pp. 63–72.

Ypma, T. J. (1995). Historical development of Newton-Raphson method, *SIAM Review* **37**, pp. 531–551.

Index